Editorial Policy

for the publication of proceedings of conferences
and other multi-authorvolumes

Lecture Notes aim to report new developments - quickly, informally and at a high level . The following describes criteria and procedures for multi-author volumes. For convenience we refer throughout to „proceedings“ irrespective of whether the papers were presented at a meeting.
The editors of a volume are strongly advised to inform contrlbutors about these points at an early stage.

§ 1. One (or more) expert participant(s) should act as the scientific editor(s) of the volume. They select the papers which are suitable (cf. §§ 2 - 5) for inclusion in the proceedings, and have them individually refereed (as for a journal). It should not be assumed that the published proceedings must reflect conference events in their entirety. The series editors will normally not interfere with the editing of a particular proceedings volume - except in fairly obvious cases, or on technical matters, such as described in §§ 2 - 5. The names of the scientific editors appear on the cover and title-page of the volume.

§ 2. The proceedings should be reasonably homogeneous i.e. concerned with a limited and welldefined area. Papers that are essentially unrelated to this central topic should be excluded. One or two longer survey articles on recent developments in the field are often very useful additions. A detailed introduction on the subject of the congress is desirable.

§ 3 . The final set of manuscripts should have at least 100 pages and preferably not exceed a total of 400 pages. Keeping the size below this bound should be achieved by stricter selection of articles and NOT by imposing an upper limit on the length of the individual papers .

§ 4. The contributions should be of a high scientific standard and of current interest. Research articles should present new material and not duplicate other papers already published or due to be published. They should contain sufficient background and motivation and they should present proofs, or at least outlines of such, in sufficient detail to enable an expert to complete them. Thus summaries and mere announcements of papers appearing elsewhere cannot be included, although more detailed versions of, for instance, a highly technical contribution may well be published elsewhere later.
Contributions in numerical mathematics may be acceptable without formal theorems/proofs provided they present new algorithms solving problems (previously unsolved or less well solved) or develop innovative qualitative methods, not yet amenable to a more formal treatment.
Surveys, if included, should cover a sufficiently broad topic, and should normally not just review the author's own recent research. In the case of surveys, exceptionally, proofs of results may not be necessary.

§ 5. „Mathematical Reviews“ and „Zentralblatt für Mathematik“ recommend that papers in proceedings volumes carry an explicit statement that they are in final form and that no similar paper has been or is being submitted elsewhere, if these papers are to be considered for a review. Normally, papers that satisfy the criteria of the Lecture Notes in Biomathematics series also satisfy this requirement, but we strongly recommend that each such paper carries the statement explicitly.

§ 6. Proceedings should appear soon after the related meeting. The publisher should therefore receive the complete manuscript (preferably in duplicate) including the Introduction and Table of Contents within nine months of the date of the meeting at the latest.

§ 7. Proposals for proceedings volumes should be sent to the Editor of the series or to Springer-Verlag Heidelberg. They should give sufficient information on the conference, and on the proposed proceedings. In particular, they should include a list of the expected contributions with their prospective length. Abstracts or early versions (drafts) of the contributions are helpful.

Further remarks and relevant addresses at the back of this book.

Lecture Notes in Biomathematics 92

S. Busenberg M. Martelli (Eds.)

Differential Equations Models in Biology, Epidemiology and Ecology

Proceedings of a Conference held in Claremont California, January 13-16, 1990

Springer-Verlag
Berlin Heidelberg New York
London Paris Tokyo
Hong Kong Barcelona
Budapest

Editors

Stavros Busenberg
Department of Mathematics
Harvey Mudd College
Claremont, CA 91711, USA

Mario Martelli
Department of Mathematics
California State University
Fullerton, CA 92634, USA

Mathematics Subject Classification (1991): 92A09, 92A15, 92A17, 43K99, 36K60

ISBN-13: 978-3-540-54283-4 e-ISBN-13: 978-3-642-45692-3
DOI: 10.1007/978-3-642-45692-3

Typesetting: Camera ready by author
46/3140-543210 - Printed on acid-free paper

Preface

The past forty years have been the stage for the maturation of mathematical biology as a scientific field. The foundations laid by the pioneers of the field during the first half of this century have been combined with advances in applied mathematics and the computational sciences to create a vibrant area of scientific research with established research journals, professional societies, deep subspecialty areas, and graduate education programs. Mathematical biology is by its very nature cross-disciplinary, and research papers appear in mathematics, biology and other scientific journals, as well as in the specialty journals devoted to mathematical and theoretical biology. Multiple author papers are common, and so are collaborations between individuals who have academic bases in different traditional departments. Those who seek to keep abreast of current trends and problems need to interact with research workers from a much broader spectrum of fields than is common in the traditional mono-culture disciplines. Consequently, it is beneficial to have occasions which bring together significant numbers of workers in this field in a forum that encourages the exchange of ideas and which leads to a timely publication of the work that is presented. Such an occasion occurred during January 13 to 16, 1990 when almost two hundred research workers participated in an international conference on Differential Equations and Applications to Biology and Population Dynamics which was held in Claremont. The occasion was particularly noteworthy because the meeting also provided a venue for the celebration of the 65th birthday of Kenneth Cooke, a seminal worker who has made a number of pioneering contributions in epidemiology, population dynamics and delay differential equations.

The mathematical and scientific work of Kenneth Cooke spans the past forty years and has strongly influenced the development of the areas of epidemic modeling and population dynamics. Its character is often recognized by the exchange and intermixing of ideas and problems from mathematical biology and delay differential equations, a style which he developed and continues to refine, and which is now widely used. The list of his publications which heads the contents of this volume serves as a tribute to this milestone in his ongoing career. It is a record of the progress of the field and a source of inspiration and ideas for future work.

This volume is devoted to papers presented at the conference in the area of mathematical biology and population dynamics. They have been selected after undergoing peer review. A companion volume in the Lecture Notes in Mathematics series of Springer contains papers devoted to delay differential equations and dynamical systems. The two areas of these volumes have had a close and fruitful interaction during this century, and Kenneth Cooke has been one of the most artful and original practitioners in this interdisciplinary research work.

The contents of this volume are collected in three groups, the first consisting of articles devoted to basic questions or new areas of mathematical biology. The second consists of research papers in epidemic modeling, an area that is now experiencing exciting and rapid development. In the third group are the research papers which deal with ecology and population dynamics. The first group of papers starts with Simon Levin's contribution discussing a fundamental question in mathematical biology, that of determining how much detail is relevant in any particular model. The other contributions in this section present and analyze the following situations: interactions between monolingual and bilingual populations; the use of time delays to model lifespans of subpopulations; and the detailed analysis of the striking dynamic physiological phenomenon of bursting electrical activity in pancreatic cells.

The second group of papers, those dealing with questions in mathematical epidemiology, includes results establishing general dynamic and structural properties of disease transmission models, as well as analyses of specific models. The notable new results in these papers include the demonstration of the possibility of destabilizing the endemic equilibrium by including age-structure in $S-I-R$ models, the axiomatic characterization of two-sex mixing interaction terms, and the deepening of our understanding of the effects of demography and control strategies on the course of epidemics.

The third group of papers deals with what is arguably the best established area of mathematical biology. The topics treated here include the study of dynamic models of phytoplankton populations, of competitive coexistence in the gradostat (a particular laboratory controlled ecosystem), of the effects of age structure, stage structure, spatial heterogeneity, delays, and density dependent growth rates in populations and competition models. The range of results and new approaches that is found in this group of papers testifies to the great vitality that marks this vast and important field.

The research results in this volume indicate several current trends in mathematical and theoretical biology. They also illustrate the crafty balance between adequate description of the biological phenomena and the inclusion of details which, while adding to the recognition of specific biological situations, may also preclude the gaining of broad insights into the dynamic behavior of general classes of population interactions. The models use differential equations, delay differential equations, and more complicated dynamical systems which can include age and stage structure. The variety of these methods points to the constant search for means of presenting more biological detail and realism in concise and symbolic representations which can be mathematically analyzed. They continue the legacy of first developing mathematical methods that fit particular population problems, and then striving to simplify them into a natural language for describing and analyzing the biological observations. Kenneth Cooke is a master craftsman of this central aspect of the field which is showing its vitality by continuing to attract outstanding young investigators.

The research conference which gave rise to this volume was supported by the National Science Foundation through grant number MCS-8912391 and by Harvey Mudd College which hosted the meeting. A large number of individuals gave invaluable help during all phases of the conference. Our special gratitude goes to those who have worked to prepare the manuscripts for publication: Sue Cook, Jeffrey McLelland, Beth Nyerges, Barbara Shade, and David Williamson. The editorial staff of Springer gave us constant support and was always ready to respond to our requests for help and information. Finally, the many researchers

who contributed time and expertise in refereeing the papers appearing in this volume and its companion provided an invaluable service not only to us the editors, but also to the authors and to all the other mathematicians and biologists who will use these volumes. Our sincere gratitude goes to all, both named and unnamed, who have helped in this endeavor.

Stavros Busenberg and Mario Martelli

Claremont, California

Contents

[illegible] 159

Publications of Kenneth Cooke

Articles

(1953): The asymptotic behavior of the solutions of linear and non-linear differential-difference equations. Trans. Amer. Math. Soc., **75**, 80-105

(1954): The rate of increase of real continuous solutions of algebraic differential-difference equations of the first order. Pacific J. Math., **4**, 483-501

(1955): A non-local existence theorem for systems of ordinary differential equations. Rend. Circ. Mat. Palermo, Ser. II, Tomo **4**, 301-308

(1955): Forced periodic solutions of a stable non-linear differential-diference equation. Annals of Math. **61**, 387-391

(1957): Hadamard matrices. With J.L. Brenner, Rev. Ci. Lima **59**, 5-13

(1958): A symbolic method for finding integrals of linear difference and differential-difference equations. Math. Magazine, 121-126

(1959): Stability theory and adjoint operators for linear differential-difference equations. With R. Bellman, Trans. Amer. Math. Soc. **92**, 470-500

(1959): On the limit of solutions of differential-difference equations as the retardation approaches zero. With R. Bellman, Proc. Nat. Acad. Sci. **45**, 1026-1028

(1959): The rate of increase of real continuous solutions of certain algebraic functional equations. Trans. Amer. Math. Soc. **92**, 106-124

(1960): Stability and asymptotic theory for linear differential-difference equations. Proc. Symposium on Ordinary Differential Equations at Mexico City, September 1959, Boletin de la Sociedad Mat. Mexicana, 277-283

(1962): On transcendental equations related to differential-difference equations. J. Math. Anal. Appl. **4**, 65-71

(1963): Differential-difference equations. International Symposium on Nonlinear Differential Equations and Nonlinear Mechanics, Academic Press, 155-171

(1965): Existence and uniqueness theorems in invariant imbedding - I: conservation principles. With R. Bellman, R. Kalaba, G.M. Wing, J. Math. Anal. Appl. **10**, 243-244

(1965): Existence and uniqueness theorems in invariant imbedding - II: convergence of a new difference algorithm. With R. Bellman, J. Math. Anal. Appl. **12**, 247-253

(1965): The condition of regular degeneration for singularly perturbed linear differential-difference equations. J. Differential Equations **1**, 39-94

(1965): Convergence of successive approximations in the shortest route problem. With D.L. Bentley, J. Math. Anal. Appl. **10**, 269-274

(1965): On the computational solution of a class of functional differential equations. With R. Bellman, J. Math. Anal. Appl. **12**, 459-500

(1966): The condition of regular degeneration for singularly perturbed systems of linear differential-difference equations. With K.R. Meyer, J. Math. Anal. Appl. **14**, 83-106

(1966): The shortest route through a network with time-dependent internodal transit times. With E. Halsey, J. Math. Anal. Appl. **14**, 493-498

(1966): Graphical solution of difficult crossing puzzles. With P. Detrick, R. Fraley, Math. Mag. **39**, 151-157

(1966): Functional differential equations close to differential equations. Bull. Amer. Math. Soc. **72**, 285-288

(1967): Functional differential equations with asymptotically vanishing lag. Rend. Circ. Mat. di Palermo, series II, XVI, 39-56

(1967): Functional differential equations: some models and perturbation problems. Address at International Symposium on Differential Equations and Dynamical Systems, Differential Equations and Dynamical Systems, J.K. Hale and J.P. LaSalle, eds., Academic Press, 167-183

(1967): Asymptotic theory for the delay-differential equation $u'(t) = -au(t - r(u(t)))$. J. Math. Anal. Appl. **19**, 160-173

(1967): Some recent work on functional-differential equations. Proc. U.S. Japan Seminar on Differential and Functional Equations, W.A. Harris, Jr., Y. Sibuya, eds., W.A. Benjamin, 27-47

(1968): Difference-differential equations and nonlinear initial-boundary value problems for linear hyperbolic partial differential equations. With D. Krumme, J. Math. Anal. Appl. **24**, 372-387

(1969): The Königsberg bridges problem generalized. With R. Bellman, J. Math. Anal. Appl. **25**, 1-7

(1970): Linear functional differential equations of asymptotically autonomous type. J. Diff. Eq. **7**, 154-174

(1971): A linear mixed problem with derivative boundary conditions. Seminar on Differential Equations and Dynamical Systems, III, Springer-Verlag

(1972): Equations modelling population growth, economic growth, and gonorrhea epidemiology. With J. Yorke, 1971 NRL-MRC Conference, Ordinary Differential Equations, L. Weiss, ed., Academic Press, New York

(1973): Some equations modelling growth processes and gonorrhea epidemics. With J. Yorke, Math. Biosciences **16**, 75-101

(1973): On a class of hereditary processes in biomechanics. With N. Distefano, B. Kashef, Math. Biosciences **16**, 359-373

(1974): A linear hyperbolic problem all of whose solutions are constant after finite time. SIAM J. Math. Anal. **5, No. 3**, 482-488

(1975): Asymptotic equivalence of an ordinary and a functional differential equation. J. Math. Anal. Appl. **51**, 187-207, Abstract in International Conference on Differential Equations, H.A. Antonsiewicz, ed., Academic Press, New York, 808-809

(1975): A discrete-time epidemic model with classes of infectives and susceptibles. Theor. Population Biol. **7**, 175-196

(1975): A discrete-time epidemic model with classes of infectives and susceptibles (Summary). Proceedings of SIMS Conference on Epidemiology, D. Ludwig, K. Cooke, eds., SIAM, Philadelphia, 132-138

(1976): An epidemic equation with immigration. Math. Biosciences **29**, 135-158

(1976): A periodicity threshold theorem for epidemics and population growth. With J.L. Kaplan, Math. Biosciences **31**, 87-104

(1977): Stability or chaos in discrete epidemic models. With D.F. Calef, E.V. Level, Nonlinear Systems and Their Applications, Academic Press, 73-93

(1978): Periodic solutions of a periodic nonlinear delay differential equation. With S. Busenberg, SIAM J. Applied Math. **35**, 704-721

(1979): Periodic solutions of delay differential equations arising in some models of epidemics. With S. Busenberg, Applied Nonlinear Analysis, V. Lakshmikantham, ed., Academic Press, New York, 67-78

(1979): Mathematical approaches to culture change. Transformations: Mathematical Approaches to Culture Change, C. Rehfrew, K. Cooke, eds., Academic Press, New York, 45-81

(1979): Stability Analysis for a vector disease model. Rocky Mountain J. of Math. **9**, 31-42

(1979): An experiment on the simulation of culture changes. With C. Renfrew, Transformations: Mathematical Approaches to Culture Change, C. Rehfrew, K. Cooke, eds., Academic Press, New York, 327-348

(1980): The effect of integral conditions in certain equations modelling epidemics and population growth. With S. Busenberg, J. Math. Biology **10**, 13-32

(1981): On the construction and evaluation of mathematical models. In Simulations in Archaeology, J.A. Sabloff, ed., Univ. New Mexico Press, Albuquerque

(1981): Stability of a functional differential equation for the motion of a radiating charged particle. With J.L. Kaplan, M. Sorg, Nonlinear Analysis: TMA **5**, 1133-1139

(1982): Vertically transmitted diseases. With S. Busenberg, Nonlinear Phenomena in the Mathematical Sciences, V. Lakshmikantham, ed., Academic Press, New York, 189-197

(1982): Models of vertically transmitted diseases with sequential continuous dynamics. With S. Busenberg, Nonlinear Phenomena in the Mathematical Sciences, V. Lakshmikantham, ed., Academic Press, New York, 179-187

(1982): Discrete delay, distributed delay and stability switches. With Z. Grossman, J. Math. Anal. Appl. **86**, 592-627

(1982): Models for endemic infections with asymptomatic cases I. one group. Mathematical Modelling **3**, 1-15

(1983): Analysis of a model of a vertically transmitted disease. With S. Busenberg, M.A. Pozio, J. Math. Biol. **17**, 305-329

(1983): Stability conditions for linear retarded functional differential equations. With J.M. Ferreira, J. Math. Anal. Appl. **96**, 480-504

(1984): Retarded differential equations with piecewise constant delays. With J. Wiener, J. Math. Anal. Appl. **99**, 265-297

(1984): Stability conditions for linear non-autonomous delay differential equations. With S. Busenberg, Quart. Appl. Math., **42**, 295-306

(1984): Mathematical models of vertical transmission of infection. Mathematical Ecology, S.A. Levin, T.G. Hallam, eds., Lecture Notes in Mathematics **54**, Springer-Verlag, 344-355

(1984): Stability of non-autonomous delay differential equations by Liapunov functionals. Infinite-Dimensional Systems, F. Kappel, W. Schappacher, eds., Lecture Notes in Mathematics **1076**, Springer-Verlag

(1985): Infection models with asymptomatics. Mathematics and Computers in Biomedical Applications, J. Eisenfeld, C. De Lisi, eds., Elsevier Science Publ., 277-282

(1985): Stability of delay differential eqations with applications in biology and medicine. Mathematics in Biology and Medicine, V. Capasso, E. Grosso, S.L. Paveri-Fontana, eds., Lecture Notes in Mathematics **57**, Springer-Verlag

(1986): Stability regions for linear equations with piecewise continuous delay. With J. Wiener, Comp. & Math. with Appl. **12A**, 695-701

(1986): One-dimensional linear and logistic harvesting models. With M. Witten, Math. Modelling **7**, 301-340

(1986): On zeroes of some transcendental equations. With P. van den Driessche, Funkcialaj Ekvacioj **29**, 77-90

(1987): An equation alternately of retarded and advanced type. With J. Wiener, Proc. Amer. Math. Soc. **99**, 72-732

(1987): Neutral differential equations with piecewise constant arguments. With J. Wiener, Bollettino UMI **7**, 321-346

(1987): Analysis of the complicated dynamics of some harvesting models. With H.E. Nusse, J. Math Biol. **25**, 521-542

(1988): Harvesting procedures with management policy in iterative density-dependent population models. With R. Elderkin, M. Witten, Natural Resource Modeling **2**, No.3, 383-420

(1988): A nonlinear equation with piecewise continuous argument. With L.A.V. Carvalho, Diff. and Integral Eqns. **1**, 359-367

(1988): The population dynamics of two vertically transmitted infections. With S. Busenberg, Theoretical Population Biology **33**, 181-198

(1988): Endemic thresholds and stability in a class of age-structured populations. With S. Busenberg, M. Iannelli, SIAM J. Applied. Math. **48** No. 6, 1379-1395

(1988): Oscillation in systems of differential equations with piecewise constant argument. With J. Wiener, J. Math. Anal. Appl. **137**, 221-239

(1989): On the role of long incubation periods in the dynamics of acquired immunodeficiency syndrome (AIDS). With C. Castillo-Chavez, W. Huang, S.A. Levin, J. Math. Biol. **27**, 373-398

(1989): Results on the dynamics for models of the sexual transmission of the human immunodeficiency virus. With C. Castillo-Chavez, W. Huang, S.A. Levin, Appl. Math. Lett. **2**, No. 4, 327-331

(1989): On dichotomic maps for a class of differential-difference equations. With L.A.V. Carvalho, submitted

(1989): On the modelling of epidemics. With C. Castillo-Chavez and S.A. Levin, High Performance Computing, J.L. Delhaye, E. Glenebe, Eds., North-Holland, Amsterdam-New York-Oxford-Tokyo, 389-402

(1989): On the role of long incubation periods in the dynamics of acquired immunodeficiency syndrome (AIDS) Part 2: Multiple group models. With C. Castillo-Chavez, W. Huang, S.A. Levin, Lecture Notes in Biomathematics **83**, C. Castillo-Chavez, ed., 200-217

(1989): The role of long periods of infectiousness in the dynamics of acquired immunodeficiency syndrome (AIDS). With C. Castillo-Chavez, W. Huang, S.A. Levin, Mathematical Approaches to Problems in Resource Management and Epidemiology, C. Castillo-Chavez, S.A. Levin, C.A. Shoemaker, eds., Lecture Notes in Biomathematics **81**, 177-189

(1990): Coexistence of Analytic and Distributional Solutions for Linear Differential Equations, I. With J. Wiener, J. Math. Anal. Appl. **148**, 390-421

Books

(1963): Differential-difference Equations. With R. Bellman, Academic Press

(1967): Differential-difference Equations. (Russian Translation), MIR, Moscow

(1968): Modern Elementary Differential Equations. With R. Bellman, Addison-Wesley

(1969): Modern Elementary Differential Equations. (Japanese Translation), Addison-Wesley, Tokyo

(1970): Algorithms, Graphs, and Computers. With R. Bellman, J. Lockett, Academic Press

(1973): Linear Algebra with Differential Equations. With D.L. Bentley, Holt, Rinehart, & Winston

(1975): Proceedings of a SIMS Conference on Epidemiology. Edited with D. Ludwig, SIAM

(1979): Transformations: Mathematical Approaches to Cultural Change. Edited with C. Renfrew, Academic Press

(1981): Differential Equations and Applications in Ecology, Epidemics, and Population Problems. Edited with S. Busenberg, Academic Press

Research Monograph

(1959): Asymptotic Behavior of Solutions of Differential-difference Equations. With R. Bellman, Mem. Amer. Math. Soc. **35**, 91

Technical Reports

(1972): The numerical solution of integro-differential equations with retardation. With S.E. List, University of Southern California Technical Report

(1972): On an epidemic equation with immigration. Universitá degli Studi di Firenze Technical Report

(1973): A model of urbanization and civilization. University of Southern California Technical Report

(1972): Delay in growth and epidemic processess. Proceedings of Symposium on Differential-Delay and Functional Equations: Control and Stability, L. Markus, ed., University of Warwick, Control Theory Centre, Report No. 12

(1960): Asymptotic behavior of solutions of linear parabolic equations. With R. Bellman, RAND Paper P-1870

(1974): Numerical experiments on equations modelling growth and epidemics. With S.E. List, Claremont Colleges Technical Report

(1975): Diffusion models for technological innovation. Dialogue Discussion Paper, The Center for the Study of Democratic Institutions, Santa Barbara

(1975): The equations of nonlinear system dynamics. With E. Level, Claremont Colleges Technical Report

(1979): Models for endemic infections with asymptotic cases, I. One group. Lefschetz Center for Dynamical Systems, Brown University

Part I

Mathematical Biology

The Problem of Relevant Detail

Simon Levin

Ecology and Systematics, Cornell University, Ithaca, New York 14853

In recent years, welcome attention has been directed to population structure in mathematical models, improving substantially upon the overly aggregated characterizations of classical interest. This conference, in honor of Kenneth Cooke and his contributions to theories involving age structure, social structure, and delays, is testament to the modern trend, and to the recognition that simplistic theories that ignore population structure often are seriously flawed in their predictions.

The need to incorporate population structure cuts across problems in ecology and evolutionary biology. In ecological community theory, the most influential theoretical result undoubtedly has been the competitive exclusion principle, which underlies the concept of the ecological niche (see the various papers in Whittaker and Levin, 1975). In its simplest form, this principle states that two species that compete for (and are limited by) the same resource cannot long coexist in the same habitat. How then do species manage to coexist in the same habitat for long periods of time, often ostensibly sharing the same resources?

In the marine intertidal, for example, space is the primary limiting resource, and is shared by a diversity of invertebrate and algal species. Since, according to the simple theory, multiple species could not coexist if they were limited by the same factor, what accounts for the observed coexistence? How are resources to be counted? Is space newly laid bare by wave action the same resource as "older" space? Does each nook and cranny in the rock surface constitute a separate resource? If different species utilize the same resource, but at different stages of their life cycles, are they being limited by the same resource? It is in these finer points regarding population and environmental structure that the interesting problems of community and population theory reside.

Population structure–the spatial, demographic, social and genetic distribution of types–holds the key to understanding the ecological issues discussed in the previous paragraph; furthermore, the same considerations apply in addressing evolutionary problems, which treat identical issues at a finer taxonomic level of resolution. The spatial distribution of a population defines the breeding neighborhoods of individuals; and the evolution of strategies to exploit the spatial and temporal structure of the environment motivates a fundamental suite of problems in life history theory.

Epidemiological models recognize a population structure of a different sort. By subdividing a population into classes according to disease status, one is able to capture the essential nature of the host-parasite relationship, the key to understanding the dynamics of infectious diseases. Alternative theories, such as those that take account only of the numbers of hosts and parasites without explicit consideration of their intimate coupling, miss the most important aspect of the tight relation between host and parasite. The subdivision of a population into susceptible and infectious classes, in particular, replaces the mass action approach with one that recognizes explicitly the linked fates of an individual host and its associated parasite population. Such a perspective is essential for explaining the evolution of reduced virulence, one of the most ubiquitous features of host-parasite interactions (Levin, 1983)

The essential detail provided by the classical epidemiological formalism often must be complicated further by the incorporation of various aspects of demographic structure. This added complexity is one of a variety of ways, including also more complicated transmission dynamics and temporal forcing, that sustained periodic and chaotic behaviour can arise (Hethcote and Levin, 1989). Of particular interest are the interplay between seasonal demography and epidemiology (Dwyer et al., 1990), and the interactions among multiple cocirculating strains (Castillo-Chavez et al., 1988, 1989).

The list of examples could be extended indefinitely. The case has been adequately made, in numerous studies, that the inclusion of the various aspects of population structure is limited only by the imagination of the modeler; can lead to fascinating mathematical questions; and most importantly, is of substantial biological import. These are not the fine details of ecological interactions; they are the essence. There can be no theory of evolution without genetic structure; variation among types is the sine qua non of evolutionary change. And little is of interest in population biology that does not make some reference to the genetic structure, the age structure, the stage structure, the spatial structure, or other features of the internal heterogeneity of the population. But if the possibilities are endless, where do we stop?

The art of modeling is in knowing how much detail to include; therefore, a fundamental research issue involves an examination of the consequences of choosing differing levels of resolution. Too little detail obscures the essential elements of an interaction. Too much detail results in overly refined, cumbersome, and counterintuitive models that are impractical in terms of the estimation of parameters, and certain to propagate errors through multitudinous pathways (see, e.g., Ludwig and Walters, 1985). The essence of good modelling of populations is to capture the essential features of aggregates in terms of the properties of individuals, while avoiding slavish attention to the irrelevant intricacies of the behaviors of those individuals.

Indeed, even if one wanted to create the most detailed model imaginable, there is no unique way to define such a model. Eliminating differences among types by creating more and more classes implies that those classes will each have fewer and fewer members. In evolutionary biology, this paradox is made clear in Dawkins' distinctions among the various concepts of fitness (Dawkins, 1982). At one extreme, one associates a fitness with the genotypic class corresponding to all individuals carrying a certain pair of genes, or a certain phenotypic trait. With increasing resolution, one includes more loci, or multiple traits. At the opposite extreme is the approach that recognizes the uniqueness of the individual, whose personal fitness can be measured, but which is a class unto itself and a unique historical event. About such individuals stories can be told, but theories are impossible.

Most mathematical theories are not about individuals, or even about groups of individuals, but about abstract entities: populations. In populations, individuals come and go; the objective of a model is to make predictions about the dynamics of such populations, projecting the expected numbers of individuals, or the numbers that will be in different classes, or the probability density functions for those classes. If the variable in question involves the number of individuals per unit area, that number can be treated as a random variable. The smaller the unit area, the less accurately can that number be predicted. In the limit, for very small areas, there either will be a single individual present, or none. Thus, as for analogous problems in physics, increasing resolution leads to increasing stochasticity, and without limit until the quantum level, where the uncertainty principle rules out all possibility of localization.

Given that there is no "correct" level of detail, it is clear that attention must be directed to the costs and benefits of differing levels of detail, to the manifestation of pattern at different scales, and hence to the consequences of changing the scale of description. Techniques from dynamical systems theory (Ruelle and Takens, 1971, Roux et al., 1983) may be used to choose the dimensionality needed to describe particular dynamic phenomena adequately, and such techniques have been applied successfully to ecological and epidemiological data (Schaffer, 1981, Schaffer and Kot, 1985, Schaffer et al., 1986). This approach seeks to examine those data and determine the dimensionality of the attractor, but gives no hint concerning the mechanisms that led to such dynamics.

The problem of aggregation (Iwasa et al., 1987, 1989) complements this approach by focusing on the search for low-dimensional descriptions that retain the signal and suppress the noise seen in data sets or in the output from higher dimensional models. In the generic approach, one starts from models that assume specific mechanisms, and seeks a lower dimensional formulation, the "best approximate aggregation," according to some criterion based on the normed difference between a full description and an approximate one. In economics, the problem is a familiar one, and attention is directed to the consequences of suppressing detail within a sector; tightly linked components may reach a quasi-equilibrium on the faster time scale, which then interacts on the slower time scale with similar quasi-equilibria reached in other sectors (Simon and Ando, 1961). Similar issues apply naturally in ecology and epidemiology.

In applied contexts, the relevant question is "How much detail leads to the most effective management?" This issue has been explored with imagination and good success by Ludwig and Walters (1985) for fisheries models. A related question is "How much detail makes sense, in terms of the capability of the data to support parameter estimation?" Large and highly detailed models may be useful as tableaux, in which all relationships are laid bare. But if the data are not sufficient to allow the relevant parameters to be estimated, such a model cannot be useful for management. Furthermore, the sensitivity of the model's behavior to certain parameters may be so great that simpler models that do not require estimation of the problematic parameter may be more practical for management. In other words, by averaging over this parameter, one obtains a model with fewer parameters, whose conclusions are less sensitive to parametric variation.

More generally, one may ask "How much detail is needed to explain the observed pattern on particular scales of observation?" The misguided assumption that the best model is one that is true to every biological detail is the enemy of the search for an understanding of what the essential interactions are, and what elements are needed to explain observations. Random walk models of movement may be criticized on the grounds that organisms do not move randomly; but neither do molecules, which are certainly governed by the laws of mechanics. The successful application of diffusion theory at the level of *populations* of molecules is based on the principle that many aspects of the behavior of statistical aggregates can be described without an accounting of the detailed interactions of individual molecules. The same philosophy underlies the application of such molecules to organisms. It is well known that such descriptions can only be accurate on particular scales; for example, the infinite speed of propagation predicted by diffusion models cannot apply to molecules or organisms. Thus, attention must be directed to determining the applicable scales of a particular description. The use of interacting particle models to examine pattern formation in ecological communities is another case in point, in which one asks what minimum level of detail is needed to explain observed pattern. This appeal to parsimony is the guiding principle in intelligent modeling, whether the goal is understanding, prediction, or management.

Since patterns differ on different scales, explanations must also differ with scale. For example, examination of the distributional patterns of krill in the Southern Ocean shows a concordance with phytoplankton and temperature on broad scales, but a quite independent distribution on fine scales (2-20 km) (Weber et al., 1986, Levin et al., 1989). The most parsimonious explanation is that physical patterns of turbulence are sufficient to explain the distribution of krill on the broader scales, moving large patches of krill about. Within these superswarms, however, a different explanation is mandated. Krill are much more patchily distributed than their resource, or than the water movements can explain. Active swimming behaviors of krill, and especially their tendency to aggregate, necessitate a different kind of model on finer scales. We (Grünbaum, Okubo, Levin, and Powell, in progress) have been developing fine scale, individual-based (Lagrangian) models of movement, with the hope that these can be related to the patterns seen at the population level.

More generally, data (often remotely collected) on distributional patterns of physical and biotic factors permit analysis of correlations that can suggest the shape of mechanistic models. This approach has been insufficiently exploited, especially for terrestrial systems; but that is changing. We have been utilizing such an approach both for the serpentine grasslands of Jasper Ridge, California (Moloney et al., in press) and for terrestrial forests of the Northeastern United

States (Levin, Moloney, Buttel, Pacala, Philpot, in progress). Both of these systems are disturbance-mediated; that is, localized disturbances create gaps in the background mosaic, and the gaps in turn present colonization opportunities for a wide variety of species. These gaps, or patches, are the dominant feature of these systems, and the principal determinants of diversity. When viewed on small scales, these systems may seem highly variable, and of relatively low diversity. As the observational window is expanded, the systems become more heterogeneous, more diverse, and (statistically) more predictable. We have modelled these systems by constructing grids of 10,000 cells, each of which is the locus of a homogeneous local dynamic.

In particular, within each cell in the grassland, for example, one posits a model for population growth and interactions. Among cells, one allows dispersal of seeds, according to a rule that can be based on empirically derived dispersal curves. Superimposed upon the entire mosaic are disturbances, which remove existing vegetation simultaneously in localized sets of contiguous cells (patches or gaps). Dispersal and disturbance both introduce spatial correlations among cells, and interact to define the correlation length in the system. Both in the model output, and in observed data, one can observe self-similarity in the distributions of key variables, at least within a correlation length determined by disturbance and dispersal patterns (Levin, 1989, Moloney et al., in press). This self-similarity is reminiscent of analogous patterns in statistical physics in the study of critical phenomena, and has stimulated us to begin investigations (with Richard Durrett) of simplified interacting particle analogues of the more complicated ecological models.

In summary, the problem of relevant detail poses a difficult dilemma. Ignoring relevant detail obscures key phenomena; appreciation of this fact has led to the current interest in population structure, the organizing theme of this conference. On the other hand, too much detail can obfuscate, burying understanding and predictability beneath a morass of irrelevance. Finding the right balance may seem a laudable goal, and is sometimes achievable if the criteria can be clearly stated. But more generally, one must recognize that there is no correct scale, and that it is essential to examine how the behavior of the system changes across scales. Pattern is not an intrinsic property of systems, but is to some extent in the eye of the beholder. Pattern is filtered through observational biases, such as the scale of description. Recognition of the multitude of scales, and of the interrelationships among them is the central problem in theoretical ecology.

Acknowledgment

I am pleased to acknowledge the very helpful suggestions of a referee. Support for this research was provided by the following grants to the author: NSF grant BSR-8806202, DOE grant DE-FG02-90ER60933, McIntire-Stennis grant NYC-183550, and Hatch grant NYC-183430. In addition, partial support was provided by the U.S. Army Research Office through the Mathematical Sciences Institute of Cornell University. This is Ecosystems Research Center Publication

ERC-233. Partial support was provided by the Ecosystems Research Center of Cornell university under Cooperative Agreement CR-812685-03 from the U.S. Environmental Protection Agency.

References

1. Castillo-Chavez, C., Hethcote, H.W., Andreasen, V., Levin, S.A., Liu, W-m. (1989): Epidemiological models with age structure, proportionate mixing, and cross immunity. J. Math. Biol. **27**, 233-258
2. Castillo-Chavez, C., Hethcote, H.W., Andreasen, V., Levin, S.A., Liu, W-m. (1988): Cross-immunity in the dynamics of homogeneous and heterogeneous populations. pp. 303-326 in Mathematical Ecology, T.G. Hallam, L.J. Gross, S.A. Levin, eds. Proc. of the Autumn Course Research Seminars, Trieste, 1986. World Scientific Publishing Co., Singapore
3. Dawkins, R. (1982): The Extended Phenotype. W.H. Freeman, Oxford and San Francisco
4. Dwyer, G., Levin, S.A., Buttel, L. (1990): A simulation model of the population dynamics and evolution of myxomatosis. Ecol. Monog. **60**, 423-447.
5. Hethcote, H.W., Levin, S.A. (1989): Periodicity in epidemiological models. In Applied Mathematical Ecology, S.A. Levin, T.G. Hallam, L.J. Gross, eds. Lecture Notes in Biomathematics **18**, 193-211, Springer-Verlag, Berlin-Heidelberg-New York
6. Iwasa, Y., Andreasen., V., Levin, S.A. (1987): Aggregation in model ecosystems. I. Perfect aggregation. Ecol. Modelling **37**, 287-302
7. Iwasa, Y., Levin, S.A., Andreasen, V. (1989): Aggregation in model ecosystems. II. Approximate aggregation. IMA J. of Math. Appl. in Med. & Biol. **6**, 1-23
8. Levin, S.A. (1989): Challenges in the development of a theory of community and ecosystem structure and function. In Perspectives in Ecological Theory, J. Roughgarden, R.M. May, S.A. Levin, eds., 242-255, Princeton University Press, Princeton, New Jersey
9. Levin, S.A. (1983): Some approaches to the modelling of coevolutionary interactions. In Coevolution, M. Nitecki, ed., 21-65, University of Chicago Press, Chicago, Illinois
10. Levin, S.A., Morin, A., Powell, T. (1989): Patterns and processes in the distribution and dynamics of Antarctic krill. In Scientific Committee for the Conservation of Antarctic Marine Living Resources Selected Scientific Papers Part I, 281-299, SC-CAMLR-SSP/5, CCAMLR, Hobart, Tasmania, Australia
11. Ludwig, D., Walters, C.J. (1985): Are age-structured models appropriate for catch-effort data? Can. J. Fish. Aquat. Sci. **42**, 1066-1072
12. Moloney, K.A., Levin, S.A., Chiariello, N.R., Buttel, L. (1991): Pattern and scale in a serpentine grassland. Theor. Pop. Biol. (in press)
13. Roux, J-C, Simoyi, R.H., Swinney, H.L. (1983): Observation of a strange attractor. Physica **8D**, 257-266
14. Ruelle, D., Takens, F. (1971): On the nature of turbulence. Commun. in Math. Phys. **20**, 167-192
15. Schaffer, W.M. (1981): Ecological abstraction: The consequences of reduced dimensionality in ecological models. Ecol. Monogr. **51**, 383-401
16. Schaffer, W.M., Ellner, S., Kot, M. (1986): Effects of noise on some dynamical models in ecology. J. Math. Biol. **24**, 479-523

17. Schaffer, W.M., Kot, M. (1985): Nearly one dimensional dynamics in an epidemic. J. Theor. Biol. **112**, 403-427
18. Simon, H.A., Ando, A. (1961): Aggregation of variables in dynamic systems. Econometrica **29**, 111-138
19. Weber, L.H., El-Sayed, S.Z., Hampton, I. (1986): The variance spectra of phytoplankton, krill and water temperature in the Antarctic Ocean south of Africa. Deep-Sea Research **33(10)**, 1327-1343
20. Whittaker, R.H., Levin, S.A. eds. (1975): Niche: Theory and Application. Benchmark Papers in Ecology/3. Dowden, Hutchinson & Ross, Inc., Stroudsburg, Pennsylvania. 448 + xv

Lifespans in Population Models: Using Time Delays

Jacques Bélair

Département de mathématiques et de statistique
Centre de recherches mathématiques Université de Montréal, C.P. 6128-A, Montréal, Québec, Canada H3C 3J7

Dedicated to Kenneth Cooke on the occasion of his 65th birthyear

Abstract

Population models with a compartment structure have to incorporate the duration of stay in each class. Using a two-compartment model of the regulation of blood cell production, we investigate two elimination mechanisms: a "random", constant rate destruction process, and a lifespan of finite duration. The limiting cases of each death mechanism alone are considered in turn, and the dynamics are compared to the full model, from the point of view of the equilibrium solutions, and their (local) asymptotic stability. The crucial quantity not to be neglected is seen to be the maturation time in the precursors' compartment.

Introduction

The behavior of the solutions of population models is known to depend on whether or not delays are incorporated into the governing equations. Indeed, the folkloric tenant that "delays destabilize" is its more widely recognized manifestation. Being folkloric, it is also known to be false in many instances: delays may have no effect on the asymptotic stability of equilibria, and increases in the values of delays may stabilize an otherwise unstable equilibrium, and numerous stability switches may occur [6].The precise way in which retardations are incorporated into the models is known to significantly alter the effect they have on the dynamics: in the example studied in [8], maturation and gestation periods may have destabilizing or even stabilizing influences.

In a class of epidemiological models [14], however, it has been shown that *average* quantities associated with the delayed terms determine the dynamics: the quantitative distribution in the delayed terms have neither qualitative nor quantitative influence on the crucial dynamical quantities (thresholds in this case).

The confusion increases even more when multiple time delays are taken into account. We use in this paper a model of the regulation of blood cell production to assess the influence of a second time lag on the stability of equilibria. We consider a two-compartment system, with both finite duration of stay and "leaky" loss of cells in each class. The limiting cases of each death mechanism alone are considered in turn, and the dynamics are compared to the full model: the existence of equilibrium solutions and their local asymptotic stability are analysed. In particular, we focus on the possibility of destabilisation of all steady states in the ruling equations.

In a context where the derivation of the equation we analyse is performed for modeling purposes, it becomes particularly important to evaluate the influence of "simplifying" assumptions which are squarely at odds with reality: for example, the assumption of constant rate of removal of infected (through recovery) in a model for measles "...does not correspond to any biological hypotheses", as pointed out in 1971 [16].This remark, in a sense, is the motivation for the present paper.

1 Derivation of the ruling equation

The equation we consider is obtained in the investigation of the regulatory mechanism for the production of platelets in mammals. A complete derivation of the equation, along with a more detailed analysis, is presented elsewhere [3].

It is generally accepted that circulating platelets are produced by an anucleated fractionation of a class of precursors, the megakaryocytes. These, in turn, come from a commitment of the pluripotential stem cells, the common ancestors, in the bone marrow, of all blood cells. It is thus natural to model this system as a two-compartment chain.

The time of stay for the cells in each class has been estimated to lie in a range of about 6 to 10 days. There is some experimental evidence for the existence of a protein, thrombopoietin, influencing the level of the number of pluripotential stem cells becoming committed to the megakaryocytic compartment: there is thus a feedback control by the number of circulating platelets of the number of cells entering the megakaryocytic compartment.

We let $n(m,t)$ denote the density of megakaroycytes at maturation level m at time t , and $p(a,t)$ stand for the density of platelets at age a at time t . We suppose that cells exit each of the compartments using one of two routes: either they leave at a fixed age, T_M for the megakaryocytes and T_s for the platelets, or they are "victims" of a random destruction mechanism. This process destroys cells at a constant rate, denoted by δ for the megakaryocytes and γ for the platelets. Letting $P(t) = \int_0^{T_s} p(a,t)da$ denote the total number of platelets at time t, and the feedback function generated by thrombopoietin written β , boundary and initial conditions can be used to derive the governing equation

$$\frac{dP}{dt} = -\gamma P(t) + e^{-\delta T_M}[\beta(P(t-T_M)) - e^{-\gamma T_s}\beta(P(t-T_M-T_s))]. \qquad (1.1)$$

A specific functional form is used for β , taken to be consistent with experimental and clinical data [3] , namely

$$\beta(P) = \beta_0 P \frac{\theta^n}{\theta^n + P^n} \tag{1.2}$$

where θ, n and β_0 are parameters. In accordance with results on the "robustness" of bifurcation sequences of discrete dynamical systems [9], we would not expect drastic qualitative changes in the behavior of solutions of eq (1.1) if a function different but "qualitatively similar" to the one given by eq. (1.2) were used.

Instead of analysing this last equation, which would take us so far afield that we would be in the middle of [3], we now consider two limiting cases, associated with the destruction mechanisms of the cells in each compartment. We thus examine, in turn, the total diseappareance of cells through "random" loss, *i.e.* elimination at a constant rate, and then a fixed-duration lifespan, in the absence of "leakage" at a constant rate. For each equation, we determine the stationary solutions and their local asymptotic stability.

1a Loss at a constant rate

When platelet deaths are incurred at a constant rate ("random destruction"), the senescence time T_s is taken to be infinite. Eq. (1.1) thus becomes

$$\frac{dP}{dt} = -\gamma P(t) + e^{-\delta T_M} \beta(P(t - T_M)), \tag{1.3}$$

which is a version of the Mackey-Glass (or Glass-Mackey) equation . It is well known to be able to display irregular ("chaotic") behavior [10] in some parameter regimes.

In our study of the equilibrium solutions of eq. (1.3) , we establish for this and the following equations, the values of the parameters β_0 and γ yielding stability of one stationary solution. Since the coefficients of the linearised equations are highly nonlinear functions of these two parameters, and may even, as above, depend on the delays, this determination is not quite straightforward.

Eq. (1.3), with β defined by eq. (1.2) , has two possible steady states, $P_1 = 0$, which always exists, and

$$P_2 = \theta \sqrt[n]{\frac{\beta_0 e^{-\delta T_M}}{\gamma} - 1},$$

which is real-valued only when $\beta_0 > \gamma e^{\delta T_M}$. The linearization of eq.(1.3) about a steady state $\underline{P}$ gives the linear equation

$$\frac{dx}{dt} = -\gamma x(t) + e^{-\gamma T_M} \beta' x(t - T_M), \tag{1.4}$$

where

$$\beta' = \frac{d\beta(\underline{P})}{dP}.$$

Since eq.(1.4) contains a single constant time delay, the values of the coefficients yielding the stability of the stationary solution can be determined (almost) explicitly. According to [12], for a fixed delay h, the values of the parameters a and b for which the null solution of the equation

$$\frac{dx}{dt} = -ax(t) - bx(t-h)$$

is stable are given by

$$-a < b < \sqrt{a^2 + r^2}, \qquad a > -\frac{1}{h}, \tag{1.5}$$

where r is the unique root of the equation $r = -a\tan(rh)$ in the interval $(0, \pi/h)$. This stability region is shown, in the (a,b) plane, in [11].

Around $P_1 = 0$, we have $\beta' = \beta_0$ and thus the stability criterion of ineqs. (1.5) becomes

$$-\gamma < -\beta_0 e^{-\delta T_M} < \sqrt{\gamma^2 + r^2}, \qquad \gamma > -\frac{1}{T_M}.$$

Since the latter two inequations are always satisfied, the region of stability for the null steady state corresponds to $\beta_0 < \gamma e^{-\delta T_M}$, *i.e.* the values of the parameters for which it is the unique equilibrium.

At the stationary point P_2, ineqs.(1.5) can be written

$$-\gamma < -\beta' e^{-\delta T_M} < \sqrt{\gamma^2 + r^2}, \qquad \gamma > -\frac{1}{T_M}, \tag{1.6}$$

where now $\beta' = \gamma(\frac{n\gamma e^{\delta T_M}}{\beta_0} - (n-1))$ and $r = -\gamma\tan(rT_M)$. Notice first that the last inequation of (1.6) is always satisfied. In addition, β' is negative when $\beta_0 > \frac{n\gamma e^{\delta T_M}}{n-1}$, and thus the second inequality in (1.6) is valid when $\beta_0 \leq n\gamma e^{\delta T_M}/(n-1)$; also, the first inequality holds when either $\beta' < 0$, or

$$\beta' > 0 \qquad \text{and} \qquad \beta_0 > \frac{n\gamma}{1 + (n-1)e^{\delta T_M}}.$$

Since P_2 exists and is real when $\beta_0 > \gamma e^{\delta T_M}$, which is a weaker condition than the one determined by the last two inequations, we see that the first inequality in (1.6) is always satisfied.

The second inequation of (1.6) fails only when $\beta' < 0$ and $(\beta')^2 e^{-2\delta T_M} < \gamma^2 + r^2$, which can be written as

$$(n(n-2) - \frac{r^2}{\gamma^2})\beta_0{}^2 - 2n\gamma e^{\delta T_M}(n-1)\beta_0 + n^2\gamma^2 e^{2\delta T_M} < 0 \quad \text{, and} \tag{1.7a}$$

$$\beta_0 > \frac{n\gamma e^{\delta T_M}}{n-1}. \tag{1.7b}$$

Write $M(\beta_0)$ for the left hand side of ineq.(1.7a). For a fixed value of γ and hence of r, $M(\beta_0)$ can be considered as a quadratic polynomial in the variable β_0. Since

$$M(\frac{n\gamma e^{\delta T_M}}{n-1}) = -1 - \frac{r^2}{\gamma^2} < 0\,,$$

and

$$\frac{d^2M}{d\beta_0^2}(0) = 2(n(n-2) - \frac{r^2}{\gamma^2})\,,$$

there is, for each fixed value of γ, and precisely when $n(n-2) - \frac{r^2}{\gamma^2} > 0$, a unique root β of $M(\beta) = 0$. Ineqs. (1.7) will then be satisfied, at a fixed value of γ, on the interval $(0, B)$, where $B > n\gamma e^{\delta T_M}/(n-1)$ is defined by $M(B) = 0$. When $n \leq 2$, there is obviously no possibility for such a B to exist.

It is easy to see that as γ takes all possible values in the interval $(0, \infty)$, there is a unique point γ_1 for which $\frac{r^2}{\gamma^2} = n(n-2)$: indeed, r is a concave function of γ, increasing from $\frac{\pi}{2T_M}$ at $\gamma = 0$ to $\frac{\pi}{T_M}$ when $\gamma = \infty$. Thus, the linear function $\gamma\sqrt{n(n-2)}$ will have exactly one point of intersection with $r(\gamma)$, at a point that we label γ_1.

From ineqs.(1.7), the value of β_0 satisfying $M(\beta_0) = 0$ can be explicitly written

$$\beta_0 = \frac{e^{-\delta T_M} n\gamma^2((n-1)\gamma - \sqrt{\gamma^2 + r^2})}{n(n-2)\gamma^2 - r^2}.$$

In addition, let $\lambda = u + iw$ be a root of the characteristic equation associated with eq (1.4), namely $\lambda = -\gamma + \beta' e^{-(\delta+\lambda)T_M}$; then at any value of β_0 for which there is a pair of imaginary roots $\lambda = \pm iw$, the derivative of the real part of a root is positive, since

$$\frac{du}{d\beta_0}(u=0) = \frac{\gamma + (\gamma^2 + w^2)T_M n\gamma^2}{-\beta'\beta_0 e^{-2\delta T_M}(w^2T_M^2 + (1+\gamma T_M)^2)} > 0.$$

It is thus clear that the stability of the stationary solution is lost, for a fixed value of γ, when β_0 is increased.

We remark that both a steep enough feedback function, one in which $n > 2$, and a large enough value of γ, namely $\gamma > \gamma_1$, are necessary for destabilization of the non-trivial steady state P_2 to be possible.

1b Fixed duration lifespan

We now consider that no loss of cells at constant rates occur, and that all cells disappear when they reach age T_s in the second compartment. Eq. (1.1), in this limit of $\gamma = \delta = 0$, becomes

$$\frac{dP}{dt} = \beta(P(t-T_M)) - \beta(P(t-T_M-T_s)). \tag{1.8}$$

Since every constant function is a solution of this last equation, it seems that a degenerate equation has been obtained. However, as pointed out in [5], a more appropriate form is the integral of eq. (1.8),

$$P(t) = c + \int_{t-T_M-T_s}^{t-T_M} \beta(P(u))du$$

where the constant c depends on the initial condition specified on the interval $[-T_M - T_s, 0)$. To have a biological meaning as a population model, this constant must be 0 , since T_s represents a life duration. The correct equation, in our context, is thus

$$P(t) = \int_{t-T_M-T_s}^{t-T_M} \beta(P(u))du. \tag{1.9}$$

The equilibrium states $\underline{P}$ of this equation must therefore satisfy $\underline{P} = T_s \beta(\underline{P})$. There are thus only two stationary solutions, $P_1 = 0$, which exists for all values of the parameters, and $P_3 = \theta \sqrt[n]{\beta_0 T_s - 1}$, which is only real-valued when $\beta_0 > 1/T_s$.

The asymptotic stability of either steady state can only be determined with respect to a subset of initial conditions: the invariance principle

$$\frac{d}{dt}[P(t) - \int_{t-T_M-T_s}^{t-T_M} \beta(P(s))ds] = 0$$

holds, and any initial function $\phi(s)$ must therefore satisfy

$$\phi(0) = \int_{t-T_M-T_s}^{t-T_M} \beta(\phi(s))ds.$$

With respect to this subset, the local stability of an equilibrium $\underline{P}$ is determined by the linearised equation

$$\frac{dx}{dt} = \beta' x(t - T_M) - \beta' x(t - T_M - T_s) \tag{1.10}$$

where β' as the same meaning as before.

Eq. (1.10) contains two time delays, and no theory provides the equivalent of Hayes' theorem to determine the stability of its null solution. We thus have to consider the characteristic equation

$$z = \beta' e^{-zT_M}(1 - e^{-zT_s}) \tag{1.11}$$

obtained by substituting $x(t) = e^{zt}$ in eq. (1.10). This equation has been considered in [4] and [7].

Elementary considerations show that there is no positive real root of eq. (1.11) whenever $\beta' < 1/T_s$, and precisely one (positive root) when $\beta' > 1/T_s$, in addition to the ever existing real root $z = 0$. Since $\frac{d\beta}{dP}(0) = \beta_0$, and the nontrivial equilibrium P_3 only exists when $\beta_0 > 1/T_s$, it is easy to see that the null solution is unstable at least when the second equilibrium is present. In addition, since

$$\frac{d\beta}{dP}(P_3) = \frac{1}{T_s}(\frac{n}{\beta_0 T_s} + 1 - n),$$

the stability of the steady state P_3 cannot be lost by a real root becoming positive.

Substituting $z = u + iw$ in eq. (1.11) and separating real and imaginary parts, we obtain

$$u = \beta' e^{-uT_M}(\cos(wT_M) - e^{-uT_s}\cos(w[T_M + T_s])) \tag{1.12a}$$

$$w = \beta' e^{-uT_M}(-\sin(wT_M) + e^{-uT_s}\sin(w[T_M + T_s])). \tag{1.12b}$$

The argument used in [3] can be applied to show that no root of eqs. (1.12) can exist when $u > 0$, as long as $\beta' > 0$. As mentioned before, the stability of the null solution is thus lost by a real root becoming positive when the second steady state appears. The equilibrium P_2 is thus stable at least when $\beta_0 < n/((n-1)T_s)$. (This argument significantly improves, and rectifies a *slight* mistake of [7]).

The pure imaginary roots of eq. (1.11) are the solutions of

$$0 = \beta'(\cos(wT_M) - \cos(w[T_M + T_s])) \tag{1.13a}$$

$$w = \beta'(-\sin(wT_M) + \sin(w[T_M + T_s])), \tag{1.13b}$$

or, equivalently,

$$0 = \sin(w[T_M + \frac{T_s}{2}])\sin(\frac{wT_s}{2}) \tag{1.14a}$$

$$w = 2\beta'\cos(w[T_M + \frac{T_s}{2}])\sin(\frac{wT_s}{2}). \tag{1.14b}$$

Since $sin(\frac{wT_s}{2}) \neq 0$, and from eq. (1.14b), eq. (1.14a) implies that

$$w = \frac{j\pi}{T_M + \frac{T_s}{2}}, \tag{1.15}$$

where j is a non-negative integer assuming, without loss of generality, that w is positive. Substituting back this last equation into eq. (1.13b), we obtain

$$\beta' \sin(\frac{j\pi}{1 + \frac{T_s}{2T_M}}) = \frac{-j\pi}{T_M + \frac{T_s}{2}}.$$

In terms of the more meaningful parameters γ and β_0 , this last equation can be written

$$\beta_0 = \frac{n}{T_s(n - 1 - \frac{2j\pi}{1+\frac{2T_M}{T_s}}\sin(\frac{j\pi}{1+\frac{T_s}{2T_M}}))}. \tag{1.16}$$

The requirement that $\beta_0 > 0$ now corresponds to

$$n - 1 - \frac{2j\pi}{1 + \frac{2T_M}{T_s}}\sin(\frac{j\pi}{1 + \frac{T_s}{2T_M}}) > 0. \tag{1.17}$$

Consider eq. (1.15) to define w as the point of intersection of the curves $y = \sin(wT_M)$ and $y = -w/\beta'$. Since for $\beta_0 > n/((n-1)T_s)$, β' is an increasing negative-valued function of β_0 , the minimum β_0 defined by eq. (1.15) will be the one corresponding to $j = 1$, namely $w = \pi/(T_m + \frac{T_s}{2})$, and will lie in the interval $(0, \pi)$.

Implicitly differentiating eq. (1.11) and evaluating at any pure imaginary root, we obtain

$$\frac{du}{d\beta_0}(u=0) = -w\frac{d\beta'}{d\beta_0}[T_M(\sin(wT_M) - \sin(w[T_m+T_s])) - T_s\sin(w[T_m+T_s])]$$
$$= \frac{2\beta'}{w}[(-1)^j\cos(wT_M)\cos(\frac{wT_s}{2}) - 1] \quad > 0.$$

There is thus a loss of stability as values of β_0 are increasing, for a fixed value of γ . In addition, due to the fact that, as a function of w, the quantity

$$1 + \frac{2\pi w}{(1+w)sin(\frac{\pi}{1+w})}$$

is monotonically increasing, from the value 3 at $w = 0$ to the value ∞ at $w = \infty$, there is no lower bound in n, defined by eq. (1.17), that is uniformly valid for all values of the ratio of the delays, T_s/T_M : stated otherwise, no value of the "steepness" parameter n in the feedback function β will yield a destabilization of the steady state for all possible values of the ratio of the delays. This is consistent with the limiting case considered heuristically in [15]: when T_s becomes very small, a destabilization of the steady state becomes "easier" at low values of n , and more roots can cross the imaginary axis from the left to the right hand side of the λ plane, leading to "chaotic" behavior of the solutions.

2 Neglecting maturation time

In this section, we consider the limiting case of eq. (1.1) when the maturation time T_M is neglected. Although this is not realistic for platelets, it is (arguably) relevant for erythrocytes, since the senescence time is more than an order of magnitude superior to the maturation time of precursors in this case. We then discuss two further limiting cases, analogous to the ones analysed in the last section. In the first one, purely random destruction of cells is supposed in either compartment, so we let $\gamma = \delta = 0$. In the second case, senescence time is taken to be so large as to make the disappearance of cells happen at the constant rate δ : in this case, we put $T_s = \infty$. There is, of course, little point in considering both of these effects together, since this would lead to an exponential growth of the population [there being no disappearance mechanism at all].

2a General case

By letting $T_M = 0$ in eq. (1.1), we obtain

$$\frac{dP}{dt} = -\gamma P(t) + \beta(P(t)) - e^{-\gamma T_s}\beta(P(t-T_s)). \tag{2.1}$$

This last equation always has the stationary solution $P_1 = 0$, and there is a second equilibrium point

$$P_4 = \theta\sqrt[n]{\frac{\beta_0(1-e^{-\gamma T_s})}{\gamma} - 1},$$

which only exists when $\beta_0 > \gamma/(1-e^{-\gamma T_s})$. The linearization of eq.(2.1) at a steady state $\underline{P}$ leads to

$$\frac{dx}{dt} = (-\gamma + \beta')x(t) - e^{-\gamma T_s}\beta' x(t - T_s), \tag{2.2}$$

where $x(t) = P(t) - \underline{P}$ and $\beta' = \frac{d\beta}{dP}(\underline{P})$. Since eq. (2.2) contains a single time delay, T_s , we can use Hayes' criterion (eq. (1.5)) to determine the stability of either steady state.

Since $\frac{d\beta}{dP}(0) = \beta_0$, the linearization about P_1 yields the region of stability

$$\beta_0 - \gamma < \beta_0 e^{-\gamma T_s} < \sqrt{(\beta_0-\gamma)^2 + r^2}, \qquad \gamma - \beta_0 > -\frac{1}{T_s}, \tag{2.3}$$

where r is the root of $r = (\beta_0 - \gamma)\tan(rT_s)$ in the interval $(0, \pi/T_s)$. The last inequation in ineqs. (2.3) imposes $\beta_0 < \gamma + 1/T_s$, whereas the first one is equivalent to $\beta_0 < \gamma/(1-e^{\gamma T_s})$ and the second one is always satisfied. Since it is equivalent to

$$0 < \beta_0^2(1-e^{-2\gamma T_s}) - 2\gamma\beta_0 + \gamma^2 + r^2,$$

which is (almost) readily seen to hold, the null solution is stable when $\beta_0 < \gamma/(1-e^{\gamma T_s})$, and becomes unstable when the second steady state P_4 appears.

To determine the stability of the latter stationary solution, we use Hayes' criterion on eq. (2.2), with

$$\beta' = \frac{d\beta}{dP}(P_4) = \frac{\gamma}{(1-e^{-\gamma T_s})^2}\left(\frac{n\gamma}{\beta_0} - (1-e^{-\gamma T_s})(n-1)\right).$$

The first and third inequalities of ineqs. (2.3) then become, respectively,

$$\beta_0 > \frac{\gamma}{1-e^{-\gamma T_s}} \quad \text{and} \quad \beta_0 > \frac{n\gamma^2 T_s}{(1-e^{-\gamma T_s})(1-e^{-\gamma T_s} + \gamma T_s(n - e^{-\gamma T_s}))}.$$

It is easy to see that the first inequation implies the latter, and that both are therefore satisfied whenever P_4 exists.

The second inequation in ineqs.(1.5) is trivially satisfied whenever $\frac{d\beta}{dP}(P_4) < 0$, which occurs when $\beta_0 > n\gamma/((n-1)(1-e^{-\gamma T_s}))$. If the latter does not hold, the stability of P_4 is characterized by

$$\begin{aligned} 0 < {} & \gamma^2(1-e^{-2\gamma T_s})\left[\frac{n\gamma}{\beta_0} - (n-1)(1-e^{-\gamma T_s})\right]^2 \\ & - 2\gamma^2(1-e^{-\gamma T_s})^2\left[\frac{n\gamma}{\beta_0} - (n-1)(1-e^{-\gamma T_s})\right] \\ & + (\gamma^2 + r^2)(1 - e{-}\gamma T_s)^4 \end{aligned} \tag{2.4}$$

where r is the root of $r = (-\gamma + \frac{d\beta}{dP}(P_4))\tan(rT_s)$ in the interval $(0, \pi/T_s)$. Considering the right hand side of ineq. (2.4) to define a quadratic polynomial in the unknown $[\frac{n\gamma}{\beta_0} - (n-1)(1-e^{-\gamma T_s})]$, it is readily seen that it has no positive root.

Eq. (2.1) , with β defined by eq. (1.2) , therefore always has a unique locally asymptotically stable equilibrium. It should be obvious at this point that the consideration of a nonnegligible maturation time in eq. (1.1) introduces for its solutions the possibility of much more complex behavior than in eq. (2.1). This is further illustrated in the two special cases which we now consider.

2b Fixed duration lifespan

In this first case, we suppose that there is no component of cell death occuring at a constant rate, so that all cells disappear when they reach age T_s. The dynamics are then regulated by

$$\frac{dP}{dt} = \beta(P(t) - \beta(P(t - T_s)). \tag{2.5}$$

As in section 1b above, the integral form

$$P(t) = \int_{t-T_s}^{t} \beta(P(u)\, du \tag{2.6}$$

is more appropriate for the investigation of the dynamics.

The stationary solutions P of eq. (2.6) satisfy $P = T_s\beta(P)$, and they are thus $P_1 = 0$, which always exits, and $P_3 = \theta\sqrt[n]{\beta_0 T_s - 1}$, which is real only when $\beta_0 > 1/T_s$. As in section 1b also, the asymptotic stability of either steady state can only be obtained with respect to a subset of functions taken as initial conditions: the linearized equation is the special case of eq. (1.10) obtained by letting $T_M = 0$. From the analysis of section 2a , it still holds that stability is switched at $\beta_0 = 1/T_s$ from the null solution to the nontrivial steady state. In the present situation, however, we can deduce global stability of an equilibrium from its local stability. From a result of [7], all solutions of eq. (2.6) asymptotically approach a constant, which must therefore be an asymptotically stable equilibrium solution. The stable equilibrium is thus globally stable, and all solutions approach zero when $\beta_0 < 1/T_s$, and the nontrivial steady state otherwise.

2c Disappearance at a constant rate

Assuming that all cells die at a constant rate, we suppose that there is no upper bound on the lifespan of cells, and therefore let T_s become infinite in eq. (2.1). Using the fact that no solution can grow faster than an exponential, we obtain as before

$$\frac{dP}{dt} = -\gamma P(t) + \beta(P(t)) \tag{2.7}$$

as the limiting governing equation (which can also be derived by putting $T_M = 0$ in eq. (1.1)).

Eq. (2.7) has two possible equilibria, $P_1 = 0$ and $P_5 = \theta\sqrt[n]{\frac{\beta_0}{\gamma} - 1}$, the latter existing when $\beta_0 > \gamma$. Since eq. (2.7) is an ordinary differential equation of dimension one, local stability implies global stability in this case also. We therefore have a behavior similar to the one detected in the previous section: the null solution is the unique steady state and is globally stable when $\beta_0 < \gamma$, and P_5 only exists when $\beta_0 > \gamma$, when it is globally stable.

Notice that in the last two cases, the nontrivial steady state can be written as $\theta \sqrt[n]{\beta_0 E - 1}$ where E is the life expectancy of cells in either of the limiting situations. This suggests, for the full equation (2.1), that the life expectancy of cells be associated with the "combined" average life duration $(1 - e^{-\gamma T_s})/\gamma$.

3 Conclusion

We have considered four limiting cases for the governing equation of a cell population. Two "extreme" mechanisms for the destruction or death of cells were studied in turn. The full basic model involves two compartments and the effect of a fixed, non-zero staying time in the initial compartment was seen to have a particularly destabilizing effect on the stationary solutions. Whereas at least one equilibrium point is always stable in his absence, the incorporation of this delay in the modeling equation is sufficient to induce the existence of a range of values, for other parameters, for which all steady states are unstable.

Depending on the precise way in which maturation times are incorporated into the ruling equations for the regulation of some animal populations, these delayed terms may have little effect, or even stabilizing influences on the stability of stationary solutions. In some epidemiological models, it has been shown that only *average* values related to memory affect the dynamical behavior of solutions.

There is some evidence, however, that the presence of non-negligible delays in dynamical equations will oftentimes yield more involved stability diagrams than if time lags were ignored. When considered in a broad perspective, retardations of *some* sort are considered to make oscillations arise in a variety of ways, "all of which involve some sort of delay" [13].

Like the cases presented here, the introduction of multiple delays can have devastating effects on the simplicity of the stability analysis. In a recent investigation of a general model for commodity prices [2], it has been found that in an equation with two time delays, one discrete and the other one distributed with an exponential kernel, the stability region, in a space of a normalized coefficient and the discrete time lag, may become multiply connected, and there may be an arbitrarily large number of stability switches.

Even more obscure at the moment is the effect of variable time lags. There are good arguments for the introduction of delays which depend on the value of the unknown function ["state-dependent-delays"] in models for commodity prices, fisheries and diseases [1]. In at least one such equation, the variability of the delay has been shown to induce destabilization of all steady states.

The analysis we have offered in this paper is further justification for a thorough investigation, in a genuine multiparameter setting, of functional differential equations containing more than one delay.

Acknowledgment

This work has been supported financially by NSERC(Canada) [OGP-8806 and Attaché de recherches] and FCAR(Québec)[EQ-2354] . We have benefitted from conversations with Michael C. Mackey and Joseph M. Mahaffy.

References

1. Bélair, J. (1991): Population models with state-dependent delays. In Proceedings Second International Conference on Mathematical Population Dynamics, Marcel Dekker, New York, 165-176
2. Bélair, J., Mackey, M. (1989): Consumer memory and price fluctuations in commodity markets: an integrodifferential model. J. Dynamics Diff. Eq. **1**, 299-325
3. Bélair, J., Mackey, M. (1990): Oscillatory pathologies of platelet control. J. Math. Biol., submitted
4. Braddock, R., van den Driessche, P. (1983): On a two lag differential delay equation. J. Austral. Math. Soc. **24**, 292-317
5. Busenberg, S., Cooke, K. (1980): The effects of integral conditions in certain equations modelling epidemics and population growth. J. Math. Biol. **10**, 13-32
6. Cooke, K., Grossman, Z. (1982): Discrete delay, distributed delay and stability switches. J. Math. Anal. Appl. **86**, 592-627
7. Cooke, K., Yorke, J. (1973): Some equations modeling growth processes and epidemics. Math. Bios. **16**, 75-101
8. Cushing, J. (1980): Model stability and instability in age structured populations. J. Theor. Biol. **86**, 709-730
9. Devaney, R. (1986): An Introduction to Chaotic Dynamical Systems. Benjamin/ Cummins, Menlo Park
10. Glass, L., Mackey, M. (1988): From Clocks to Chaos: the Rhythms of Life. Princeton University Press, Princeton
11. Hale, J. (1977): Functional Differential Equations. Springer-Verlag, Berlin-Heidelberg-New York
12. Hayes, N. (1950): Roots of the transcendental equation associated with a certain difference-differential equation. J. London Math. Soc. **25**, 226-232
13. Hethcote, H., Levin, S. (1989): Periodicity in epidemiological models. In Applied Mathematical Ecology, Lecture Notes in Biomathematics **18**, Springer-Verlag, Berlin-Heidelberg-New York, 193-211
14. Hethcote, H., Tudor, D. (1980): Integral equations models for endemic infectious diseases. J. Math Biol. **9**, 37-47
15. Kaplan, J., Marotto, F. (1977): Chaotic behavior in dynamical systems. In Nonlinear Sciences and Applications to Life Sciences, Academic Press, New York, 199-210
16. Yorke, J. (1971): Selected topics in differential delay equations. In Japan-U.S. Seminar on Ordinary and Functional Differential Equations, Lecture Notes in Mathematics **243**, Springer-Verlag, Berlin-Heidelberg-New York, 16-28

Convergence to Equilibria in General Models of Unilingual-Bilingual Interactions

H. I. Freedman[1] *and I. Baggs*[2]

[1] Department of Mathematics, Applied Mathematics Institute, University of Alberta, Edmonton, Canada T6G 2G1
[2] Department of Mathematics, University of Alberta, Edmonton, Canada T6G 2G1

Abstract

We derive general models for the interaction of unilingual and bilingual components of a population, and classify those models for which the dynamics is trivial by obtaining criteria for the nonexistence of periodic solutions.

1 Introduction

In countries throughout the world the decline and possible extinction of a minority language can be an issue which generates considerable concern, Whether the threatened language is French in Canada, Spanish in the United States, Welsh in Wales or Swahili in Kenya, the people who identify with that language will also feel that their identity is threatened. The survival and promotion of a particular language can become a serious political issue as has been seen in Canada, parts of the United States and many other countries (see [3], [4], [8] and [9]).

There are many factors which can eventually determine whether a particular language fluorishes and grows or declines and eventually becomes extinct (see Wardrough [9] Ch. 1). However, if a language is to expand, it must gain speakers faster than speakers are lost. In this paper we model the dynamics of interaction between two language groups in a simple setting. Namely, we model the case where a bilingual population interacts with a unilingual population in a closed environment. In Canada and the United States for example, there are numerous communities where there is one population which speaks only English and a second population which is bilingual in English and French or English and German, or English and Spanish or English and Italian or English and some other language. It is well known that in an English speaking country a considerable amount of pressure exists for children to be fluent in English. Often English is seen as the language of opportunity. As a result the minority language may be in a struggle for survival.

In [2], we have attempted to model the interaction between unilingual and bilingual components of a population by utilizing population models which include logistic growth and very specific assumptions on the dynamics which determine the proportion of populations which learn the second language. The assumption of logistic growth poses no problems, since such growth dynamics for a population is commonly used. In this paper we can assume a more general growth law which includes the logistic as a special case, and this does not affect our analysis.

However, the models considered in [2] may be criticized on other grounds. One of these is the fact that the dynamics leading to learning the second language are not known, and hence the specific function representing these dynamics may be incorrect. Hence we propose a more general function for our model.

A second area of criticism with the models in [2] may be that we were not able to eliminate the possibility of nontrivial periodic solutions in all cases. It is important for planning purposes that the eventual population ratios be predictable. This means, mathematically that one would like to have a class of models where solutions approach a stable steady state. It will be one of the main purposes of this paper to describe such a class of models.

In the next section we describe our models and the equilibria. In section 3, we derive criteria for the desired trivial dynamics.

2 Models and equilibria

Let $x_1(t)$ and $x_2(t)$ denote the concentrations of the unilingual and bilingual components of the population, respectively, at time t. Then we propose as a set of models of their growth and interaction the system of autonomous ordinary differential equations

$$\begin{aligned}\dot{x}_1(t) &= B_1(x_1) - D_1(x_1) - x_1x_2f(x_1,x_2) + P_1B_2(x_2)\\ \dot{x}_2(t) &= P_2B_2(x_2) - D_2(x_2) + x_1x_2f(x_1,x_2), \\ x_1(0) &\geq 0,\ x_2(0) \geq 0,\ \cdot = \frac{d}{dt}.\end{aligned} \tag{2.1}$$

The assumptions on, and interpretations of the various components of these models are given below. However, a priori, we assume that all functions are sufficiently smooth so that solutions of (2.1) exist, are unique and are continuable for all $t \geq 0$.

$B_i(x_i)$ and $D_i(x_i)$ are the birth and death rates, respectively of x_i. We assume (see [5], [6]) that

$$(H1): B_i(0) = D_i(0) = 0,\ B_i(x_i) > 0 \text{ and } D_i(x_i) > 0 \text{ for } x_i > 0,$$

$$B_i'(0) > D_i'(0),\ \exists K_i > 0 \ni B_i(K_i) = D_i(K_i),\ B_i'(K_i) < D_i'(K_i).$$

Allowing for different K_i (carrying capacities) for the different language groups recognizes the possibility of different cultural backgrounds.

Note that it is well known that under assumptions (H1), solutions of

$$\dot{x} = B(x) - D(x), \; x(0) > 0$$

approach K as $t \to \infty$, even if $B(x(t))$ is replaced by $B(x(t-\tau))$ (see [5]).

$f(x_1, x_2)$ is a function, related to the rate at which the unilingual population learns the second language. It is reasonable to expect that $f(x_1, x_2)$ is a decreasing function of its arguments, since as populations become large, the need to become bilingual in order to function in society diminishes. However, for mathematical purposes, such an assumption is not necessary. All we require at this time is

$$(H2) : f(x_1, x_2) > 0 \text{ for } x_1, x_2 > 0.$$

Finally, P_1, $0 < P_1 < 1$, is the average probability that children born to bilingual parents will enter the speaking population as unilingual. $P_2 = 1 - P_1$.

In [2], $B_i(x_i) = B_i x_i$, $D_i(x_i) = D_i x_i + L_i x_i^2$ (hence $K_i = (B_i - D_i)/L_i$), $f(x_1, x_2) = \alpha(1 + x_1)^{-1}$. The model in [2] is, therefore, included among the set of models (2.1).

System (2.1) always has an equilibrium at the origin, $E_0(0,0)$, and on the x_1-axis, $E_1(K_1, 0)$. It may or may not have one or more interior (positive) equilibria of the form $\hat{E}(\hat{x}_1, \hat{x}_2)$, $\hat{x}_1 > 0, \hat{x}_2 > 0$. These are the only possible equilibria.

It is of interest to know when $\hat{E}$ exists. $\hat{E}$ will exist if the system of algebraic equations

$$B_1(x_1) - D_1(x_1) - x_1 x_2 f(x_1, x_2) + P_1 B_2(x_2) = 0 \tag{2.2}$$

$$P_2 B_2(x_2) - D_2(x_2) + x_1 x_2 f(x_1, x_2) = 0.$$

has a positive solution. Instead of (2.2), it is more convenient to consider the system consisting of the second equation in (2.2) and the equation obtained by adding the equations of (2.2), giving

$$B_1(x_1) - D_1(x_1) + B_2(x_2) - D_2(x_2) = 0 \tag{2.3}$$

$$P_2 B_2(x_2) - D_2(x_2) + x_1 x_2 f(x_1, x_2) = 0. \tag{2.4}$$

Equations (2.3) and (2.4) may be graphed in the $x_1 - x_2$ plane. Let Γ and Γ_2, respectively, be their graphs. Then positive intersections of Γ and Γ_2 represent positive equilibria.

From (H1), Γ is a closed curve (an ellipse in [2]) which passes through $(0,0)$, $(K_1, 0)$, $(0, K_2)$ and (K_1, K_2).

Now consider Γ_2 when $x_2 = 0$. Let $\hat{x}_1$, if it exists, be such that $(\hat{x}_1, 0)$ lies on Γ_2.

Then

$$\hat{x}_1 f(\hat{x}_1, 0) = D_2'(0) - P_2 B_2'(0) \tag{2.5}$$

by l'Hospital's rule. Now let $\hat{x}_2$ be such that $(0, \hat{x}_2)$ lies on Γ_2. Then $\hat{x}_2$ satisfies $P_2 B_2(\hat{x}_2) = D_2(\hat{x}_2)$ and since $P_2 < 1$, $\hat{x}_2 < K_2$. In fact, it may be that $\hat{x}_2 < 0$.

With the above, we can now obtain criteria which guarantee the existence of at least one $\hat{E}$.

Theorem 2.1 *If either of*

(i) $\hat{x}_2 \geq 0$ *and equation (2.5) has no solution* $\hat{x}_1$ *such that* $0 < \hat{x}_1 < K_1$*;*
(ii) equation (2.5) has a unique solution $\hat{x}_1$ *such that* $0 < \hat{x}_1 < K_1$*,*
then $\hat{E}$ *exists.*

Proof. If (i) holds, then clearly Γ and Γ_2 must intersect. If (ii) holds then Γ_2 "enters" the positive quadrant at $x_1 = \hat{x}_1$ and cannot "exit" until after K_1, and so intersects Γ. □

Note that in [2], it was sufficient for either $\hat{x}_2 \geq 0$ or $\hat{x}_1 < K_1$ to hold to guarantee that $\hat{E}$ exists, since there Γ_2 was an increasing curve.

Finally we mention some elementary results on the stability of the two boundary equilibria.

E_0 is unstable. E_1 has a stable manifold along the x_1 axis. It will be stable or unstable locally in the x_2 direction as $P_2B_2'(0) - D_2'(0) + K_1 f(K_1, 0)$ is negative or positive, respectively.

3 Criteria for trivial dynamics

In this section we describe a class of functions $f(x_1, x_2)$ for which no nontrivial periodic solutions exists. We first assume that additional hypotheses on the growth laws hold. These are completely consistent with logistic-type growth.
(H3):

$$\frac{d}{dx_1}\left[\frac{B_1(x_1) - D_1(x_1)}{x_1}\right] = -\beta_1(x_1) < 0$$

$$\frac{d}{dx_2}\left[\frac{P_2B_2(x_2) - D_2(x_2)}{x_2}\right] = -\beta_2(x_2) < 0.$$

In establishing our criteria, we utilize a theorem called Dulac's theorem (see [1], p. 205) which is stated below.

Lemma 3.1 *(Dulac's Theorem). Consider the system*

$$\dot{x}_1 = F_1(x_1, x_2)$$
$$\dot{x}_2 = F_2(x_1, x_2), \tag{3.1}$$

where F_1 *and* F_1 *are* C^1 *functions in a simply connected region* $\mathcal{R}$*. Let* $A(x_1, x_2) \neq 0$ *be* C^1 *in* $\mathcal{R}$ *as well. Then if*

$$\Delta(x, y) = \frac{\partial}{\partial x_1}[A(x_1, x_2)F_1(x_1, x_2)] + \frac{\partial}{\partial x_2}[A(x_1, x_2)F_1(x_1, x_2)]$$

does not change sign (or is identically zero) in $\mathcal{R}$*, there are no nontrivial periodic solutions lying entirely within* $\mathcal{R}$*.*

Theorem 3.2 *Let* $f(x_1, x_2) \in C^1$*. Let hypotheses (H2) and (H3) hold. Then if* $f(x_1, x_2)$ *has the form*

$$f(x_1, x_2) = G(x_1 + x_2) + H_1(x_1) + H_2(x_2) + \delta_1 x_1 + \delta_2 x_2 \tag{3.2}$$

where

$$H_1'(x_1) \geq 0, \ H_2'(x_2) \leq 0, \ \delta_1 \geq \delta_2 \tag{3.3}$$

for all $x_1 \geq 0$, $x_2 \geq 0$, then system (2.1) has no nontrivial periodic solutions in $\{(x_1, x_2) \,|0 < x_1, \ 0 < x_2\}$.

Proof. Since the x_1 axis is invariant, and $\dot{x}_1|_{x1=0} = B_1(0) - D_1(0) + P_1 B_2(x_2) > 0$ for $x_2 \geq 0$, any periodic solution must lie entirely in the interior of the positive quadrant. Hence if we choose $\mathcal{R} = \{(x_1, x_2)|x_1 \geq \varepsilon, \ x_2 \geq \varepsilon\}$ for arbitrary $\varepsilon > 0$, and show that there are no periodic solutions in $\mathcal{R}$, we will have established the theorem.

Choose $A(x_1, x_2) = x_1^{-1} x_2^{-1}$. Then $A(x_1, x_2) \in C^1$ over $\mathcal{R}$. We now compute $\Delta(x_1, x_2)$ as given in Dulac's theorem for system (2.1) and get

$$\Delta(x_1, x_2) = -\frac{\beta_1(x_1)}{x_2} - \frac{\beta_2(x_2)}{x_1} - \frac{P_1 B_2(x_2)}{x_1^2 x_2} + f_{x_2}(x_1, x_2) - f_{x_1}(x_1, x_2). \tag{3.4}$$

From (3.2), $f_{x_2}(x_1, x_2) - f_{x_1}(x_1, x_2) = H_2'(x_2) - H_1'(x_1) + \delta_2 - \delta_1$, which by (3.3) is nonpositive. Hence by (H3), $\Delta(x_1, x_2) < 0$ and the result follows from Dulac's theorem. □

Corollary 3.3 *If the hypotheses of Theorem 3.2 hold and $\hat{E}$ does not exist, then* $\lim_{t\to\infty}(x_1(t), x_2(t)) = (K_1, 0)$.

Proof. Since $\dot{x}_1(t) + \dot{x}_2(t) = B_1(x_1) - D_1(x_1) + B_2(x_2) - D_2(x_2)$, the total population corresponding to solutions "outside" of Γ is decreasing. Hence the total population is bounded. Then all solutions must approach an equilibrium and the only one available is E_1. □

Remark 1 Note that the form of $f(x_1, x_2)$ given by (3.2) was arrived at in attempts to solve the equation $f_{x_2}(x_1, x_2) - f_{x_1}(x_1, x_2) \leq 0$ (see [7]).

Remark 2 In [2], we used $f(x_1, x_2) = \alpha(1 + x_1)^{-1}$. This function does not satisfy (3.2). A reasonable function which would satisfy (3.2) is $f(x_1, x_2) = \alpha(1 + x_1 + x_2)^{-k}$, $k \geq 1$.

Acknowledgment

Research partially supported by the Natural Sciences and Engineering Research Council of Canada, Grant No. A4823.

References

1. Andronov A. A., Leontovich E. A., Gordon, I.I., Maier, A. G. (1973): Qualitative Theory of Second-order Dynamic Systems. John Wiley and Sons, New York
2. Baggs, I., Freedman, H. I. (1990): A mathematical model for the dynamics of interactions between a uniligual and bilingual population: persistence versus extinction. J. Math. Sociology, **16**, 51-75
3. Bourhis, R. Y. (1984): Conflict and Language Planning in Quebec. Clevedon, Multilingual Matters
4. Fishman, J. A. (ed.) (1978): Advances in the Study of Societal Multilingualism. Mouton Press, The Hague
5. Freedman, H. I., Gopalsamy, K. (1986): Global stability in time-delayed single species dynamics. Bull. Math. Biol. **48**, 485-492
6. Freedman, H. I., Waltman, P. (1978): Predator influence on the growth of a population with three genotypes. J. Math. Biol. **6**, 367-374
7. Rektorys, K. (1969): Survey of Applicable Mathematics. M.I.T. Press, Cambridge
8. Wagner, S. T. (1981): The historical background of bilingualism and biculturalism in the United States. In The New Bilingualism: An American Dilemma (M. Ridge, ed.), USC Press, Los Angeles, 29-52
9. Wardhough, R. (1987): Languages in Competition: Dominance, Diversity and Decline. Basil Blackwell, New York

The Sherman-Rinzel-Keizer Model for Bursting Electrical Activity in the Pancreatic β-Cell

Mark Pernarowski[1], *Robert M. Miura*[2] *and J. Kevorkian*[1]

[1] Department of Applied Mathematics, University of Washington, Seattle, Washington 98195, USA

[2] Departments of Mathematics and Pharmacology & Therapeutics and Institute of Applied Mathematics, University of British Columbia, Vancouver, British Columbia, Canada V6T 1Z2

Abstract

Pancreatic β-cells exhibit periodic bursting electrical activity (BEA) consisting of active and silent phases. The Sherman-Rinzel-Keizer (SRK) model of this phenomenon consists of three coupled first-order nonlinear differential equations which describe the dynamics of the membrane potential, the activation parameter for the voltage-gated potassium channel, and the intracellular calcium concentration. These equations are nondimensionalized and transformed into a Liénard differential equation coupled to a single first-order differential equation for the slowly changing nondimensional calcium concentration. Leading-order perturbation problems are derived for the silent and active phases of the BEA on slow and fast time scales. Numerical solutions of these leading-order problems are compared with those for the exact equation in their respective regions. The leading-order solution in the active phase has a limit cycle behavior with a slowly varying frequency. It is observed that the "damping term" in the Liénard equation is small numerically.

Keywords: Bursting electrical activity — Nonlinear oscillators — Limit cycles — Perturbation problems — β-cells — Sherman-Rinzel-Keizer model

1 Introduction

Bursting electrical activity in excitable cells is a dynamical phenomenon in which the membrane potential undergoes a succession of alternating active and silent phases. The active phase is characterized by a sequence of rapid oscillations and the silent phase is characterized by a slowly changing membrane potential. The overall phenomenon consisting of active and silent phases appears periodic and, in the case of the β-cell, is associated with a putative slowly oscillating intracellular calcium concentration.

As a biological phenomenon, bursting in the β-cell is important because it is related to insulin release. In particular, the ratio of the duration of the active phase to the overall period, i.e., the plateau fraction, is proportional to the rate at which the β-cell releases insulin in response to given glucose concentrations (Ozawa and Sand 1986, Henquin and Meissner 1984, Meissner and Schmelz 1974). These β-cells are located within the pancreas and occur in clusters called the Islets of Langerhans. Early electrophysiological experiments (Dean and Matthews 1970) were performed *in vitro* on cells within the islets of the mouse. In these experiments, bursting electrical activity (BEA) was observed when isolated mouse islets were bathed in a D-glucose solution. Higher concentrations of glucose result in longer active phases, and for sufficiently high concentrations, only the active phase is present, i.e., there is a complete absence of the silent phase.

Through the use of ionic channel blocking agents and externally applied currents, a variety of ionic channels in the membrane have been identified (Dean and Matthews 1970a,b, Meissner and Schmelz 1974, Atwater et al. 1978a,b,c, Meissner and Preissler 1980). Based on an accumulation of such data, Atwater et al. (1980) devised a biophysical model of BEA in β-cells which led Chay and Keizer (1983) to develop a "minimal" mathematical model. Their model is "minimal" in the sense that it includes the fewest number of ionic transport mechanisms necessary to explain the phenomenon. Since voltage clamp data for β-cells were not available at that time, the channel activation functions in the Chay-Keizer (CK) model were specified by modifying those in the Hodgkin-Huxley (1952) model for electrical activity in the squid giant axon. Other models for BEA in β-cells have been proposed (cf. Keizer, 1988).

Rorsman and Trube (1986) performed voltage clamp experiments on β-cells and curve fitted these channel activation functions to the data. Sherman et al. (1988) obtained an improved fit to Rorsman and Trube's experimental I-V curve and have incorporated this into their "minimal" model. Although the maximal channel conductances in the Sherman-Rinzel-Keizer (SRK) model differ substantially from those in the earlier CK model, the membrane potentials in both models exhibit the same qualitative behaviors. This apparent contradiction will be resolved in this paper.

As a mathematical phenomenon, BEA depends on fast and slow processes with distinctly different time scales. Comparison of the short time scale of one oscillation in the active phase to the long time scale of the overall active-silent phase period leads naturally to the use of perturbation methods. The "small" parameter on which to base a perturbation analysis is the fraction of free to bound calcium ions, f. However, the main difficulty in performing such an analysis on the CK and SRK models is that they contain nonlinearities which make analytical work difficult and numerical calculations necessary.

From this mathematical point of view, the SRK model is more suitable for analytical treatment since it is linear in the potassium channel activation variable, n. The 20 parameters (10 parameters define the channel activation functions and 10 biological parameters describe the cell membrane electrical properties) contained in the SRK model make numerical computations essential. However, extensive numerical parametric studies are limited by this large number of parameters. Two purposes of applying perturbation techniques to the SRK model are to identify the important parameters in each phase and to determine their effects on the bursting phenomenon.

In this paper, the SRK equations are nondimensionalized, yielding 15 dimensionless parameters, which is still a large number of parameters. The potassium activation variable, n, is scaled so that its values in the active phase are near unity. Also, a dimensionless small parameter, ε, is defined. Then the resulting system of three first-order nonlinear ordinary differential equations is transformed into a system containing a second-order differential equation and a first-order equation. The second-order equation is linear in its derivatives and in the active phase is weakly coupled to the first-order equation.

From these equations, the leading-order problems for the active and silent phases are obtained in separate perturbation analyses involving "fast" and "slow" times. For the active phase, the "fast" time problem has a leading- order solution which is a limit cycle. The period of this limit cycle slowly increases as the calcium concentration increases. A multiple scales procedure applied to the calcium equation yields this slow evolution. As the calcium concentration approaches its maximum value, the rapidly oscillating active phase trajectory passes through a separatrix and the local instability of the solution near the system's leading-order nullcline causes a transition back to the silent phase. The leading-order equation in the active phase can be put in Lienard form with a "damping" term which is $O(1)$ in ε so the multiple scales procedures of Kuzmak (1959) and Luke (1966) for "strongly nonlinear" oscillators are not strictly applicable. However, it is found that the damping term is numerically small and this leads to the possibility of treating it formally as $o(1)$ in ε.

In Sect. 2, the SRK model is introduced and assumptions leading to the equations are described. Nondimensionalization of the SRK model and scaling considerations are discussed in Sect. 3. Here, the small parameter, ε, is defined. The transformation of the system of three first-order ordinary differential equations to a second-order equation coupled to a first-order equation is given in Sect. 4. Detailed numerical studies of the phenomenon have been carried out and some of the results are presented in Sect. 5. These computations are used to motivate the derivations of the perturbation problems for the active and silent phases presented in Sect. 6. For the leading-order problems, the solution in the active phase is compared to a numerical solution of the exact problem and the trajectory in the silent phase is shown to lie on a leading-order nullcline.

2 The SRK model

The cellular membrane is a phospholipid bilayer and ions can flow between intracellular and extracellular regions through macromolecular pores (channels) imbedded in this bilayer. Generally, intracellular and extracellular concentrations of a given ion are very different and the resulting net charge differences across the membrane produce a potential difference, V, normally negative inside relative to the outside. The channels in the bilayer facilitate or impede ionic transport depending on various influences, e.g., the membrane potential. Unlike classical diffusion processes, the rate of ionic transport through the membrane is not proportional to the difference of intracellular and extracellular concentrations.

The SRK model of β-cell electrical activity (Sherman et al. 1988) is a set of ionic current and concentration balance equations where the Ca^{2+} and K^+ ionic transport rates through the membrane are modelled using voltage-, time-, and concentration-dependent channel conductances. The intracellular calcium concentration is small and for the range of V observed during the BEA, calcium ions flow from the extracellular space into the intracellular compartment. Simultaneously the large intracellular potassium concentration drives K^+ ions up the electrical potential gradient into the extracellular space through their respective channels.

Spatial effects are neglected by assuming the membrane is an homogenous medium with uniformly distributed channels. Cell parameters such as cell radius, cellular volume, and total membrane capacitance, C_m, are each assumed constant. Membrane ionic currents increase intracellular Ca^{2+} concentrations, and uptake of Ca^{2+} (e.g., by the endoplasmic reticulum, mitochondria, or membrane pumps) tends to decrease the intracellular calcium concentration, Ca_i, at a rate proportional to Ca_i. Only the calcium and potassium ionic currents are assumed essential for BEA in β-cells. The channel conductances associated with these currents were curve fitted by Sherman et al. (1988) to the experimental results on β-cells obtained by Rorsman and Trube (1986). Thus the SRK model consists of the following set of equations:

$$C_m\frac{dV}{d\tau} = \bar{g}_{Ca}m_\infty(V)\tilde{h}(V)(V_{Ca}-V) - \bar{g}_k n(V-V_K) \tag{2.1}$$

$$-g_{K-Ca}(Ca_i)(V-V_K)\ ,$$

$$\frac{dn}{d\tau} = \frac{n_\infty(V)-n}{\tau_n(V)}\ , \tag{2.2}$$

$$\frac{dCa_i}{d\tau} = f[\alpha\bar{g}_{Ca}m_\infty(V)\tilde{h}(V)(V_{Ca}-V) - k_{Ca}Ca_i]\ , \tag{2.3}$$

where

$$\alpha = \frac{1}{2V_{Cell}F},\quad F = 96485 \text{ coul/mole is Faraday's constant,} \tag{2.4}$$

and,

$$g_{K-Ca}(Ca_i) = \bar{g}_{K-Ca}\frac{Ca_i}{K_d+Ca_i}\ , \tag{2.5}$$

$$m_\infty(V) = \frac{1}{1+\exp[(V_m-V)/S_m]}\ , \tag{2.6}$$

$$\tilde{h}(V) = \frac{1}{1+\exp[(V-V_h)/S_h]}\ , \tag{2.7}$$

$$n_\infty = \frac{1}{1+\exp[(V_n-V)/S_n]}\ , \tag{2.8}$$

$$\tau_n(V) = \frac{\bar{\tau}_n}{\exp[(V-V_b)/S_a]+\exp[-(V-V_b)/S_b]}\ . \tag{2.9}$$

The equation for the membrane potential (2.1) is a current balance equation. The term on the left side represents the membrane capacitive current with capacitance, C_m, while the terms on the right side represent the ionic currents through the membrane due to voltage-gated calcium channels, voltage-gated potassium channels, and calcium-activated potassium channels, respectively. The activation variable, n, associated with the voltage-gated potassium channel is assumed to obey the relaxation equation (2.2) with channel activation function, $n_\infty(V)$, and relaxation time, $\tau_n(V)$, given by (2.8) and (2.9), respectively. (Equations (2.2) and (2.9) differ from those in the SRK model in that the SRK parameter λ has been absorbed into $\bar{\tau}_n$.) Since the relaxation time for the calcium channel's activation variable, m, is much shorter than $\tau_n(V)$, this variable is replaced by its respective channel activation function, $m_\infty(V)$, given by (2.6).

Table 1. Left table shows comparison of parameter values in SRK and CK models with CK values taken from Chay and Keizer (1985). Right table gives parameter values used to define $\tilde{h}(V), m_\infty(V), n_\infty(V)$, and $\tau_n(V)$ in the SRK model.

Parameter	SRK	CK
V_{Ca} (mV)	110	100
V_K (mV)	-75	-75
$\bar{g}_{Ca}$ (pS)	1400	8000
$\bar{g}_K$ (pS)	2500	7500
$\bar{g}_{K-Ca}$ (pS)	30000	50
V_{Cell} (μm^3)	1150	905
C_m (fF)	5310	4524
K_d (μM)	100	1
k_{Ca} (ms^{-1})	.03	.02
κ_d (μM)	.472	2.591

Parameter	SRK
V_b (mV)	-75
V_h (mV)	-10
V_m (mV)	4
V_n (mV)	-15
S_a (mV)	65
S_b (mV)	20
S_h (mV)	10
S_m (mV)	14
S_n (mV)	5.6
$\bar{\tau}_n$ (ms)	37.5

Equation (2.3) represents a balance equation for the intracellular calcium concentration where f is the ratio of free to bound calcium ions, assumed to be constant. The first term on the right side of (2.3) represents a calcium concentration increase due to an influx of Ca^{2+} through the voltage-gated calcium channels where α is a geometrical factor which accounts for cell shape and volume. The second term represents a decrease in the cytoplasmic Ca^{2+} concentration due to intracellular buffering. Although many mechanisms for calcium buffering have been observed, the decrease caused by mitochondria, via a glucose-dependent process stimulated by oxidative phosphorylation, is thought to be the most prominent (Chay and Keizer, 1983). With this interpretation of calcium buffering, the rate constant, k_{Ca}, increases with glucose concentration.

A summary of values for the maximal channel conductances, $\bar{g}_{Ca}$, $\bar{g}_K$, and $\bar{g}_{K-Ca}$, the Nernst potentials for calcium and potassium, V_{Ca} and V_K, and the remaining parameters defining the channel activation functions and relaxation time can be found in Table 1.

3 Nondimensionalization and scaling

In order to carry out a systematic perturbation analysis, it is essential to first nondimensionalize and scale the variables. As the choices of dimensionless variables are not unique, the dimensional equations were integrated numerically to help provide some insight into making such choices. Details and some results of the numerical computations are given in Sect. 5.

The membrane potential exhibits BEA and the corresponding calcium concentration has a characteristic "sawtooth" shape (cf. Fig. 1). The approximate ranges of each dimensional dependent variable in the BEA are: 1) membrane potential (mV): $-66.8 < V < -23.1$; 2) activation variable: $0.00011 < n < 0.107$; and 3) calcium concentration (μM): $0.532 < Ca_i < 0.611$.

The activation variable, n, is replaced by a scaled variable, w, with numerical values near one in the active phase. Time is nondimensionalized so that the maximum relaxation time associated with the K^+-channel is approximately equal to one. To nondimensionalize Ca_i, a scaled dissociation constant, κ_d with units of μM, was used. The choices of nondimensionalized (and scaled) membrane potential, activation variable, calcium concentration, and time are:

$$v \equiv -V/V_K \ , \tag{3.1}$$

$$w \equiv \gamma_K \, n, \quad \gamma_K \equiv \frac{\bar{g}_K \bar{\tau}_n}{C_m} \ , \tag{3.2}$$

$$c \equiv Ca_i/\kappa_d, \quad \kappa_d \equiv \frac{C_m K_d}{\bar{g}_{K-Ca}\, \bar{\tau}_n} \ , \tag{3.3}$$

$$t \equiv \tau/\bar{\tau}_n \ . \tag{3.4}$$

Note that v has been chosen to take on negative values although this is not the natural mathematical choice. However, this choice maintains the physiological intuition associated with V. Denoting differentiations with respect to t by ($\dot{}$), the corresponding nondimensional ordinary differential equations are:

$$\dot{v} = i_{Ca}(v) - w(v+1) - g(c)(v+1) \ , \tag{3.5}$$

$$\dot{w} = \frac{w_\infty(v) - w}{\tau_w(v)} \ , \tag{3.6}$$

$$\dot{c} = \varepsilon[\beta i_{Ca}(v) - c] \ , \tag{3.7}$$

where

$$i_{Ca}(v) = \frac{\gamma_{Ca}(v_{Ca} - v)}{(1+\exp[(v_m - v)/s_m])\,(1+\exp[(v - v_h)/s_h])} \ , \tag{3.8}$$

$$g(c) = \frac{c}{1 + c/\gamma_{K-Ca}} \ , \tag{3.9}$$

$$w_\infty(v) = \frac{\gamma_K}{1+\exp[(v_n - v)/s_n]} \ , \tag{3.10}$$

$$\tau_w(v) = \frac{1}{\exp[(v - v_b)/s_a] + \exp[-(v - v_b)/s_b]} \ , \tag{3.11}$$

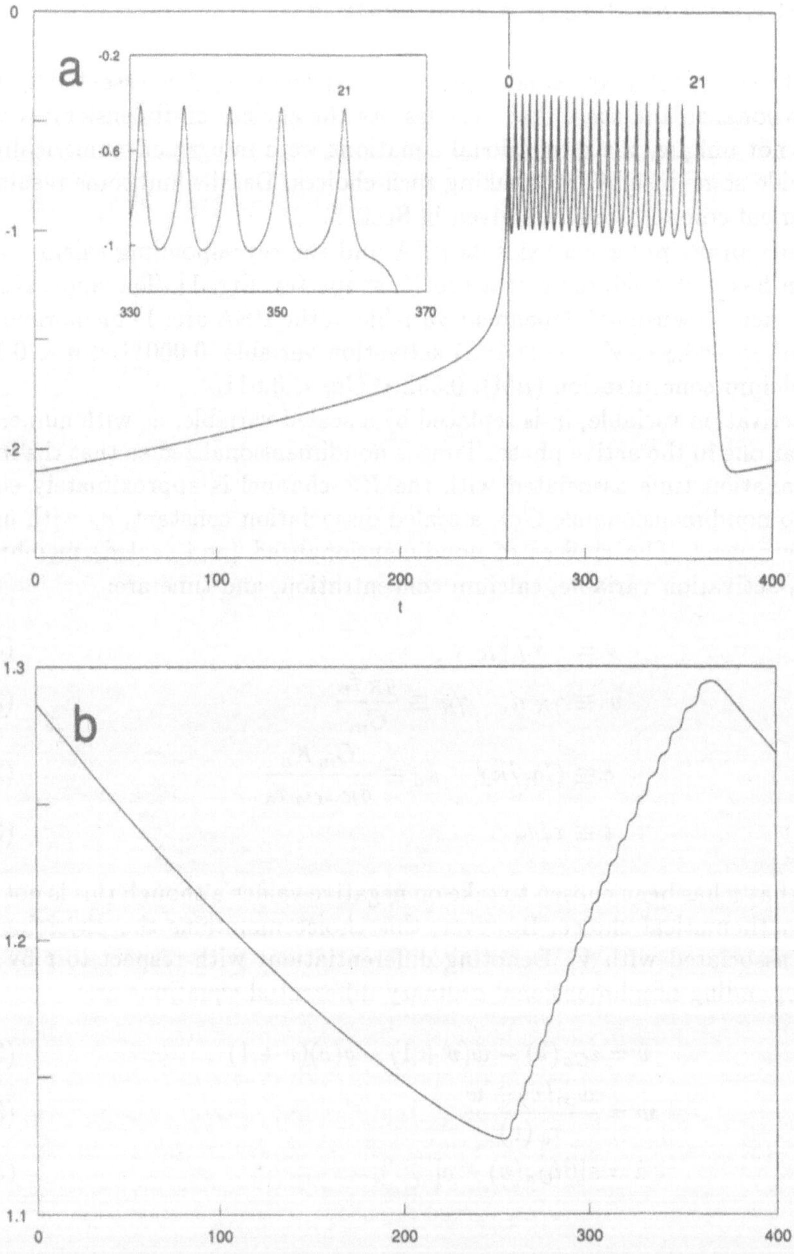

Fig. 1. One cycle of BEA in transformed variables. a shows the transformed membrane potential u. The active phase spikes are numbered from 0 to 21 for comparison purposes in later figures. The insert details the slow frequency decrease of these oscillations. b shows the dimensionless calcium concentration c.

are the dimensionless calcium current, calcium-activated potassium conductance, activation function, and the relaxation time, respectively. The notational changes made in the dimensionless SRK equations permit the 15 dimensionless parameters to be written in the compact form:

$$v_i = -V_i/v_K \quad \text{where} \quad i = \text{Ca,b,h,m,n} \ , \tag{3.12}$$

$$s_i = -S_i/V_K \quad \text{where} \quad i = \text{a,b,h,m,n} \ , \tag{3.13}$$

$$\gamma_i = \frac{\bar{g}_i \bar{\tau}_n}{C_m} \quad \text{where} \quad i = \text{Ca,K,K-Ca} \ , \tag{3.14}$$

$$\varepsilon = f \bar{\tau}_n k_{Ca} \ , \tag{3.15}$$

$$\beta = \frac{-\alpha V_K \bar{g}_{K-Ca}}{k_{Ca} K_d} \ . \tag{3.16}$$

The values of these parameters are summarized in Table 2. Since $\varepsilon/f = 1.125$ it is evident that the smallness of ε is a consequence of the small ratio, f, of free to bound intracellular calcium ions.

Table 2. Dimensionless parameter values for SRK model

Parameter	SRK	Parameter	SRK
f	0.001	v_h	-0.133
ε	0.001125	v_m	0.0533
β	3.380	v_n	-0.200
γ_K	17.655	s_a	0.866
γ_{Ca}	9.887	s_b	0.266
γ_{K-Ca}	211.864	s_h	0.133
v_{Ca}	1.467	s_m	0.186
v_b	-1.000	s_n	0.0746

The approximate ranges of each dimensionless dependent variable in the BEA are: 1) membrane potential: $-0.878 < v < -0.308$; 2) activation variable: $0.00202 < w < 1.88$; and 3) calcium concentration: $1.13 < c < 1.30$.

Both $i_{Ca}(v)$ and $w_\infty(v)$ are positive and monotonically increasing functions of v during the BEA. Their values range from 0.1 to about 1.7 and from near 0.0 to about 3.5, respectively. The relaxation time, $\tau_w(v)$, is concave down and its values range from about 0.44 to 0.58. Also, since $\gamma_{K-Ca} \gg 1, g(c) \cong c$. Spike periods in the active phase, measured from peak to peak, range from about 3.8 to about 8.9, and are considerably less than the overall active-silent phase period, $T \cong 380$.

4 Transformation to a perturbed Liénard form

The nondimensional equation (3.5) can be simplified by introducing the change of variable

$$u = \ell n(v+1) \tag{4.1}$$

which is defined for $v > -1$ and whose inverse is given by $v = e^u - 1$. Then (3.5) becomes

$$\dot{u} = e^{-u} i_{Ca}(v(u)) - w - g(c) \ , \tag{4.2}$$

where each term on the right side corresponds to a dimensionless conductance. In particular, the variable w should be re-interpreted as the dimensionless conductance associated with the voltage-gated potassium channel.

By eliminating w, the two first-order differential equations (3.6) and (4.2) can be combined into a single second-order differential equation which is in a perturbed Liénard form (Minorsky, 1962). This is accomplished by differentiating (4.2) with respect to t and using equations (3.6), (3.7), and (4.2). The resulting system of differential equations for u and c is:

$$\ddot{u} + F(u)\dot{u} + G(u,c) = \varepsilon H(u,c) \ , \tag{4.3}$$

$$\dot{c} = \varepsilon[\beta h(u) - c] \ , \tag{4.4}$$

where by defining

$$\Gamma(u) \equiv e^{-u} i_{ca}(v(u)), \ T_n(u) \equiv \tau_w(v(u)), \ N(u) \equiv w_\infty(v(u)), \tag{4.5}$$

the functions F, G, H, and h are given by:

$$F(u) = \frac{1}{T_n(u)} - \frac{d\Gamma}{du}(u) \ , \tag{4.6}$$

$$G(u,c) = \frac{N(u) + g(c) - \Gamma(u)}{T_n(u)} \ , \tag{4.7}$$

$$H(u,c) = \frac{dg}{dc}(c)[c - \beta h(u)] \ , \tag{4.8}$$

$$h(u) = i_{Ca}(v(u)) \ . \tag{4.9}$$

On the t time scale, the coupling between (4.3) and (4.4) is weak since $\dot{c} = O(\varepsilon)$ with $\varepsilon \ll 1$. More specifically, when ε is set equal to zero in the system (4.3)-(4.4), c acts as a parameter in (4.3) which then is in the form of a Liénard differential equation.

Throughout most of the BEA, $\Gamma(u)$ is concave up and $h(u)$ is a monotonically increasing function of u. The functions $\Gamma(u), T_n(u), N(u)$, and $h(u)$ are each positive, and $F(u), G(u,c)$, and $H(u,c)$ have signs which depend on the values of u and c. In particular, the sign of $F(u)$ during the active phase is important since it determines whether the damping in (4.3) is positive or negative.

Two advantages in studying the system (4.3)-(4.4) instead of the system (3.5)-(3.7) are: 1) the transformation (4.1) makes (4.3) quasi-linear, i.e., the resulting second-order equation for v would have been quadratic in $\dot{v}$ and 2) perturbed second-order differential equations for oscillatory systems are commonly treated in the literature on perturbation theory (Bender and Orszag 1978, Kevorkian and Cole 1981).

5 Numerical results

Computations on the SRK model were carried out on a VAX 3500 using the CMLIB ordinary differential equation solving routine DDRIV3. The routine was set for a Gear integrator and all computations were carried out using double precision. The relative error for most runs was fixed at 10^{-7} and the time step was chosen to be sufficiently small so that the details of individual spikes could be resolved graphically.

The solutions $u(t)$ and $c(t)$ were obtained by numerical integration of (4.3)-(4.4) and are represented in Fig. 1. The calcium cycle shown in Figure 1b has values of c which increase during the active phase and decrease during the silent phase. The oscillations of $u(t)$ in the active phase approximately begin and end at those times where $\dot{c} = 0$. The weak coupling between u and c during the active phase is manifest by noting the small amplitude of the oscillations in c.

Two-dimensional projections of the 3-dimensional BEA cycle in $(u, \dot{u}, c)$-space are shown in Fig. 2. Figure 2a is the projection onto the (u, c)-plane and Fig. 2b is the projection onto the (u, c)-plane. During the silent phase, the solution lies close to the $\dot{u} = 0$ plane and values of c decrease. When c becomes small a transition to the active phase occurs. During the active phase the solution slowly spirals upwards as c increases. When values of c become large, the oscillation in the active phase terminates, a transition back to the silent phase occurs, and the entire cycle then repeats itself.

For numerical experiments with durations much longer than those shown in Fig. 1, there is a persistence of the bursting and sawtooth patterns exhibited by u and c, respectively. Although the accuracy of such experiments may be questioned, it is believed that the phenomenon is periodic. Numerical experiments, where a variety of different initial conditions were used, support the conjecture that the repetitive BEA represents an attractive 3-dimensional limit cycle in $(u, \dot{u}, c)$-space. If an initial condition is chosen which has a value of $c(0)$ larger than the minimum value of $c(t)$ in Fig. 1b, the resulting trajectory approaches the numerical solution in Fig. 1 within one BEA cycle. If $c(0)$ is chosen smaller than the minimum $c(t)$ in Fig. 1b, the initial part of the first active phase can contain extra spikes. Subsequent active phases, however, all contain the same number of spikes.

The existence of two separate time scales is very evident in Fig. 1a. During the silent phase $\dot{u}$ nearly equals zero for a time span of $\Delta t \cong 250$. In contrast, each oscillation of $u(t)$ in the active phase has an amplitude of $O(1)$ and occurs in a time span $\Delta t \cong 5$. Furthermore, there is a slow increase in the frequency of these oscillations which is reminiscent of frequency changes found in other nonlinear oscillators. Also, it is worth noting that the oscillations in the active phase are approximately symmetric about the $\dot{u} = 0$ plane (see Fig. 2a).

The solution crosses the $\dot{c} = 0$ nullsurface (i.e., the surface in the $(u, \dot{u}, c)$-space defined by the right-hand-side of (4.4) equal to zero) in the transition regions connecting the active and silent phases. The position of the middle portion of this nullsurface is between the active and silent phases and explains the increasing/decreasing behavior of c with the active/silent phases of u.

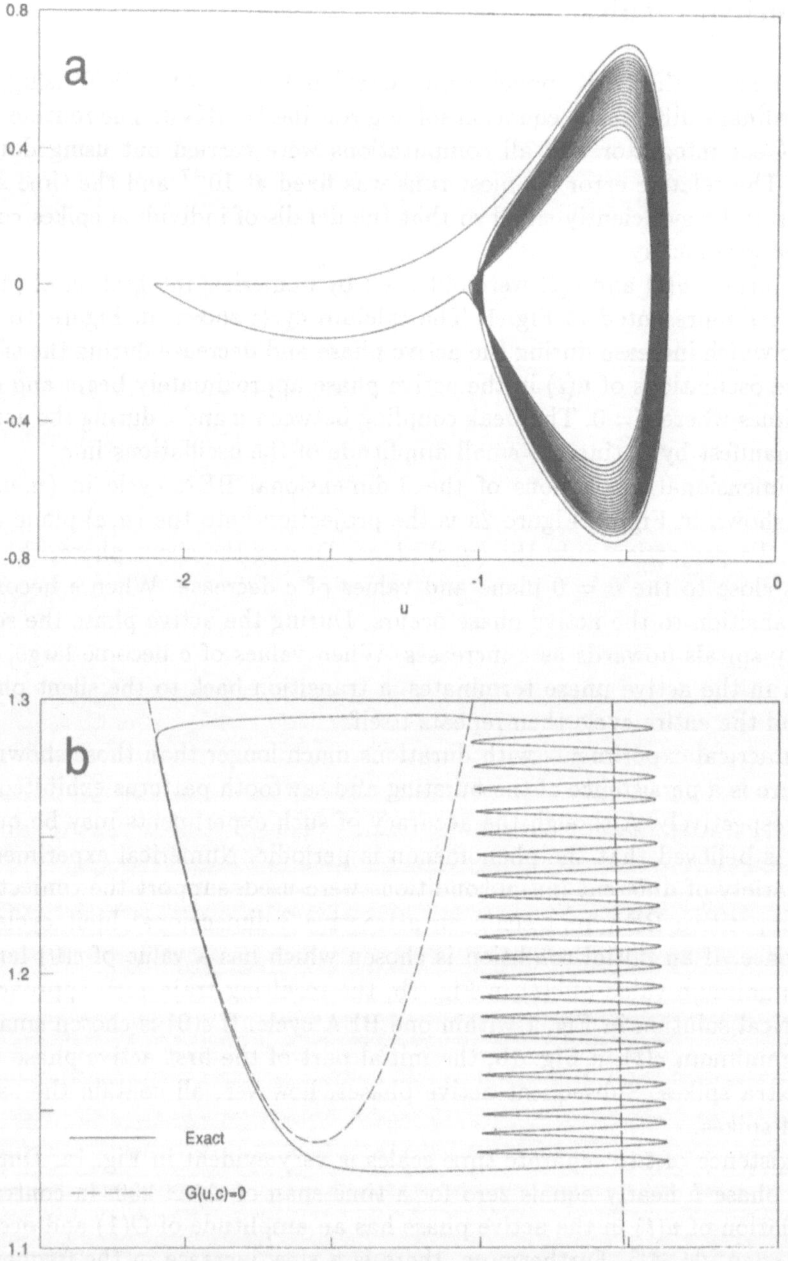

Fig. 2. **Projections of one complete BEA cycle onto a the $(u, \dot{u})$-plane and b the (u, c)-plane. Dashed curve corresponds to nullcline for Liénard equation.**

6 Leading-order perturbation problems

The silent-active phase cycle governed by (4.3)-(4.4) consists of a silent phase, a transition to the active phase, an active phase, and a transition back to the silent phase. A complete perturbation analysis of the BEA cycle requires separate treatments of each part of this cycle coupled with a set of matching conditions. In this section, the leading-order perturbation problems for the active and silent phases are determined systematically from (4.3)-(4.4) for $\varepsilon \to 0$. Then the closeness of the solutions of these approximate problems to the exact solution will be shown by comparing the corresponding numerical solutions.

The silent phase is characterized by slow changes of both u and c on the t scale. More specifically, du/dc is $O(1)$ in the silent phase (see Fig. 2b) and since $\dot{c}$ is $O(\varepsilon)$, then $\dot{u}$ must be $O(\varepsilon)$. A slow time, $\tilde{t}$, and new dependent variables are defined by

$$\tilde{t} \equiv \varepsilon t, \quad U(\tilde{t};\varepsilon) \equiv u(\varepsilon^{-1}\tilde{t};\varepsilon), \quad C(\tilde{t};\varepsilon) \equiv c(\varepsilon^{-1}\tilde{t};\varepsilon) \ . \tag{6.1}$$

Both U and C undergo $O(1)$ changes on $\tilde{t}$ time intervals of $O(1)$. If $()'$ denotes differentiation with respect to $\tilde{t}$, (4.3)-(4.4) become:

$$\varepsilon^2 U'' + \varepsilon F(U)U' + G(U,C) = \varepsilon H(U,C) \ , \tag{6.2}$$

$$C' = \beta h(U) - C \ . \tag{6.3}$$

Formally, the leading-order silent phase problem is determined by substituting

$$U(\tilde{t};\varepsilon) \sim U_0(\tilde{t}) + \varepsilon U_1(\tilde{t}) + \cdots \ , \tag{6.4}$$

$$C(\tilde{t};\varepsilon) \sim C_0(\tilde{t}) + \varepsilon C_1(\tilde{t}) + \cdots \ , \tag{6.5}$$

into (6.2)-(6.3), expanding the functions F, G, and H in Taylor series, and then setting ε equal to zero in the resulting equations. The solution for (U_0, C_0) is subsequently determined from the algebraic equation

$$G(U_0, C_0) = 0 \ , \tag{6.6}$$

and the first-order differential equation

$$C_0' + C_0 = \beta h(U_0) \ . \tag{6.7}$$

The algebraic equation (6.6) can be solved explicitly for C_0 obtaining

$$C_0 = \gamma_0(U_0) \equiv \frac{\Gamma(U_0) - N(U_0)}{1 + \gamma_{K-Ca}^{-1}[N(U_0) - \Gamma(U_0)]} \ . \tag{6.8}$$

A confirmation of (6.6) as part of the correct leading-order problem for the silent phase is shown in Fig. 2b [3] where the curve defined by (6.8) has been superimposed onto the numerical solution of (4.3)-(4.4). From (6.2) the

[3] Since this paper does not treat the SRK model as a bifurcation problem, it is more suitable to re-orient the usual "Z" curve (see Fig. 3, Rinzel and Lee 1986) to a backwards "N" curve in the (u, c)-plane.

$G(u,c) = 0$ surface, S, is seen to be the leading-order U' nullsurface i.e., with $\varepsilon \to 0$. The surface S divides the $(u, \dot{u}, c)$-space into two regions with $G > 0$ above S and $G < 0$ below S. The left branch of S is defined as the points $(u, \dot{u}, c)$ which have values of u less than that at the local minimum of $\gamma_0(u)$ (see Fig. 2b). The middle branch of S has values of u between those at the local minimum and maximum of $\gamma_0(u)$. Finally, the right branch of S has values of u greater than that at the local maximum of $\gamma_0(u)$.

Figure 2b clearly demonstrates that the relationship between u and c defined by the exact equations (4.3)-(4.4) is closely approximated by (6.8) throughout the silent phase. Although this confirms the leading-order trajectory in $(u, \dot{u}, c)$-space, a confirmation of the leading-order dynamics requires a comparison of the dynamics for the approximate and exact solutions. Eliminating C_0 in (6.7) using (6.8), the dynamics of the leading-order problem are determined by

$$U_0' = Z(U_0) \equiv \frac{\gamma_0(U_0) - \beta h(U_0)}{[N'(U_0) - \Gamma'(U_0)][1 + \gamma_{K-Ca}^{-1}\gamma_0(U_0)]^2} \ . \tag{6.9}$$

The numerical solutions of (6.9) and (6.2)-(6.3) are compared in Fig. 3. The closeness of $U_0(\tilde{t})$ to $U(\tilde{t})$ confirms that the dynamics of the leading-order silent phase solution approximate the dynamics of the exact solution. Also, the leading-order approximation $U_0(\tilde{t})$ breaks down as it approaches the local minimum of $\gamma_0(U_0)$ where $Z(U_0)$ becomes undefined.

Since $C > \beta h(U)$ along the left branch (cf. Fig. 2b), $C' < 0$ from (6.3) and C must decrease as the solution moves along $G = 0$ towards the local minimum of $\gamma_0(U)$. The solution cannot move up the lower part of the middle branch since $C' < 0$, hence it moves away from S into a transition region separating the silent and active phases. As the solution moves through this region, the value of G decreases until the leading-order result (6.6) is no longer valid. The assumption that du/dc is $O(1)$ breaks down and the slow time $\tilde{t}$ in (6.1) is no longer an appropriate variable for a perturbation analysis. From Fig. 2b, the value of u is observed to increase rapidly and a transition to the active phase occurs.

Once the solution enters the active phase, the appropriately scaled equations are (4.3)-(4.4). However, it is clear from Fig. 2b that whereas the rapid oscillations occur on the fast time scale t, the slow evolution of c occurs on the slower time scale, $\tilde{t} = \varepsilon t$. Applying the method of multiple scales, let

$$u(t;\varepsilon) = U(\bar{t}, \tilde{t}; \varepsilon) \sim u_0(\bar{t}, \tilde{t}) + \varepsilon u_1(\bar{t}, \tilde{t}) + \cdots \ , \tag{6.10}$$

$$c(t;\varepsilon) = C(\bar{t}, \tilde{t}; \varepsilon) \sim c_0(\bar{t}, \tilde{t}) + \varepsilon c_1(\bar{t}, \tilde{t}) + \cdots \ , \tag{6.11}$$

where $\bar{t}$ defined by $d\bar{t}/dt = \omega(\tilde{t}; \varepsilon)$ is a strained fast time and the $u_i(\bar{t}, \tilde{t})$ and $c_i(\bar{t}, \tilde{t})$, $i = 0, 1, 2, \ldots$, are assumed to be strictly periodic in $\bar{t}$ with fixed period, T_p. Thus (4.3)-(4.4) yield the following leading-order problem for the active phase:

$$\omega^2 \frac{\partial^2 u_0}{\partial \bar{t}^2} + \omega F(u_0) \frac{\partial u_0}{\partial \bar{t}} + G(u_0, c_0) = 0 \ , \tag{6.12}$$

$$\frac{\partial c_0}{\partial \bar{t}} = 0 \ . \tag{6.13}$$

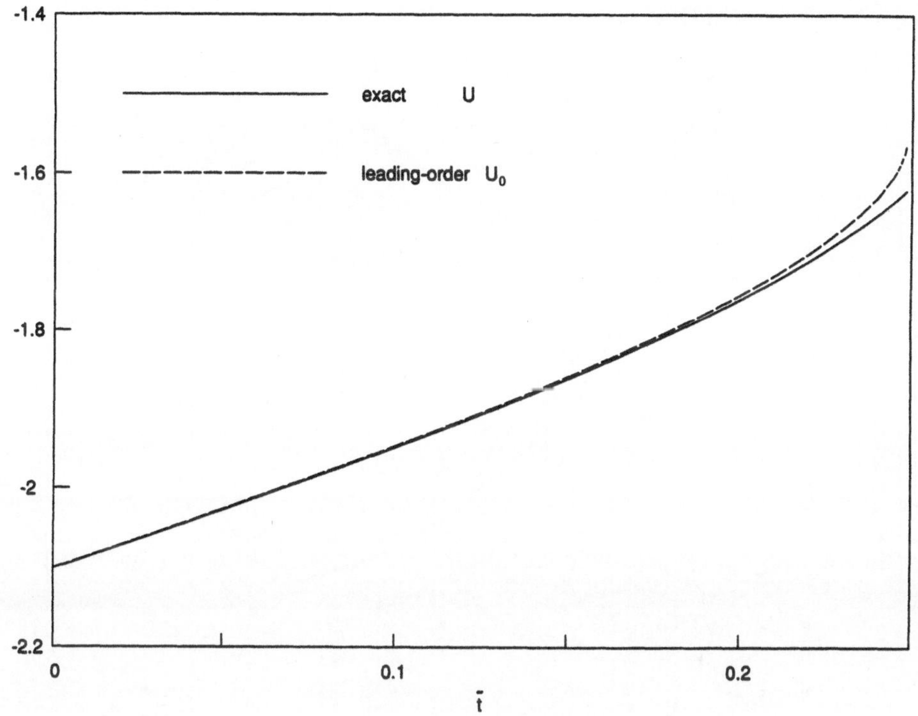

Fig. 3. Comparison of exact and leading-order solutions in the silent phase.

For a fixed value of c_0 in the range observed during the active phase, (6.12) has a limit cycle solution (see Fig. 4). Therefore, equations (6.12)-(6.13) imply that for time intervals in $\bar{t}$ of $O(1)$, each cycle in the active phase can be approximated by the single periodic solution, $u_0 = \Omega(\bar{t} - \phi; c_0)$, of the strictly nonlinear Lienard equation (6.12) in which c_0 is a parameter and ϕ is the phase. A standard linear stability analysis of (6.12) yields critical points located on the $G = 0$ surface, with the critical points on the middle branch classified as saddle points and on the right branch as unstable spirals.

Equation (6.13) implies that c_0 is a function of $\tilde{t}$ only. The evolution of $c_0(\tilde{t})$ is obtained by averaging the equation for c_1 in $\bar{t}$ over the period T_p of the periodic solution of (6.12). Thus the evolution of c_0 is determined by

$$\frac{dc_0}{d\tilde{t}} + c_0 = \beta \bar{h}(c_0) \equiv \frac{\beta}{T_p} \int_0^{T_p} h(\Omega(\bar{t}; c_0)) d\bar{t} \ , \tag{6.14}$$

where ϕ has been omitted because Ω is periodic in $\bar{t}$. Since the solution $\Omega(\bar{t}; c_0)$ is not analytically available in general, the function $\bar{h}(c)$ was computed numerically and is plotted in Fig. 5a. In Fig. 5b the numerical solution of the leading-order active phase problem (6.12)-(6.14) is compared to the numerical solution of the exact equations (4.3)-(4.4). Note that this solution of the leading-order active

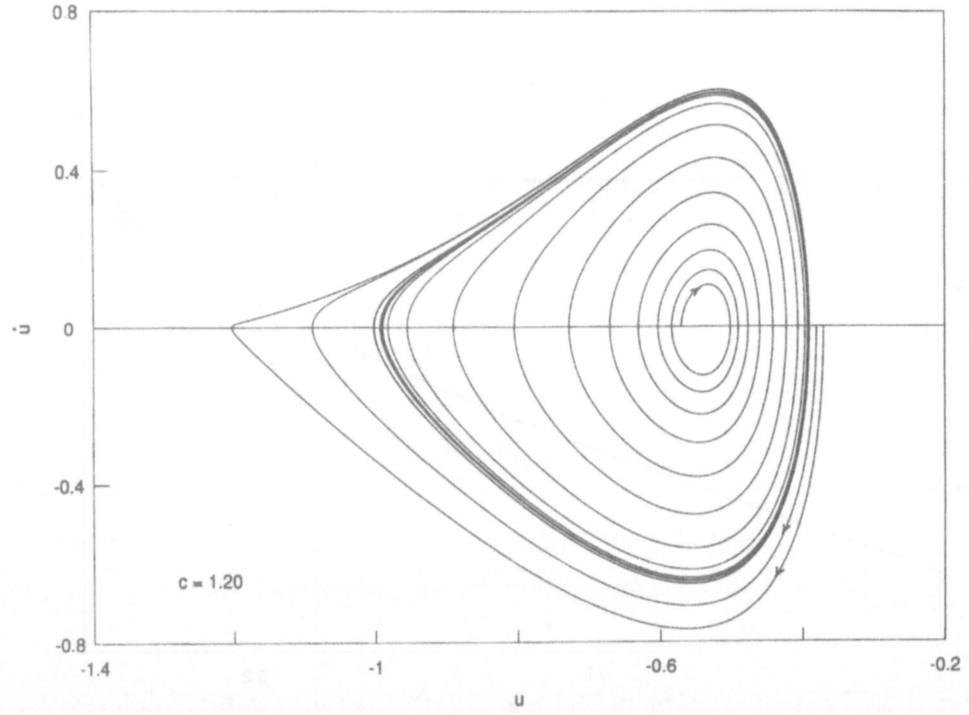

Fig. 4. Demonstrates the existence of an attractive limit cycle for the leading-order active phase equation.

phase problem closely tracks the exact solution, however, the two solutions become phase shifted because the slow time dependence of ϕ has not been taken into account.

For "strongly nonlinear" oscillators, Kuzmak (1959) and Luke (1966) have developed a multiple scales perturbation procedure for determining solutions which are valid for times of $O(1/\varepsilon)$. Unfortunately, the Kuzmak-Luke multiple scales procedures are not directly applicable to (4.3)-(4.4) since they do not take into account the cumulative effect of the $O(1)$ damping term. This damping term, $F(u)\dot{u}$, and its changes in sign during the active phase are necessary for a limit cycle solution of (6.12) to exist. However, numerical computations reveal that the damping term is numerically small relative to $\ddot{u}$ and $G(u,c)$ in the leading-order active phase equation. In Fig. 6 is plotted $\int^t G(u,c)dt$ versus $\dot{u}$ and it is noted that the curve deviates little from the straight line with slope -1. Thus the cumulative effect of $F(u)\dot{u}$ is small and makes the Lienard equation a strongly nonlinear oscillator for which the Kuzmak-Luke method and the near-identity averaging method (Kevorkian 1987) are formally applicable. This approach to the leading-order perturbation problem in the active phase will be explored in a subsequent paper.

Since $0 < \dot{c} = O(\varepsilon)$ in the active phase, the exact solution of (4.3)-(4.4)

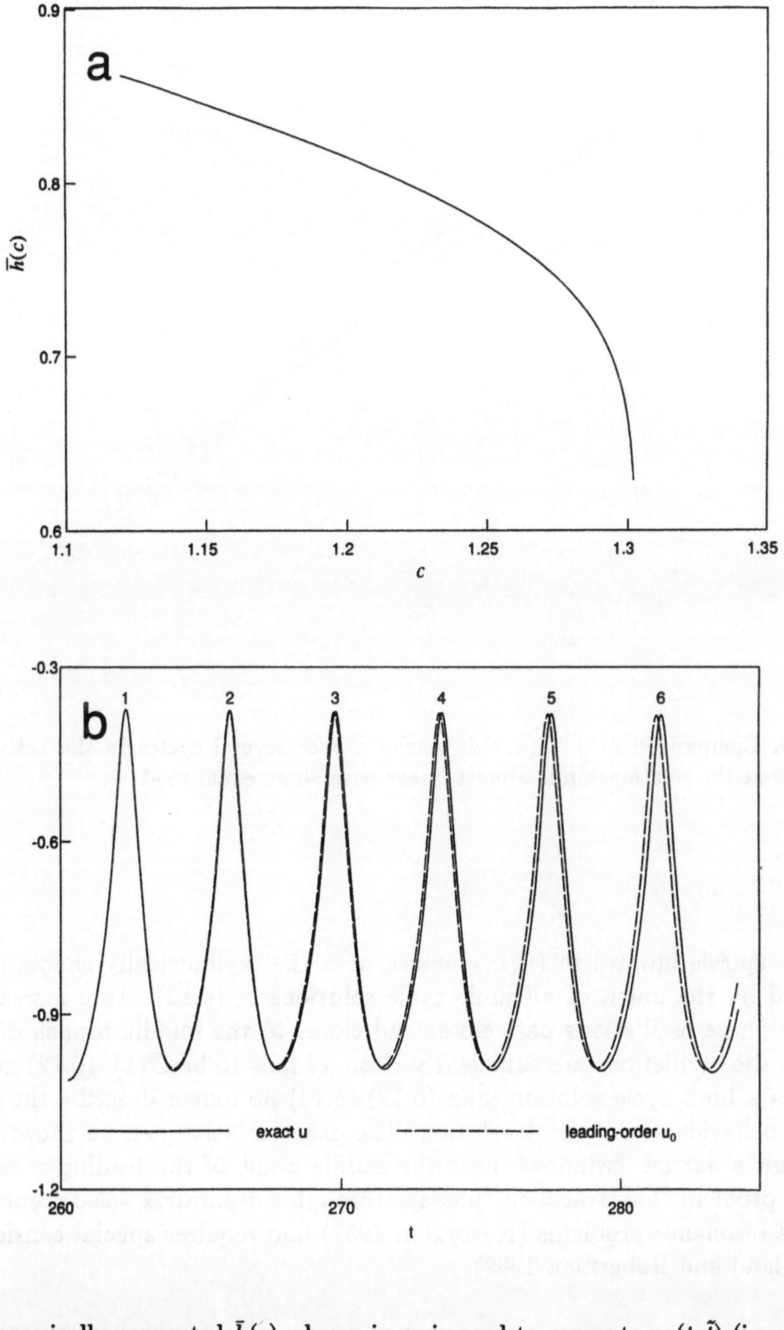

Fig. 5. Numerically computed $\bar{h}(c)$, shown in **a**, is used to compute $u_0(t, \tilde{t})$ (i.e. with ω=1 and ϕ constant). Comparison of exact and leading-order solutions for u in the active phase are shown in **b**.

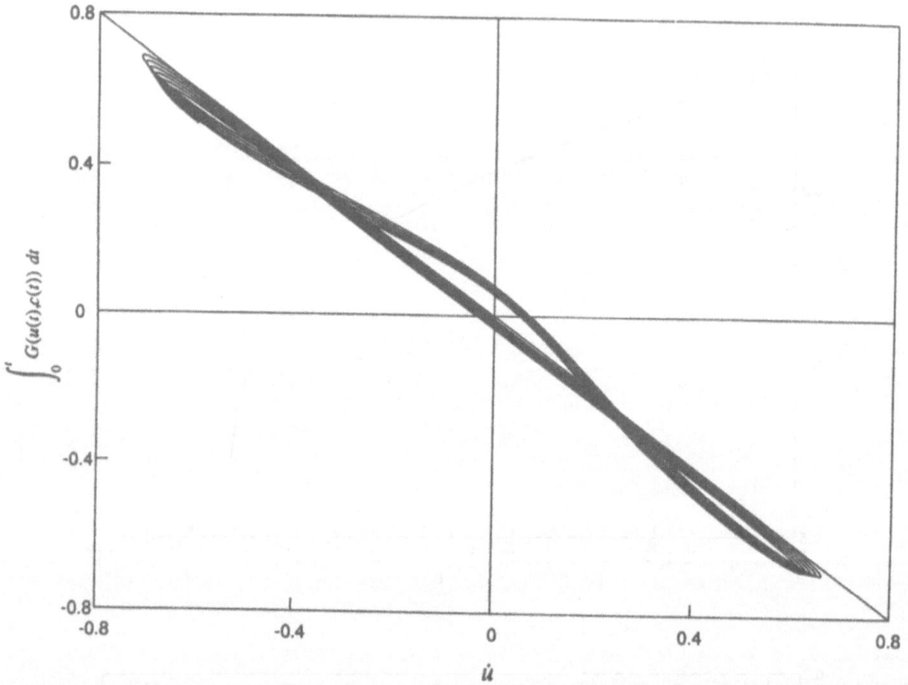

Fig. 6. Comparison of $\int^t G(u,c)dt$ versus $\dot{u}$ over several cycles in the active phase. Note that the relationship is almost linear with slope equal to -1.

slowly spirals upward in $(u, \dot{u}, c)$-space near the (cylindrically shaped) surface formed by the union of all limit cycle solutions of (6.12). As c increases, the active phase oscillations pass closer and closer to the middle branch of $G = 0$. When the oscillations are sufficiently close, G fails to be $O(1)$, (6.12) no longer admits a limit cycle solution, and (6.12)-(6.14) no longer describe the leading-order behavior of the exact solution. The exact solution can be shown to pass through a narrow "window" near the saddle point of the leading-order active phase problem. This so-called "passage through a separatrix" also occurs in sustained resonance problems (Kevorkian 1987) and requires special consideration (Bourland and Haberman 1989).

Once outside the separatrix, the exact solution makes a transition back to the silent phase (cf. Fig. 2b). As it does, it crosses the $\dot{c}$ nullcline and the values of c begin to decrease slowly. When the solution nears the left branch of $G = 0$, the leading-order behavior of the exact solution is governed again by (6.6)-(6.7). The silent-active phase cycle then repeats itself creating the bursting phenomenon.

7 Discussion

The goal of the analysis presented in this paper has been to develop a simplified framework for the mathematical analysis of the SRK equations. It has not been the authors' aim to speculate on the validity of the assumptions leading to these equations. The three new results in this paper are nondimensionalization of the SRK equations, transformation of the dimensionless model equations into an equation in perturbed Lienard form which is weakly coupled to a first-order equation, and formulation of the leading-order problems (in ε) in the silent and active phases of the BEA.

Parameter studies made without the use of analytical techniques are impractical since the total number of parameters defining the SRK system is large (10 biological parameters and 10 artificial parameters to define the functions $n_\infty(V), h(V), m_\infty(V)$, and $\tau_n(V)$). The nondimensionalized equations contain 15 parameters which, although still large, is a significant improvement.

The nondimensionalization also explains (more precisely) how different biological parameters affect the BEA. (The CK equations can be nondimensionalized in a completely analogous way to the nondimensionalization presented in Sect. 3.) In both the CK and SRK models, the intracellular calcium concentration, Ca_i, is nondimensionalized by a scaled dissociation constant which is inversely proportional to the ratio $\bar{g}_{K-Ca}/K_d$. The values of this ratio (50 and 300 for the CK and SRK models, respectively) differ less than the values of $\bar{g}_{K-Ca}$ (50 and 30000 for the CK and SRK models, respectively). Consequently, the values of the scaled dissociation constants, $\kappa_d = 0.472$ for the SRK model and $\kappa_d^{(CK)} = 2.591$ for the CK model, are not that different. Therefore, it is evident that the ratio $\bar{g}_{K-Ca}/K_d$ is more important than the parameter values $\bar{g}_{K-Ca}$ and K_d for maintaining the BEA. The primary effect of this ratio is to change the range of the intracellular calcium concentration, Ca_i.

The transformation (4.1) provides a simpler formulation of (2.1)-(2.3) from which to derive the active and silent phase leading-order perturbation problems of Sect. 6. The leading-order silent phase equations (6.6)-(6.7) are mathematically equivalent to the dimensional silent phase equations presented by Rinzel (1987) but were derived here using a perturbation analysis which involves an explicitly defined slow time $\tilde{t}$. Furthermore, the silent phase analysis presented in Sect. 6 differs from Rinzel's treatment since c is eliminated in (6.9).

As a result of the transformation (4.1), the limit cycle structure of the leading-order active phase problem is expressed explicitly in a single Liénard differential equation. The observation that the damping term is numerically small during each oscillation is important because the Liénard equation then describes a strongly nonlinear oscillator.

Acknowledgment

We are grateful to J. Rinzel and A. Sherman for bringing this subject and their model to our attention. This work was supported in part by the Department of

Energy under Grant No. DE-FG06-86ER25019 (M.P., J.K.) and by the Natural Sciences and Engineering Research Council of Canada under Grant No. A-4559 (R.M.M.)

References

1. Atwater, I., Ribalet, B., Rojas, E. (1978a): Cyclic changes in potential and resistance of the β-cell membrane induced by glucose in islets of Langerhans from mouse. J. Physiology (Lond.). **278**, 117–139
2. ——. (1978b): Mouse pancreatic β-cells: tetraethylammonium blockage of the potassium permeability increase induced by depolarization. J. Physiology (Lond.). **288**, 561-574
3. Atwater, I., Dawson, C.M., Ribalet, B., Rojas, E. (1978c): Potassium permeability activated by intracellular calcium ion concentration in the pancreatic β-cell. J. Physiology (Lond.). **288**, 575-588
4. Atwater, I., Dawson, C.M., Scott, A., Eddlestone, G., Rojas, E. (1980): The nature of the oscillatory behaviour in electrical activity from pancreatic β-cell. Hormone and Metabolic Research, supp.10, 100-107
5. Bender, C.M., Orszag, S.A. (1978): Advanced Mathematical Methods for Scientists and Engineers. McGraw-Hill, New York
6. Bourland, F.J., Haberman, R. (1990): Separatrix crossing: time-invariant potentials with dissipation. SIAM J. Appl. Math. **50**, 1716-1744
7. Chay, T.R., Keizer, J. (1983): Minimal model for membrane oscillations in the pancreatic β-cell. Biophys. J. **42**, 181-189
8. ——.(1985): Theory of the effect of extracellular potassium on oscillations in the pancreatic β-cell. Biophys. J. **48**, 815-827
9. Dean, P.M., Matthews, E.K. (1970a): Glucose-induced electrical activity in pancreatic islet cells. J. Physiology (Lond.). **210**, 255-264
10. ——. (1970b): Electrical activity in pancreatic islet cells: effects of ions. J. Physiology (Lond.). **210**, 265-275
11. Henquin, J.C., Meissner, H.P. (1984): Effects of theophylline and dibutyryl cyclic adenosine monophosphate on the membrane-potential of mouse pancreatic β-cells. J. Physiology (Lond.). **351**, 595-612
12. Hodgkin, A.L., Huxley, A.F. (1952): A quantitative description of membrane current and its application to conduction and excitation in nerve. J. Physiology (Lond.). **117**, 500-544
13. Keizer, J. (1988): Electrical activity and insulin release in pancreatic beta cells. Math. Biosci. **90**, 127-138
14. Kevorkian, J., Cole, J.D. (1981): Perturbation Methods in Applied Mathematics. (Applied Mathematical Sciences, Vol. 34). Springer-Verlag, Berlin-Heidelberg-New York
15. Kevorkian, J. (1987): Perturbation techniques for oscillatory systems with slowly varying coefficients. SIAM Rev. **29**, 391-461
16. Kuzmak, G.E. (1959): Asymptotic solutions of nonlinear second order differential equations with variable coefficients. P.M.M. **23**, 730-744
17. Luke, J.C. (1966): A perturbation method for nonlinear dispersive wave problems. Proc. Roy. Soc., Ser. A. **292**, 403-412

18. Meissner, H.P., Preissler, M. (1980): Ionic mechanisms of the glucose-induced membrane potential changes in β-cells. Hormone and Metabolic Research, supp.10, 91-99
19. Meissner, H.P., Schmelz, H. (1974): Membrane potential of beta-cells in pancreatic islets. Pfluegers Arch. **351**, 195-206
20. Minorsky, N. (1962): Nonlinear Oscillations. Robert E. Krieger Publishing, Malabar, Florida
21. Ozawa, S., Sand, O. (1986): Electrophysiology of endocrine cells. Phys. Reviews, **66**, 887-952
22. Rinzel, J. (1987): A formal classification of bursting mechanisms in excitable systems. In: Teramato, E., Yamaguti, M. (eds.) Mathematical Topics in Population Biology, Morphogenesis, and Neurosciences. Lecture Notes in Biomath. Vol. 71 267-281, Springer-Verlag, Berlin-Heidelberg-New York
23. ——, (1987): Dissection of a model for neuronal parabolic bursting. J. Math. Biol. **25**, 653-675
24. Rinzel, J., Lee, Y.S. (1986): On different mechanisms for membrane potential bursting. In: Othmer, H.G. (ed.) Nonlinear Oscillations in Biology and Chemistry, Lecture Notes Biomath. **66**, 19-83, Springer-Verlag, Berlin-Heidelberg-New York
25. Rorsman, P., Trube, G. (1986): Calcium and delayed potassium currents in mouse pancreatic β-cells under voltage-clamp conditions. J. Physiology. **374**, 531-550
26. Sherman, A., Rinzel, J., Keizer, J. (1988): Emergence of organized bursting in clusters of pancreatic β-cells by channel sharing. Biophys. J. **54**, 411-425

Part II

Epidemiology

Models for the Spread of Universally Fatal Diseases II

Fred Brauer

Department of Mathematics, University of Wisconsin, Madison, Wisconsin 53706

Abstract

We consider a simple model for a universally fatal disease with an infective period long enough to allow natural deaths during the infective period. The analysis of this model is considerably more complicated than the analysis of a model with an infective period short enough that the population dynamics are confined to the susceptible class. However, the basic result that in some circumstances the stability of an endemic equilibrium may depend on the distribution of infective periods is shared by both models.

1 Introduction

There are simple classical models for the spread of infectious diseases due to Soper (1929) and Wilson-Burke (1942) which can be interpreted as models for universally fatal diseases in a population which would grow exponentially in the absence of disease. For a disease with recovery it is possible to incorporate births and deaths into the model but to keep the total population size constant. This is not possible for a universally fatal disease; in order to incorporate births and deaths into such a model and keep the total population size constant it is necessary to assume nonlinear population dynamics. Recently, a start has been made on the study of disease models in populations of varying sizes see for example Pugliese (1990) and Busenberg and Van den Driessche (1990). The classical models for infectious diseases assume either an exponential distribution of infective periods as in the Soper model or an infective period of fixed length as in the Wilson-Burke model. The first study including an arbitrary distribution of infective periods is the model of Cooke and Yorke (1973), which also describes age-structured population dynamics and a variety of other applications. In the model of Cooke and Yorke, for a disease with recovery with no immunity against re-infection, the behavior of the model depends on the mean infective period but not on the distribution of infective periods. A model for a universally fatal disease with an arbitary distribution of infective periods has been studied by Brauer (1990a), and it has been shown that in some circumstances the stability of an endemic equilibrium may depend on the distribution of infective periods. This analysis was carried out under the assumption that only the susceptible members of the populations contribute to the population dynamics, except for deaths fo infectives from the disease. Such an assumption is appropriate for rapidly debilitating animal diseases such as rabies but not for diseases with long infective periods such as AIDS. In this paper we formulate a model for a universally fatal disease with birth and death rates depending on total population size and with deaths other than from the disease distributed proportionally between the susceptible and infective classes. This model is of interest in its own right and is also a step towards a model for diseases which are fatal to some victims but from which others recover. There are diseases such as measles which are rarely fatal in developed countries but from which there is substantial mortality in less-developed countries. Another possible direction of extension would be towards a model for a disease with infectivity depending on age of infection but without a full age structure. Such a model may be of use in describing AIDS in a simple manner.

2 Basic model

The model studied in Brauer (1990a) is

$$\begin{aligned} S'(t) &= gS(t) - \hat{C}\{S(t)+I(t)\}S(t)I(t) \\ I(t) &= \int_0^t \hat{C}\{S(x)+I(x)\}S(x)I(x)\}P(t-x)dx, \end{aligned} \tag{1}$$

for values of t large enough that members who were infective at $t = O$ have been removed. The hypotheses which led to this model are:
(H1) The rate of change of population size in the absence of infection is a function g of population size. All births are in the susceptible class, all deaths other than from disease are in the susceptible class, and infective members do not contribute to the birth rate. The population has a carrying capacity K, with

$$g(K) = 0,\ g'(K) < 0,\ g(N) \le 0 \text{ for } N \ge K \tag{2}$$

(H2) The number of contacts per infective in unit time is a function $C(N)$ of total population size $N = S + I$, with

$$C(N) > 0,\ C'(N) \ge 0,\ [C(N)/N]' \le 0. \tag{3}$$

The rate of new infections is then $\hat{C}(N)SI$. It is convenient to define

$$\hat{C}(N) = C(N)/N$$

so that

$$\hat{C}'(N) \le 0 \tag{4}$$

and the rate of new infections is $\hat{C}(N)SI$. It is also convenient to define

$$\beta = \hat{C}(K). \tag{5}$$

(H3) The fraction of infectives remaining infective a time s after becoming infective is a function $P(s)$ with

$$P(0) = 1,\ P(s) \geq 0,\ \int_0^\infty P(s)ds = \tau < \infty,\ P \text{ non} - \text{increasing.} \tag{6}$$

It has been shown [Brauer (1990a)] that for the model (1) there is a contact number

$$K\hat{C}(K)\int_0^\infty P(s)ds = \beta\tau K.$$

If the contact number is less than 1 the system (1) has a single asymptotic equilibrium, namely the disease-free equilibrium $S = K$, $I = O$, and this equilibrium is asymptotically stable. If the contact number exceeds 1 the disease-free equilibrium is unstable, but there is also an endemic asymptotic equilibrium (S, I) with $S < K$, $I > O$. To analyze the stability of this equilibrium we form the characteristic equation, which has the form

$$b\hat{P}(\lambda) = \frac{\lambda + a}{\lambda + c} \tag{7}$$

where $\hat{P}(\lambda)$ denotes the Laplace transform of P,

$$\hat{P}(\lambda) = \int_0^\infty e^{-\lambda s}P(s)ds$$

and

$$a = I\hat{C}(S + I) + SI\hat{C}'(S + I) - g'(S)$$

$$b = s\hat{C}(S + I) + SI\hat{C}'(S + I)$$

$$c = -g'(S).$$

Then the following result is applicable.

Theorem 1 [Hethcote, Stech, and van den Driessche (1981)] *If $a > |c|$ and $0 < b\tau \leq 1$, then all roots of (7) have negative reeal part, but if $a > c$ with c negative and $|c|$ sufficiently large, there may be roots with positive real part.*

It is not difficult to verify with the aid of (3) and (4) that $a > c$, $0 < b\tau \leq 1$. Thus if $g'(S) < 0$ all roots of (7) have negative real part but if $g'(S) > 0$ there is a possibility of roots of (7) with positive real part. However if $P(s) = e^{-s/\tau}$ all roots of (7) have negative real part regardless of the value of $g'(S)$ [Brauer (1990a)], while if $P(s) = 1$ $(0 \leq s \leq \tau)$, $P(s) = 0$ $(s > \tau)$, there can indeed be roots with positive real part [Hethcote, Stech, and van den Driessche (1981)].

The significance of this result is that for a fatal disease modelled by the system (1) the stability of the endemic equilibrium may depend on the distribution of infective periods if the contact number is high enough and if the population dynamics permit an equilibrium with $g'(S) > 0$.

3 Long infective period

In order to model a disease with an infective period long enough to allow natural deaths during the infective period, we replace the hypothesis (H1) by a different assumption on the population dynamics. We continue to assume (H2) and (H3) but instead of (H1) we assume: (H1*) There is a birth rate $B(S)$ per susceptible and a death rate $D(N)$ per member of the population, $N = S + I$. All births are in the susceptible class and the death rate in each class is proportional to the size of the class. The population has a carrying capacity K,

$$B(K) = D(K), \ B'(K) < D'(K), \ B(N) \leq D(N) \ \text{if} \ N \geq K. \tag{8}$$

In addition, we assume

$$\begin{aligned} B'(S) \leq 0, \ [SB(S)]' = B(S) + SB'(S) \geq 0 \\ 0 \leq D'(N) \leq D(N)/N. \end{aligned} \tag{9}$$

The assumption (H1*) implies that in unit time there are $SD(N)$ deaths in the susceptible class and $ID(N)$ (natural) deaths in the infective class. If $z(t)$ denotes the number of members who became infective at time x who have not died of natural causes by time t, then $z'(t) = -z(t)D\{N(t)\}$, and this implies

$$z(t) = z(x)e^{-\int_x^t D\{N(y)\}dy}.$$

Thus the fraction of the members who became infective at time x and who have not died either of natural causes or from disease is

$$P(t-x)\exp\left(-\int_x^t D\{N(y)\}dy\right).$$

This leads us to the model

$$\begin{aligned} S'(t) &= S(t)B\{S(t)\} - S(t)D\{N(t)\} - \hat{C}\{N(t)\}S(t)I(t) \\ I(t) &= \int_0^t \hat{C}\{N(x)\}S(x)I(x)e^{-\int_x^t D\{N(y)\}dy}P(t-x)dx. \end{aligned} \tag{10}$$

Here it is convenient to use N in the model to denote $S + I$.

The conditions for an asymptotic equilibrium (S, I) of (10), with $N = S + I$, are

$$SB(S) = SD(N) + \hat{C}(N)SI$$

$$I = \hat{C}(N)SI \int_0^\infty e^{-D(N)S} P(s)ds.$$

Then either $I = 0$, which implies $SB(S) = SD(N)$, so that $S = N = K$, or

$$\hat{C}(N)S \int_0^\infty e^{-D(N)s} P(s)ds = 1.$$

We define

$$Q(s) = e^{-D(N)s} P(s)$$

and

$$\hat{Q}(\lambda = \int_0^\infty e^{-\lambda s} Q(s)ds,$$

so that

$$\hat{Q}(0) = \int_0^\infty Q(s)ds.$$

Then the conditions for an endemic equilibrium are

$$\begin{aligned} B(S) &= D(N) + \hat{C}(N)I \\ S\hat{C}(N)\hat{Q}(0) &= 1. \end{aligned} \tag{11}$$

The existence of an endemic equilibrium requires $S < K$, or $K\hat{C}(N)\hat{Q}(0) > 1$. If there is an endemic equilibrium we have

$$\frac{1}{\hat{Q}(0)} = S\hat{C}(N) < N\hat{C}(N) = C(N) \leq C(K) = \beta K$$

by (3) and (5). Thus the existence of an endemic equilibrium requires $\beta K\hat{Q}(0) > 1$. The same argument as that used for the model of Section 2 [Brauer (1990a)] shows that the disease-free equilibrium $(K, 0)$ of (10) is asymptotically stable if and only if

$$\beta K\hat{Q}(0) < 1.$$

In particular, an endemic equilibrium exists if and only if the disease-free equilibrium is unstable.

The linearization of the system (10) about an endemic equilibrium (S, I) is

$$\begin{aligned} u' &= [B(S) + SB'(S) - D(N) - SD'(N) - SI\hat{C}'(N) - I\hat{C}(N)]u \\ &\quad - [SD'(N) + SI\hat{C}'(N) + S\hat{C}(N)]v, \end{aligned}$$

$$\begin{aligned} v(t) &= \int_0^t [I\hat{C}(N) + SI\hat{C}'(N)]Q(t-x)u(x)dx \\ &\quad + \int_0^t [S\hat{C}(N) + SI\hat{C}'(N)]Q(t-x)v(x)dx \\ &\quad - \int_0^t SID'(N)\hat{C}(N)Q(t-x)\left[\int_x^t \{u(y) + v(y)\}dy\right] dx. \end{aligned}$$

A complicated calculation gives the characteristic equation at the endemic equilibrium

$$\hat{Q}(\lambda) = \frac{\lambda^2 + d\lambda + \hat{Q}(0)c}{a\lambda^2 + b\lambda + c} \tag{12}$$

with

$$\begin{aligned} a &= \nu S \\ b &= [D(N) + SD'(N) - B(S) - SB'(S)]\nu S - \mu ISD'(N) + \gamma \\ c &= \gamma[D(N) + \mu I - \nu S - B(s) - SB'(S)] \\ d &= \mu I + D(N) + SD'(N) - B(S) - SB'(S) + \gamma\hat{Q}(0) \end{aligned} \tag{13}$$

where

$$\begin{aligned} \mu &= \hat{C}(N) + S\hat{C}'(N) \\ \nu &= \hat{C}(N) + I\hat{C}'(N) \\ \gamma &= \hat{C}(N)SID'(N). \end{aligned} \tag{14}$$

From the condition (3) it is easy to deduce that

$$\begin{gathered} 0 \le -I\hat{C}'(N) \le \mu \le \hat{C}(N) \\ 0 \le -S\hat{C}'(N) \le \nu \le \hat{C}(N). \end{gathered} \tag{15}$$

In the particular case when $D(N)$ is a constant, so that $D'(N) = 0$, we have $\gamma = 0$ and $c = 0$. In this case the characteristic equation (12) reduces to

$$a\hat{Q}(\lambda) = \frac{\lambda + d}{\lambda + \frac{b}{a}},$$

which is of the form (7) and can be analyzed by Theorem 1. It is easy to show using (15) that

$$0 < a\hat{Q}(0) \le 1 \tag{16}$$

and that

$$d - \frac{b}{a} = \beta I + ID'(N) = [\beta + D'(N)]I > 0,$$

where $\beta = \hat{C}(K) = \hat{C}(N)$, so that $d > b/a$. If the contact number exceeds 1, so that there is an endemic equilibrium, but is close to 1, then I is close to zero and N and S are close to K. For such contact numbers

$$\begin{aligned} b &\approx \beta K[D(K) + KD'(K) - B(K) - KB'(K)] \\ &= \beta K^2[D'(K) - B'(K)] > 0, \end{aligned}$$

using (8). Then Theorem 1 shows that the endemic equilibrium is asymptotically stable for contact numbers close enough to 1, but may become unstable for large contact numbers with some choices of $P(s)$. In other words, the qualitative behavior of the model (10) is the same as that of the model (1) in the special case $D'(N) = 0$.

4 Analysis of the general case

In the general case $D'(N) \neq 0$, the analysis of the characteristic equation (12) is considerably more complicated. The hypothesis (9) implies $\gamma > 0$. We have the following result, whose proof may be found in the appendix.

Theorem 2 *Under the conditions*

$$c < 0,\ 0 < a\hat{Q}(0) \leq 1,\ 0 < b\hat{Q}(0) \leq d \tag{17}$$

and

$$\int_0^\infty sQ(s)ds < \frac{d - b\hat{Q}(0)}{-c} \tag{18}$$

all roots of the characteristic equation (12) *have negative real part.*

For a general Q we can analyze the stability of the endemic equilibrium only for contact numbers close to 1. If the contact number is close to 1, so that $S \approx K$, $N \approx K$, $I \approx 0$, we have

$$\hat{C}(N) \approx \beta,\ \ \hat{Q}(0) \approx \frac{1}{\beta K} \tag{19}$$

because of (5) and (11). For such contact numbers we also have

$$\mu \approx \beta + K\hat{C}'(K) > 0,\ \nu \approx \beta,\ \frac{\gamma}{I} \approx \beta K D'(K). \tag{20}$$

Using (19) and (20) we expand in powers of I to obtain

$$\begin{aligned} a &\approx \beta K > 0 \\ b &\approx K[D'(K) - B'(K)]\beta K - K^2\hat{C}'(K)D'(K)I > 0 \\ c &\approx -\beta K D'(K)I[\beta K + KB'(K)] < 0 \\ d &\approx K[D'(K) - B'(K)] + [\beta + K\hat{C}'(K) + D'(K)]I > 0. \end{aligned} \tag{21}$$

In (21) we have retained the terms in I in the approximations for b, c, d because they will be needed in the application of Theorem 2 to the model (10). Then

$$\begin{aligned} d - b\hat{Q}(0) &\approx \left[\beta + K\hat{C}'(K) + D'(K) + \frac{K\hat{C}'(K)D'(K)}{\beta}\right] I \\ &= [K\hat{C}'(K) + \beta]\left[1 + \frac{D'(K)}{\beta}\right] I \geq 0. \end{aligned} \tag{22}$$

Because of (16) we have now established that all conditions in (17) are satisfied for contact numbers close to 1.

Again using (21) along with (22) we have

$$\frac{d - b\hat{Q}(0)}{-c} \approx \frac{[K\hat{C}'(K) + \beta][1 + \frac{D'(K)}{\beta}]}{\beta K D'(K)[\beta K + KB'(K)]}. \tag{23}$$

We can not hope to establish the inequality (18) for arbitrary $\hat{C}(N)$ because while $K\hat{C}'(K)+\beta \geq 0$ in general, the choice $C(N)=\lambda$, so that $\hat{C}(N)=\frac{\lambda}{N}$ and $\lambda = \beta K$, gives $K\hat{C}'(K)+\beta = 0$ and

$$\frac{d-b\hat{Q}(0)}{-c} \approx 0.$$

The choice $\hat{C}(N)=\beta$ gives

$$\frac{d-b\hat{Q}(0)}{-c} \approx \frac{D'(K)+\beta}{\beta K D'(K)[\beta K + KB'(K)]}. \tag{24}$$

Integration by parts gives

$$\begin{aligned}\int_0^\infty sQ(s)ds &= \int_0^\infty se^{-D(N)s}P(s)ds \\ &= -\frac{1}{D(N)}[sP(s)e^{-D(N)s}]_0^\infty + \frac{1}{D(N)}\int_0^\infty e^{D(N)s}[sP'(s)+P(s)]ds \\ &= \frac{1}{D(N)}\int_0^\infty e^{-D(N)s}[sP'(s)+P(s)]ds \\ &\leq \frac{1}{D(N)}\int_0^\infty e^{-D(N)s}P(s)ds = \hat{Q}(0)/D(N),\end{aligned}$$

using $P'(s) \leq 0 \quad (0 \leq s < \infty)$. Now, for contact numbers close to 1, we can estimate $\int_0^\infty sQ(s)ds$ by $1/\beta K D(K)$. Since $B'(K) \geq 0, D'(K) \leq D(K)/K$ by the hypothesis (8), (24) gives

$$\frac{d-b\hat{Q}(0)}{-c} > \frac{\beta}{\beta^2 K^2 D'(K)} > \frac{1}{\beta K D(K)}.$$

From this we see that (18) is satisfied and thus that Theorem 2 is applicable. We now have the following result.

Theorem 3. *If the function $\hat{C}(N)$ is constant, then the endemic equilibrium of the model* (10) *is asymptotically stable at least for contact numbers sufficiently close to 1.*

The question of stability of the endemic equilibrium for more general $\hat{C}(N)$ is open. It is reasonable to conjecture that while the choice $\hat{C}(N)=\beta K/N$ must be excluded there is a class of non-constant functions $\hat{C}(N)$ for which stability can be established.

5 A fox rabies model

A model for fox rabies has been proposed [Anderson et al (1981)] which is of the form (10) except for the incorporation of an exposed period and has an exponential distribution of infective periods. This model exhibits instability of the endemic equilibrium for high contact numbers, but if the model did not include an exposed period the endemic equilibrium would be asymptotically stable for all contact numbers.

We shall examine the special case of (10) with $P(s) = e^{-s/\tau}$ for which the model (10) reduces to the system of ordinary differential equations

$$\begin{aligned} S' &= SB(S) - SD(N) - \hat{C}(N)SI \\ I' &= \hat{C}(N)SI - ID(N) - \frac{1}{\tau}I. \end{aligned} \tag{25}$$

The linearization of (25) about an equilibrium (S, I) has coefficient matrix

$$M = \begin{bmatrix} B(S) + SB'(S) - D(N) - SD'(N) - \mu I & -SD'(N) - \nu S \\ [\mu - D'(N)]I & \nu S - D(N) - \frac{1}{\tau} - ID'(N) \end{bmatrix}$$

The endemic equilibrium satisfies

$$S\hat{C}(N) = D(N) + \frac{1}{\tau}, \quad B(S) = D(N) + \hat{C}(N)I$$

and this enables us to rewrite this coefficient matrix as

$$M = \begin{bmatrix} SB'(S) - SD'(N) - SI\hat{C}'(N) & -DS'(N) - \nu S \\ [\mu - D'(N)]I & [S\hat{C}'(N) - D'(N)]I \end{bmatrix}.$$

The endemic equilibrium is asymptotically stable if and only if $\operatorname{tr} M < 0$, $\det M > 0$. We have

$$\begin{aligned} \operatorname{tr} M &= SB'(S) - SD'(N) - SI\hat{C}'(N) + SI\hat{C}'(N) - D'(N)I \\ &= SB'(S) - ND'(N) < 0, \end{aligned}$$

because of (9). Also,

$$\begin{aligned} \frac{\det M}{SI} &= [B'(S) - D'(N) - I\hat{C}'(N)][S\hat{C}'(N) - D'(N)] \\ &\quad + [D'(N) + \nu][\mu - D'(N)] \\ &= B'(s)S\hat{C}'(N) - B'(S)D'(N) \\ &\quad + [\hat{C}(N)]^2 + S\hat{C}(N)\hat{C}'(N) + I\hat{C}(N)\hat{C}'(N) \\ &= B'(S)[S\hat{C}'(N) - D'(N)] + \hat{C}(N)[\hat{C}(N) + N\hat{C}'(N)] \end{aligned}$$

and this is positive because of the assumptions (3), (4), (9). This establishes the asymptotic stability of the endemic equilibrium of (25) for all contact numbers. We thus have the same situation observed for the simpler model (1): For some choices of birth and death rates, destabilization of the endemic equilibrium may depend on the distribution of infective periods.

Another possible cause of destabilization of the endemic equilibrium is an exposed period, as in the rabies model of Anderson et al (1981). Models with nonlinear population dynamics and with exposed and infective periods of fixed length have been formulated as delay equations with two delays [Brauer (1989)]. The formulation of models with arbitrarily distributed exposed and infective periods leads to integral equations whose kernel is the convolution of the exposed and infective kernels [Hethcote and Tudor (1980)]. Models with nonlinear population dynamics and arbitrarily distributed exposed and infective periods remain to be formulated and analyzed.

Another direction of generalization would be a model for a disease from which a fraction of infectives recover. Such a model would have to generalize the model (10) of Section 3 by allowing natural deaths in each class rather than the model (1) of Section 2. It should also have a birth rate of susceptibles depending on the recovered class size as well as the susceptible class size. If the fraction of infectives who recover is p $(0 \leq p \leq 1)$, a model would be

$$S' = (S+R)B(S+R) - SD(N) - \hat{C}(N)SI$$

$$I(t) = \int_0^t \hat{C}\{N(x)\}S(x)I(x)e^{-\int_x^t D\{N(y)\}dy}P(t-x)dx \tag{26}$$

$$R'(t) = -p\int_0^t \hat{C}\{N(x)\}S(x)I(x)e^{-\int_x^t D\{N(y)\}dy}P'(t-x)dx - R(t)D\{N(t)\}.$$

The model (10) is the special case $p = 0$ of (26). The case $p = 1$ of (26) would be an S-I-R model with recovery, for which the endemic equilibrium is always asymptotically stable [Brauer (1990b)]. Thus as p varies from 0 to 1 there may be a transition from instability and oscillation about the endemic equilibrium to stability, and the dependence of the behavior on the recovery fraction p is of interest.

Appendix

In order to prove Theorem 2, we begin with a general but simple lemma.

Lemma a *Suppose that f and g are analytic in an open set containing the right half plane $Re\lambda \geq 0$ with $f(\overline{z}) = \overline{f(z)}$, $g(\overline{z}) = \overline{g(z)}$ and assume that*

(i) $f(0) = g(0) > 0$

(ii) $|f(iy)| < |g(iy)|, \quad 0 < y < \infty$

(iii) g *has a single zero in* $Re\lambda > 0$

(iv) $f'(0) > g'(0)$.

Then, except for a simple root at $\lambda = 0$, all roots of $f(\lambda) = g(\lambda)$ satisfy $Re\lambda < 0$.

Proof. We consider the equation $rf(\lambda) = g(\lambda)$ with r varying from 0 to 1. For $r = 0$ there is a single root in $Re\lambda > 0$ and roots depend continuously on r. No root crosses the imaginary axis for $0 \leq r < 1$ because a crossing would require that either $\lambda = 0$ or $\lambda = iy$ is a root for some value of r, impossible since

$$|rf(0)| < |f(0)| = |g(0)|$$

$$|rf(iy)| < |f(iy)| < |g(iy)|.$$

There is a root $\lambda(r)$ with $\lambda(1) = 0$ because of (i). Implicit differentiation of $rf\{\lambda(r)\} = g\{\lambda(r)\}$ gives

$$\lambda'(r) = \frac{f\{\lambda(r)\}}{-rf'\{\lambda(r)\} + g'\{\lambda(r)\}}$$

and letting $r \to 1-$ we obtain

$$\lambda'(1) = \frac{f(0)}{g'(0) - f'(0)} < 0.$$

Thus the root $\lambda(r)$ approaches zero from the right half plane and $\lambda(0)$ must be the zero of g in the right half plane. This leaves no roots of $f(\lambda) = g(\lambda)$ in $Re\lambda \geq 0$.

To prove Theorem 2 we apply this lemma with

$$f(\lambda) = \hat{Q}(\lambda), \quad g(\lambda) = \frac{\lambda^2 + d\lambda + \hat{Q}(0)c}{a\lambda^2 + b\lambda + c}.$$

Then $f(0) = g(0) = \hat{Q}(0) > 0$ if $c \neq 0$ and g has a single zero with $Re\lambda > 0$ if $a > 0, c < 0$. Because

$$f'(0) = -\int_0^\infty sQ(s)ds, \quad g'(0) = \frac{d - b\hat{Q}(0)}{c},$$

the condition (iv) is satisfied if

$$-\int_0^\infty sQ(s)ds > \frac{d - b\hat{Q}(0)}{c}.$$

If $d > 0, c < 0$, this is equivalent to

$$\int_0^\infty sQ(s)ds < \frac{d - b\hat{Q}(0)}{-c}.$$

The verification of the hypothesis (ii) is more complicated. It is easy to calculate $|f(iy)| \leq \hat{Q}(0)$ and

$$|g(iy)|^2 = \frac{y^4 + \{d^2 - 2\hat{Q}(0)c\}y^2 + \{\hat{Q}(0)\}^2c^2}{a^2y^4 + (b^2 - 2ac)y^2 + c^2}.$$

We wish to minimize $|g(iy)|^2$ over $0 \leq y < \infty$. The sign of the derivative of $|g(iy)|^2$ with respect to y for $y = 0$ is the same as the sign of $\{d^2 - b^2[\hat{Q}(0)]^2\} - 2\hat{Q}(0)c\{1 - a\hat{Q}(0)\}$ and this is positive if $b > 0$, $c < 0$, $d > 0$, $b\hat{Q}(0) \leq d$, $a\hat{Q}(0) \leq 1$. If the function $|g(iy)|^2$ has no critical points, the minimum of $|g(iy)|^2$ must be $[\hat{Q}(0)]^2$, attained for $y = 0$. If $|g(iy)|^2$ has only one critical point, this critical point is a relative maximum and the minimum of $|g(iy)|^2$ is the smaller of $|g(0)|^2 = [\hat{Q}(0)]^2$ and $\lim_{y\to\infty} |g(iy)|^2 = \frac{1}{a^2}$. If $a\hat{Q}(0) \leq 1$, this minimum is again $[\hat{Q}(0)]^2$. In either case, $|f(iy)| < |g(iy)|$ for $0 < y < \infty$, and thus it remains to show that $|g(iy)|^2$ has at most one critical point.

We let $z = y^2$, $h(z) = |g(i\sqrt{2})|^2$ for $0 < z < \infty$, so that

$$h(z) = \frac{z^2 + \{d^2 - 2\hat{Q}(0)c\}z + [\hat{Q}(0)]^2c^2}{a^2z^2 + (b^2 - 2ac)z + c^2}.$$

If $a > 0$, $c < 0$, the denominator does not vanish for $0 \leq z < \infty$. The derivative of $h(z)$ has the sign of

$$(s - ar^2)z^2 + 2c^2\{1 - a^2[\hat{Q}(0)]^2\}z + \{rc^2 - [\hat{Q}(0)]^2c^2s\}$$

with

$$r = d^2 - 2\hat{Q}(0)c, \quad s = b^2 - 2ac.$$

Both r and s are nonnegative if $a > 0$, $c < 0$. If $a\hat{Q}(0) \leq 1$ and $|b|\hat{Q}(0) \leq d$, we have

$$rc^2 - [\hat{Q}(0)]^2c^2s = \{d^2 - b^2[\hat{Q}(0)]^2\} - 2\hat{Q}(0)c\{1 - a\hat{Q}(0)\} \geq 0.$$

Thus if $s - ar^2 \geq 0$ the derivative of $h(z)$ has no positive zero and if $s - ar^2 < 0$ the derivative of $h(z)$ has a single positive zero; in either case $|g(iy)|^2$ has at most one positive critical point.

We have now completed the verification of the hypotheses (i)-(iv) and may apply the lemma to yield the desired result:

Theorem 2 *If $c < 0$, $0 < a\hat{Q}(0) \leq 1$, $0 < |b|\hat{Q}(0) \leq d$ and if*

$$\int_0^\infty sQ(s)ds < \frac{d - b\hat{Q}(0)}{-c},$$

then all roots of the equation

$$\hat{Q}(\lambda) = \int_0^\infty e^{-\lambda s}Q(s)ds = \frac{\lambda^2 + d\lambda + \hat{Q}(0)c}{a\lambda^2 + b\lambda + c},$$

where $Q(0) = 1$, Q is non-increasing and $\int_0^\infty Q(s)ds < \infty$, have negative real part.

References

1. Anderson, R.M., Jackson, H.C., May, R.M., Smith, A.M. (1981): Population dynamics of fox rabies in Europe. Nature **289**, 765–771
2. Brauer, F. (1989): Epidemic models in populations of varying size. In "Mathematical Approaches to Problems in Resource Management and Epidemiology", C. Castillo-Chavez, S.A. Levin, and C. Shoemaker (eds.), Lecture Notes in Biomathematics **81**, Springer-Verlag, Berlin–Heidelberg–New York, 109–123
3. Brauer, F. (1990a): Models for the spread of universally fatal diseases. J. Math. Biology **28**, 451–462
4. Brauer, F. (1990b): Some infectious disease models with population dynamics and general contact rates. Differential and Integral Equations **5**, 827-836
5. Busenberg, S., van den Driessche, P. (1990): Analysis of a disease transmission model in a population with varying size. J. Math. Biol. **28**, 257–270
6. Cooke, K.L., Yorke, J.A. (1973): Some equations modelling growth processes and gonorrhea epidemics. Math. Biosc. **16**, 75–101
8. Hethcote, H.W., Stech, H.W., van den Driessche P. (1981): Stability analysis for models of diseases without immunity. J. Math. Biol. **13**, 185–198
9. Hethcote, H.W., Tudor, D.W. (1980): Integral equation models for endemic infectious diseases. J. Math. Biol. **9**, 37–47
10. Pugliese, A. (1990): Population models for diseases with no recovery. J. Math. Biol. **28**, 65–82
11. Soper, H.E. (1929): Interpretation of periodicity in disease prevalence. J. Royal. Statistical Soc. **92** , 34–73
12. Wilson, E.B., Burke, M.H. (1942): The epidemic curve. Proc. Nat. Acad. Sci. **28** , 361–367

Nonexistence of Periodic Solutions for a Class of Epidemiological Models

Stavros Busenberg[1] *and P. van den Driessche*[2]

[1] Department of Mathematics, Harvey Mudd College, Claremont, CA 91711 USA
[2] Department of Mathematics, University of Victoria, Victoria, B.C. V8W 3P4 Canada

Dedicated, on the occasion of his 65th birthday, to Kenneth L. Cooke who inspired us both to work in this area.

Abstract

Disease transmission models are formulated under assumptions that the size of the population varies and the force of infection is of the proportionate mixing type. Conditions are given that rule out the possibility of periodic solutions for such models. Examples are considered and sharp thresholds identified.

1 Introduction

The behavior of a general $SAIS$ (S = susceptible, A = asymptomatic, I = infective,) epidemiological model in a population of varying size is governed by a differential equation in $\mathbf{R}^3_+$. When the force of infection is of the proportionate mixing type, the nonlinear terms of this equation are homogeneous of degree one and satisfy a balance condition. Moreover, in the linear part, the off-diagonal terms are nonnegative.

We consider a general equation of this form, and give conditions which rule out periodic solutions, including limit cycles, homoclinic orbits and oriented phase polygons. Our method involves a new technique and extends results of Busenberg and van den Driessche (1990) for an $SIRS$ (R = recovered) model, in which a generalization of the Bendixson-Dulac criterion is proved and used in the analysis.

In certain special cases, the nonexistence of periodic solutions, in combination with analysis of the existence and stability of equilibrium points, provides a complete global analysis of the model.

2 Mathematical formulation

We consider a differential equation in $\mathbf{R}^3_+$ of the form

$$x' = Mx + f(x) \tag{2.1}$$

where $'$ denotes the derivative d/dt. The 3×3 constant matrix $M = [m_{ij}]$ is assumed to be ***essentially nonnegative***, that is the off-diagonal entries of M are nonnegative, and it is also assumed that at least one is positive. The nonlinear function $f : \mathbf{R}^3_+ \to \mathbf{R}^3_+$ is assumed to be continuously differentiable and ***homogeneous of degree one***, that is $f(ax) = af(x)$ for $a > 0$, and to obey the balance condition $\sum_{i=1}^{3} f_i = 0$. In the disease transmission models, we are concerned with solutions having nonnegative components $x_i(t) \geq 0$, and we also assume that $(Mx)_i + f_i(x) \geq 0$ when evaluated on $x_i = 0$. For $\sum_{i=1}^{3} x_i \neq 0$, we introduce the normalized variable (see, e.g., Hahn (1967, Section 57), Hadeler et al. (1988), Hofbauer and Sigmund (1988), Busenberg and Hadeler (1990), Busenberg and van den Driessche (1990))

$$y = x/\sum_{i=1}^{3} x_i, \tag{2.2}$$

with $\sum_{i=1}^{3} y_i = 1$. The normalized variable satisfies

$$y' = My - (\mathbf{1} \cdot My)y + f(y), \tag{2.3}$$

where $\mathbf{1}$ denotes the vector in $\mathbf{R}^3$ with every entry equal to one. Clearly, the hyperplane $S \equiv \left\{ y : \sum_{i=1}^{3} y_i = 1 \right\}$ is invariant under the flow induced by (2.3), since

$$\left(\sum_{i=1}^{3} y_i \right)' = (\mathbf{1} \cdot My) \left(1 - \sum_{i=1}^{3} y_i \right) + \sum_{i=1}^{3} f_i(y) = 0.$$

We now write (2.3) in component form for $i = 1, 2, 3$:

$$y_i' = (My)_i - (\mathbf{1} \cdot My)y_i + f_i(y_1, y_2, y_3). \tag{2.4}$$

On S we can write the first equation of (2.4) in the following alternate forms:

$$y_1' = f_{12}(y_1, y_2) + f_1(y_1, y_2, 1-y_1-y_2) = f_{13}(y_1, y_3) + f_1(y_1, 1-y_1-y_3, y_3), \tag{2.5}$$

where

$$\begin{aligned} f_{12} \equiv\ & (m_{11} - m_{13})y_1 + (m_{12} - m_{13})y_2 + m_{13} \\ & - y_1[(m_{11} + m_{21} + m_{31} - m_{13} - m_{23} - m_{33})y_1 \\ & + (m_{12} + m_{22} + m_{32} - m_{13} - m_{23} - m_{33})y_2 + m_{13} + m_{23} + m_{33}]. \end{aligned} \tag{2.6}$$

The function f_{13} is defined from f_{12} by interchanging subscripts 2 and 3. In (2.5) the first equation is in terms of y_1 and y_2 only, whereas the second equation is in terms of y_1 and y_3 only, and f_1 contains the homogeneous terms. Similarly we can write

$$y_2' = f_{23}(y_2, y_3) + f_2(1 - y_2 - y_3, y_2, y_3) = f_{21}(y_1, y_2) + f_2(y_1, y_2, 1 - y_1 - y_2) \tag{2.7}$$

where

$$\begin{aligned} f_{23} \equiv\ & (m_{22} - m_{21})y_2 + (m_{23} - m_{21})y_3 + m_{21} \\ & - y_2[(m_{12} + m_{22} + m_{32} - m_{11} - m_{21} - m_{31})y_2 \\ & + (m_{13} + m_{23} + m_{33} - m_{11} - m_{21} - m_{31})y_3 + m_{11} + m_{21} + m_{31}], \end{aligned} \tag{2.8}$$

and f_{21} is defined from f_{23} by interchanging subscripts 3 and 1;

$$y_3' = f_{31}(y_1, y_3) + f_3(y_1, 1 - y_1 - y_3, y_3) = f_{32}(y_2, y_3) + f_3(1 - y_2 - y_3, y_2, y_3) \tag{2.9}$$

where

$$\begin{aligned} f_{31} \equiv\ & (m_{31} - m_{32})y_1 + (m_{33} - m_{32})y_3 + m_{32} \\ & - y_3[(m_{11} + m_{21} + m_{31} - m_{12} - m_{22} - m_{32})y_1 \\ & + (m_{13} + m_{23} + m_{33} - m_{12} - m_{22} - m_{32})y_3 + m_{12} + m_{22} + m_{32}], \end{aligned} \tag{2.10}$$

and f_{32} is defined from f_{31} by interchanging subscripts 1 and 2.

Now let $S^+ \equiv S \cap \mathbf{R}^3_+$, $S^0 \equiv S^+ - \partial S^+$, and define g on S^0 so its transpose g^T is given by

$$\begin{aligned} g^T \equiv (\ & \frac{1}{y_1y_3} f_{31}(y_1, y_3) + \frac{1}{y_1y_3} f_3(y_1, 1 - y_1 - y_3, y_3) - \frac{1}{y_1y_2} f_{21}(y_1, y_2) \\ & - \frac{1}{y_1y_2} f_2(y_1, y_2, 1 - y_1 - y_2), \quad \frac{1}{y_1y_2} f_{12}(y_1, y_2) \\ & + \frac{1}{y_1y_2} f_1(y_1, y_2, 1 - y_1 - y_2) - \frac{1}{y_2y_3} f_{32}(y_2, y_3) \\ & - \frac{1}{y_2y_3} f_3(1 - y_2 - y_3, y_2, y_3), \quad \frac{1}{y_2y_3} f_{23}(y_2, y_3) \\ & + \frac{1}{y_2y_3} f_2(1 - y_2 - y_3, y_2, y_3) - \frac{1}{y_1y_3} f_{13}(y_1, y_3) \\ & - \frac{1}{y_1y_3} f_1(y_1, 1 - y_1 - y_3, y_3)\ ). \end{aligned} \tag{2.11}$$

Note that g on S is equal to $y \times (My - (1 \cdot My)y + f(y))/(y_1y_2y_3)$. Then $\nabla \times g$ is composed of two terms, one involving the functions f_{ij} coming from the linear part of (2.1) and the other involving the functions f_i coming from the nonlinear homogeneous part of (2.1). We call these $\nabla \times g_M$ and $\nabla \times g_f$, respectively, and have the following results, the first of which simply collects two relations we need in the sequel.

Lemma 2.1 *Using the definitions and assumptions above,*

$$\nabla \times g_M \cdot \mathbf{1} = -\left(\frac{m_{13}}{y_1^2 y_2} + \frac{m_{23}}{y_1 y_2^2} + \frac{m_{32}}{y_1 y_3^2} + \frac{m_{12}}{y_1^2 y_3} + \frac{m_{21}}{y_2^2 y_3} + \frac{m_{31}}{y_2 y_3^2}\right), \tag{2.12}$$

and

$$\nabla \times g_f \cdot \mathbf{1} = \frac{f_2 y_3(y_2 - y_1) + f_3 y_2(y_3 - y_1)}{y_1^2 y_2^2 y_3^2} + \frac{\frac{\partial f_2}{\partial y_2} - \frac{\partial f_2}{\partial y_1} + \frac{\partial f_3}{\partial y_3} - \frac{\partial f_3}{\partial y_1}}{y_1 y_2 y_3}. \tag{2.13}$$

Proof. As $\mathbf{1} = (1,1,1)$, the term $\nabla \times g_M \cdot \mathbf{1}$ is computed by adding the terms of $\nabla \times g$ involving the functions f_{ij} from (2.11). Performing these simple but lengthy computations and collecting terms gives $\nabla \times g_M \cdot \mathbf{1}$ as in (2.12). Similarly for the nonlinear terms, and using the fact that $f_1 = -f_2 - f_3$, gives (2.13). □

Theorem 2.2 *Assume that $M = [m_{ij}]$ has $m_{ij} \geq 0$ for $i \neq j$, with at least one strict inequality; and $f : R_+^3 \to R_+^3$ is continuously differentiable, homogeneous of degree 1 with $f_1 + f_2 + f_3 = 0$. If on S^0*

$$f_2 y_3(y_2 - y_1) + f_3 y_2(y_3 - y_1) + y_1 y_2 y_3 \left(\frac{\partial f_2}{\partial y_2} - \frac{\partial f_2}{\partial y_1} + \frac{\partial f_3}{\partial y_3} - \frac{\partial f_3}{\partial y_1}\right) \leq 0, \tag{2.14}$$

then there are no periodic solutions of the system (2.3), namely $y' = My - (\mathbf{1} \cdot My)y + f(y)$, in S^0. Moreover, if $(My)_i - (\mathbf{1} \cdot My)y_i + f_i(y)$ evaluated at $y_i = 0$ is nonnegative for $i = 1,2,3$, and is positive for at least one i, then this result holds in the invariant region S^+.

Proof. The nonexistence of periodic solutions (including closed orbits, homoclinic loops and oriented phase polygons) follows from Theorem 4.1 of Busenberg and van den Driessche (1990). We check the conditions there, namely

$$g \cdot (My - (\mathbf{1} \cdot My)y + f(y)) = 0$$

and (by Lemma (2.1), (2.14) and the assumptions on M) we have

$$\nabla \times g \cdot \mathbf{1} = \nabla \times g_M \cdot \mathbf{1} + \nabla \times g_f \cdot \mathbf{1} < 0.$$

The fact that S is invariant has already been established, while the invariance of $\mathbf{R}_+^3$ follows from the fact that (2.3) and assumptions on M and f ensure that y_i' evaluated at $y_i = 0$ is nonnegative. If, in addition, one of these terms is strictly positive, then the boundary ∂S^+ is not a phase polygon for the system, and the nonexistence of closed orbits in S^+ is established. □

The balance condition, $\sum_{i=1}^{3} f_i = 0$, which occurs naturally in the epidemiological models we consider, is needed only to obtain a simpler form (2.13) for

$\nabla \times g_f \cdot \mathbf{1}$. Conditions other than (2.14) are needed when this does not hold. As M is assumed to be essentially nonnegative with at least one negative off-diagonal entry, $\nabla \times g_M \cdot \mathbf{1}$ is strictly negative by (2.12).

3 A general epidemiological model

We consider a model of disease transmission in a nonconstant population N divided into three groups, susceptibles, asymptomatics (infectives *without* symptoms) and infectives (infectives *with* symptoms), the numbers in each class being given by S, A, I, respectively, thus $N = S + A + I$. For a discussion of the role of asymptomatic individuals in a constant population model, see Cooke (1982). Our model is given schematically in figure 1, and all parameters are assumed nonnegative.

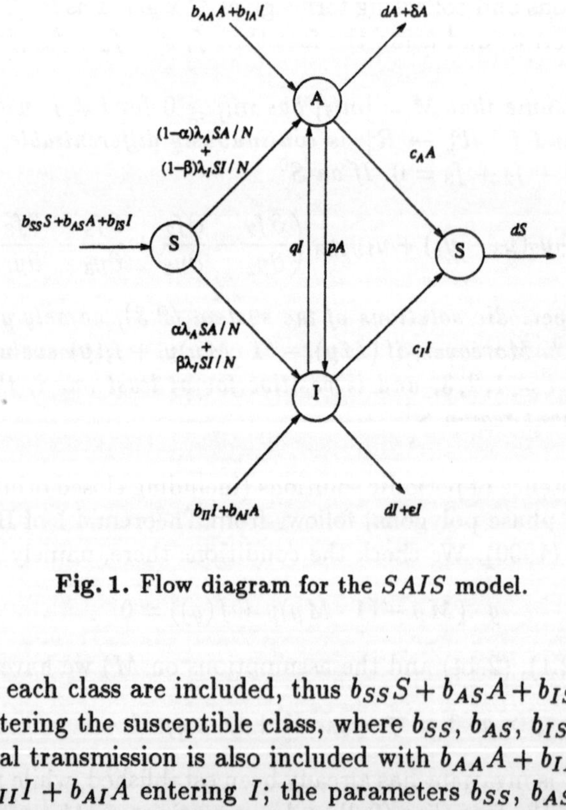

Fig. 1. Flow diagram for the *SAIS* model.

Births from each class are included, thus $b_{SS}S + b_{AS}A + b_{IS}I$ is the number of newborns entering the susceptible class, where b_{SS}, b_{AS}, b_{IS} are all assumed positive. Vertical transmission is also included with $b_{AA}A + b_{IA}I$ newborns entering A and $b_{II}I + b_{AI}A$ entering I; the parameters b_{SS}, b_{AS}, b_{IS}, b_{AA}, b_{IA}, b_{II}, b_{AI}, are birth rates, the first subscript denoting the parent class and the second subscript denoting the class of the offspring. The disease free death rate is d, and excess per capita death rates δ, ϵ are assumed in A, I, respectively. Recoveries from A, I are at rates c_A, c_I respectively. There is also the possibility of passing from asymptomatic to infective at rate p, and in the opposite direction at a rate q. The force of infection is of the proportionate mixing type, with λ_A, λ_I the effective per capita contact rate of infective individuals in classes

A and I, respectively. The rate at which susceptibles become infected is thus $\lambda_A SA/N + \lambda_I SI/N$. Proportions $\alpha, \beta \in [0,1]$ of those infected by asymptomatics, infectives, respectively, pass directly into the infective class. This model includes many epidemiological models previously studied in the literature, for example it includes the SIRS model of Busenberg and van den Driessche (1990); some special cases and previous work will be discussed in the next section.

The above hypotheses lead to the following model equations, an example of system (2.1), for the nonnegative variables:

$$S' = b_{SS}S + b_{AS}A + b_{IS}I - dS - \lambda_A SA/N - \lambda_I SI/N + c_A A + c_I I, \tag{3.1}$$

$$\begin{aligned} A' =& b_{AA}A + b_{IA}I + (1-\alpha)\lambda_A SA/N + (1-\beta)\lambda_I SI/N \\ &- (d+\delta+c_A+p)A + qI, \end{aligned} \tag{3.2}$$

$$I' = b_{AI}A + b_{II}I + \alpha\lambda_A SA/N + \beta\lambda_I SI/N - (d+\epsilon+c_I+q)I + pA, \tag{3.3}$$

giving

$$N' = b_{SS}S + (b_{AS} + b_{AA} + b_{AI} - \delta)A + (b_{IS} + b_{IA} + b_{II} - \varepsilon)I - dN. \tag{3.4}$$

Given initial data which are nonnegative, we can easily show that nonnegative solutions are defined for all time $t \geq 0$, thus the model is well posed.

We are interested in solutions with the total population N varying, thus we work with proportions in the three epidemiological classes, namely the normalized variables of Section 2, $s \equiv S/N = y_1$, $a \equiv A/N = y_2$, $i \equiv I/N = y_3$. The feasibility region is $S^+ \equiv \{s \geq 0, a \geq 0, i \geq 0, s+a+i=1\}$. We note that the linear part of these equations have nonnegative off-diagonal entries, with $m_{12} = b_{AS} + c_A$ and $m_{13} = b_{IS} + c_I$ both positive, as assumed for matrix M in Section 2. Also the nonlinear terms satisfy the assumptions there, in particular, the homogeneity and balance conditions. The proportion equations are:

$$\begin{aligned} s' =& b_{SS}s + b_{AS}a + b_{IS}i - \lambda_A sa - \lambda_I si + c_A a + c_I i \\ &- s[b_{SS}s + (b_{AS} + b_{AA} + b_{AI} - \delta)a + (b_{IS} + b_{IA} + b_{II} - \epsilon)i], \end{aligned} \tag{3.5}$$

$$\begin{aligned} a' =& b_{AA}a + b_{IA}i + (1-\alpha)\lambda_A sa + (1-\beta)\lambda_I si - (\delta + c_A)a - pa + qi \\ &- a[b_{SS}s + (b_{AS} + b_{AA} + b_{AI} - \delta)a + (b_{IS} + b_{IA} + b_{II} - \epsilon)i], \end{aligned} \tag{3.6}$$

$$\begin{aligned} i' =& b_{AI}a + b_{II}i + \alpha\lambda_A sa + \beta\lambda_I si - (\epsilon + c_I)i + pa - qi \\ &- i[b_{SS}s + (b_{AS} + b_{AA} + b_{AI} - \delta)a + (b_{IS} + b_{IA} + b_{II} - \epsilon)i]. \end{aligned} \tag{3.7}$$

Note that the disease free death rate d does not occur in these. We use our previous theorem to show the nonexistence of periodic solution for this disease transmission model.

Theorem 3.1 *The model system* (3.5)-(3.7) *has no periodic solutions in* S^+.

Proof. The condition (2.14) of theorem 2.2 reduces to

$$sai[-\alpha\lambda_A sa/i - (1-\beta)\lambda_I si/a]$$

which is certainly ≥ 0 in S^0 as all parameters are nonnegative and $\alpha, \beta \in [0,1]$. The conditions for invariance of S^+ are easily checked. We need to consider the boundary ∂S^+. But $s'(0,a,i) = (b_{AS} + c_A)a + (b_{IS} + c_I)i$ which is positive by our assumptions on the birth rates. Similarly, $a'(s,0,i)$ and $i'(s,a,0)$ are nonnegative. Thus the boundary cannot be a phase polygon for the system, and we have eliminated the possibility of periodic solutions. □

We now investigate the disease free equilibrium (DFE), namely $(s,a,i) = (1,0,0)$. We find that the following threshold parameters are important:

$$R_{0A} = \frac{(1-\alpha)\lambda_A}{c_A + p + \delta + b_{SS} - b_{AA}}, \quad R_{0I} = \frac{\beta\lambda_I}{c_I + q + \epsilon + b_{SS} - b_{II}}. \tag{3.8}$$

We make the biologically reasonable assumption tht $b_{SS} > \max\ \{b_{AA}, b_{II}\}$ so that the denominators do not vanish. For the A class, R_{0A} gives a measure of the relative strength of the disease transmission via contacts versus dilution through recovery, transfer into I, excess death, or relative increase in the S population by births and is thus a *disease reproduction number*; whereas R_{0I} gives a similar measure for the I class. These parameters play an important role in the following result.

Theorem 3.2 *The system* (3.5)-(3.7) *always has the DFE. If* $b_{AI} + \alpha\lambda_A + p > 0$ *and* $b_{IA} + (1-\beta)\lambda_I + q > 0$, *then it is the only equilibrium with either a or i equal to* 0. *In this case, the DFE is locally asymptotically stable in* S^+ *if* $R_{0A} < 1$, $R_{0I} < 1$, *and*

$$(1 - R_{0A})(1 - R_{0I}) > \frac{(b_{AI} + \alpha\lambda_A + p)(b_{IA} + (1-\beta)\lambda_I + q)}{(c_A + p + \delta + b_{SS} - b_{AA})(c_I + q + \epsilon + b_{SS} - b_{II})}. \tag{3.9}$$

Proof. Setting the derivatives in (3.5)-(3.7) to zero, gives the three nonlinear equations which must hold at an equilibrium point. From (3.7), when $i = 0$, $(b_{AI} + \alpha\lambda_A s + p)a = 0$, and from (3.6) when $a = 0$, $(b_{IA} + (1-\beta)\lambda_I s + q)i = 0$. Thus when both parameter sums are positive, that is, at least one of each set $\{b_{IA}, \alpha\lambda_A, p\}$ and $\{b_{IA}, (1-\beta)\lambda_I, q\}$ is strictly positive, we see that, at an equilibrium, $a = 0$ implies $i = 0$ and conversely.

Local stability of the DFE is governed by the eigenvalues of the Jacobian matrix

$$\begin{pmatrix} b_{AA} + (1-\alpha)\lambda_A & b_{IA} + (1-\beta)\lambda_I + q \\ -(\delta + c_A + p + b_{SS}) & \\ & \\ b_{AI} + \alpha\lambda_A + p & b_{II} + \beta\lambda_I \\ & -(\epsilon + c_I + q + b_{SS}) \end{pmatrix}, \tag{3.10}$$

since the third eigenvalue is $-b_{SS}$ which is negative. Thus, using (3.8), if $R_{0A} < 1$ and $R_{0I} < 1$, the main diagonal entries, and hence the trace of the matrix, are negative. Condition (3.9) is then equivalent to the determinant being positive. These trace and determinant conditions are necessary and sufficient for the matrix to be stable, hence, the DFE is locally asymptotically stable under the stated conditions. □

The first two conditions of the theorem are simply that the disease reproduction number for each class is less than one, whereas inequality (3.9) is a coupled condition.

We have already ruled out periodic solutions, so to establish global stability of the DFE, we need to rule out the possibility of an endemic equilibrium (s^*, a^*, i^*) in S^+ with $i^* > 0$. We consider now some important special cases.

4 Examples

Example 4.1 The SIRS model. Identifying the class A with a recovered class R, and setting $b_{SS} = b_{AS} = b_{IS} = b > 0$, $b_{AA} = b_{AI} = b_{II} = b_{IA} = 0$, $p = c_I = \lambda_A = 0$, $\beta = 1$, λ_I, c_A, $q > 0$, we obtain an SIRS model with no vertical transmission. A complete global analysis is given in Busenberg and van den Driessche (1990). Here it suffices to state that, in that case, $R_{0A} = 0$, and the determinant condition (3.9) reduces to $R_{0I} = \lambda_I/(q + \epsilon + b) < 1$. Although $b_{AI} + \alpha\lambda_A + p = 0$, the equilibrium point $(\overline{s}, 0, \overline{a})$ with $\overline{a} = (\delta + c_A + b)/\delta > 1$ is outside S. We are able to rule out an endemic equilibrium, and to prove that the DFE is globally asymptotically stable in S^+ iff $R_{0I} \leq 1$. When $R_{0I} > 1$ there is a unique endemic proportion equilibrium, which is globally asymptotically stable in $S^+ - \{(1, 0, 0)\}$; thus the reproduction number R_{0I} gives a sharp threshold.

Busenberg and Hadeler (1990) consider a generalization of this SIRS model which includes vertical transmission in the infective class ($b_{II} > 0$), with the biologically reasonable assumptions that the birth rates satisfy $b_{SS} \geq b_{AS} \geq b_{IS} + b_{II}$. They also assume that disease related death in the infective class is at least as large as that in the removed class, that is $\epsilon \geq \delta$. The above sharp threshold result continues to hold, with the parameter R_{0I} defined in (3.8) modified by the vertical transmission term.

For an SIR model with permanent immunity the parameter c_A is zero, and our results, so modified, hold in this special case.

Example 4.2 SEIS model. We can obtain an SEIS model by identifying A with an exposed (but not yet infectious) class E, and setting $q = c_A = \lambda_A = \beta = 0$, $\lambda_I, c_I, p > 0$. As in the SIRS model, we neglect vertical transmission and assume that the birth rate in each class is equal to $b > 0$. For this model, which was shown in Busenberg and van den Driessche (1990) to have no periodic solutions, we see that both parameters in (3.8) are zero. Theorem 3.2 gives that the DFE, which is the only equilibrium on ∂S^+, is locally asymptotically stable in S^+ if

$$1 > \frac{\lambda_I}{(b + c_I + \epsilon)} \frac{p}{(b + p + \delta)}.$$

If this inequality is reversed, then the DFE is unstable. If $p \to \infty$, that is the average exposed period $1/p \to 0$, the model reduces to an SIS model, with the above inequality then becoming $1 > \lambda_I/(b + c_I + \epsilon)$.

The corresponding SEI model, in which there is no cycling back into the susceptible class, is obtained from the SEIS model by setting c_I equal to zero. This is analysed by Mena Lorca (1988, section 5.7) under the additional assumption that there is no disease related death in the exposed class ($\delta = 0$). The sharp threshold result is obtained with reproduction number $\lambda_I p/[(b+\epsilon)(b+p)]$.

For diseases with no recovery and no latent class the correct model is an SI one. This can be obtained from our SEIS model by letting $p \to \infty$ and $c_I = 0$. Pugliese (1990) considers this model with a general shape of density dependent natural mortality and incidence rate and some vertical transmission. The sharp threshold result is also obtained in this case with reproduction number $(\lambda_I + b_{II})/(b+\epsilon)$. Brauer (1990) formulates a model for universally fatal diseases including nonlinear population dynamics and a distribution of infective periods which generalizes our average infective period $1/\epsilon$. A linear analysis again gives a sharp threshold.

Example 4.3 Simplified SIAS model. Consider equations (3.1)-(3.3) with no vertical transmission, no possibility of moving directly between groups A and I (i.e., $b_{AA} = b_{AI} = b_{II} = b_{IA} = p = q = 0$) and $\alpha = \beta$. The resulting model corresponds to the SIAS model in a population of constant size considered by Cooke (1982). Making the additional assumption that the birth rate in each class is equal to $b > 0$, we obtain from (3.8) the threshold parameters

$$R_{0A} = \frac{(1-\alpha)\lambda_A}{c_A + \delta + b}, \quad R_{0I} = \frac{\alpha\lambda_I}{c_I + \epsilon + b}.$$

Thus, for $\alpha \in (0,1)$ and $\lambda_A, \lambda_I > 0$, theorem 3.2 shows that the DFE is locally asymptotically stable in S^+ if $R_{0A} + R_{0I} < 1$. This sum corresponds to the sharp threshold found by Cooke (1982).

Acknowledgment

The research of the first author was partially supported by the National Science Foundation through Grant No. DMS-8902712, and that of the second author by NSERC A-8965 and the University of Victoria Committee on Faculty Research and Travel.

References

1. Brauer, F. (1990): Models of the spread of universally fatal diseases. J. Math. Biol. 28, 451-462
2. Busenberg, S., Hadeler, K. (1990): Demography and Epidemics. Math. Bios. 101, 63-74
3. Busenberg, S., van den Driessche, P. (1990): Analysis of a disease transmission model in a population with varying size. J. Math. Biol. 28, 257-270

4. Cooke, K.L. (1982): Models for endemic infections with asymptomatic cases I. One group. Math. Modelling **3**, 1-15
5. Hadeler, K., Waldstätter, R., Wörz-Busekros, A. (1988): Models for pair formation in bisexual populations. J. Math. Biol. **26**, 635-649
6. Hahn, W. (1967): Stability of Motion. Springer-Verlag, Berlin-Heidelberg-New York
7. Hofbauer, J., and Sigmund, K. (1988): The Theory of Evolution and Dynamical Systems. Cambridge University Press, Cambridge
8. Mena Lorca, J. (1988): Periodicity and stability in epidemiological models with disease-related deaths. Ph.D. Thesis, University of Iowa
9. Pugliese, A. (1990): Population models for diseases with no recovery. J. Math. Biol. **28**, 65-82

On the Solution of the Two-Sex Mixing Problem

C. Castillo-Chavez[1] *and S. Busenberg*[2]

[1] Biometrics Unit/Center for Applied Mathematics, 341 Warren Hall, Cornell University, Ithaca, NY 14853-7801
[2] Department of Mathematics, Harvey Mudd College, Claremont, CA 91711

The work described in this paper has been motivated by work with Kenneth Cooke. Ken has used his considerable experience in the modeling and analysis of disease transmission, and most recently in the development of models that may help our fight against AIDS. Many of the ideas discussed in this article arose out of our study of Ken's work, our discussions with him, and our collaborative efforts with Ken over the years. We dedicate this paper to him as we celebrate his 65th birthday.

Abstract

In this paper we describe an axiomatic framework that allows for the general incorporation of sexual structure into two-sex pair-formation models for sexually-transmitted diseases. A representation theorem describing all solutions to this mixing framework as perturbations of particular solutions is proved. Two-sex age structured demographic and age-structured epidemiological models that make use of our framework, and are therefore capable of describing the dynamics of individuals and/or pairs of individuals, are formulated.

1 Introduction

The modeling of sexual transmission of diseases can be said to have its genesis in the work of Sir Ronald Ross. Several ideas introduced in his modeling work on malaria have proved to be very useful in the development of a mixing framework for social/sexual interactions as well as in the development of models for the spread of venereal diseases. For example, the recognition that there must be a conservation of the number of interactions between individuals involved in a disease transmission process, a fact often ignored by modelers, was already clearly articulated in Ross' work on malaria. For malaria, this meant that the number of bites on humans must equal the number of humans bitten (Ross 1911, p. 666-7). In sexually transmitted diseases (STD's) we recognize this constraint requiring the equality of the number of sexual partnerships formed between individual human interacting groups (a kind of group reversibility property or a conservation law.) The consequences of this constraint will be further discussed later in this paper. Ross also observed that models with fixed and variable size populations must be treated differently, and may have radically different properties (Ross 1916, pp. 212, 215, 222). The fact that in the study of the dynamics of malaria the size of the host and vector populations play a key role in transmission forced him to introduce a special mixing structure given by a linear function of the ratio of the vector to host population sizes. We will show later that all solutions to our two-sex mixing framework are given by multiplicative perturbations of these special solutions.

Models for the spread of STD's were not systematically studied for over fifty years. In 1973, Cooke and Yorke analyzed and developed the first models for the spread of gonorrhea. This and subsequent papers re-opened this important area of research which reached a significant plateau with the application of these new adavances to the problem of gonorrhea dynamics and control. A description of these applications to U.S. data is clearly detailed in the monograph by Hethcote and Yorke (1984).

This paper is organized as follows: In Section 2, we formulate a general two-sex model for the spread of gonorrhea. This model allows us to discuss the problem of pair-formation or mixing. In Section 3, we discuss some special mixing solutions and provide a representation theorem for all possible two-sex mixing (pair-formation) solutions. In Section 4 we formulate a two-sex-structured demographic model and a two-sex age-structured epidemiological model that follow pairs of individuals. Models of this type have been formulated earlier by Fredrickson (1971), McFarland (1972), Dietz (1988), Dietz and Hadeler (1988), Castillo-Chavez (1989), Hadeler (1989a, b), and Castillo-Chavez et al. (1991). Section 4 begins with an axiomatic description theorem for the two-sex mixing problem in an age-structured population, and illustrates the role of age–dependent mixing in contact and pair formation models.

2 Two-sex gonorrhea model with variable population size

In order to provide a context for the sexual interactions of a heterosexual population, we introduce a two-sex model with variable population size for the transmission dynamics of gonorrhea. Traditional gonorrhea models (see Hethcote and Yorke, 1984) have assumed that the mixing subpopulations have constant size. This assumption may be very useful when we deal with the relative evaluation of control strategies (loc. cit.). However, this assumption is not appropriate in situations in which we wish to evaluate the impact of different mixing patterns in disease dynamics., The assumption of interacting populations of constant size leads to time-independent mixing probabilities (i.e. constant contact matrices) and hence to mixing patterns that are valid only for populations that have already reached a steady state.

We consider a population of heterosexually active individuals. This population is divided into classes or subpopulations. Classes can be indentified by sex, race, socio-economic background, average degree of sexual activity, etc. Models that incorporate factors such as chronological age, age of infection, variable infectivity, and partnership duration can be found in our earlier work (see Busenberg and Castillo-Chavez, 1989, 1991). An example of such a model is given in Section 4. We consider N-sexually active populations of females and L-sexually active populations of males. Each population is divided into two epidemiological classes: $S_j^f(t)$ and $S_i^m(t)$ (susceptible females and males, i.e., unifected and sexually-active, at time t); $I_j^f(t)$ and $I_i^m(t)$ (infected females and males, at time t); for $j = 1,...,N$ and $i = 1,...,L$. Hence, the sexually-active individuals of each sex and each subpopulation at time t are represented by $T_j^f(t) = S_j^f(t) + I_j^f(t)$ and $T_i^m(t) = S_i^m(t) + I_i^m(t)$.

$B_j^f(t)$ and $B_i^m(t)$ denotes the jth and i^{th} incidence rates for females in group j and males in group i at time t, that is, the number of new infective cases in each subpopulation per unit time. $B_j^f(t)$ and $B_i^m(t)$ are complicated functions that depend on the frequency and type of sexual interactions that susceptible females of group j and susceptible males of group i have with all other sexually-active individuals, in this case, of the opposite sex (although this condition can be easily relaxed).

If A_j^f and A_i^m denote the "recruitment" rates (assumed constant), μ_j^m and μ_i^m denote the (contant) removal rates from sexual activity, and γ_j^f and γ_i^m denote the (constant) recovery rates from gonorrhea infection, then we can write the following contact model for the transmission dynamics of gonorrhea:

$$\frac{dS_j^f(t)}{dt} = A_j^f - B_j^f(t) - \mu_j^f S_j^f(t) + \gamma_j^f I_j^f(t), \tag{1}$$

$$\frac{dI_j^f(t)}{dt} = B_j^f(t) - (\gamma_j^f + \mu_j^f) I_j^f(t), \tag{2}$$

$$\frac{dS_i^m(t)}{dt} = A_i^m - B_i^m(t) - \mu_i^m S_i^m(t) + \gamma_i^m I_i^m(t), \tag{3}$$

$$\frac{dI_i^m(t)}{dt} = B_i^m(t) - (\gamma_i^m + \mu_i^m) I_i^m(t), \tag{4}$$

$i = 1,...,L$ and $j = 1,...,N$.

Of course, this model is not fully described until we provide explicit expressions for $B_j^f(t)$ and $B_i^m(t)$. The formulae for the incidences will be provided in two steps: first we will provide expressions for the incidences in terms of the set of mixing probabilities $\{p_{ij}(t)$ and $q_{ji}(t)$: $i = 1,...,L$ and $j = 1,...,N\}$; and secondly, these mixing probabilities will be described (in the next section) in terms of an axiomatic system for sexual interactions.

To describe the formulae for the female and male incidences we need the following definitions:

$p_{ij}(t)$: the fraction of partnerships of males in group i with females in group j at time t,
$q_{ji}(t)$: the fraction of partnerships of females in group j with males in group i at time t,
$T_i^m(t)$: male population size of group i at time t,
$T_j^f(t)$: female population size of group j at time t.
c_i: average (constant) number of female partners per unit time of males in group i, or the ith-group rate of (male) pair-formation,
b_j: average (constant) number of male partners per unit time of females in group j, or the jth-group rate of (female) pair-formation,
β_i^m: disease transmission coefficient (constant) of males in group i,
β_j^f: disease transmission coefficient (constant) of females in group j.

Using these definitions we obtain the following expressions for the incidence rates:

$$B_i^m(t) = c_i S_i^m(t) \sum_{j=1}^{N} \beta_j^f p_{ij}(t) \frac{I_j^f(t)}{T_j^f(t)}, \tag{5}$$

and

$$B_j^f(t) = b_j S_j^f(t) \sum_{i=1}^{L} \beta_i^m q_{ji}(t) \frac{I_i^m(t)}{T_i^m(t)}. \tag{6}$$

3 Two-sex mixing framework

Special solutions for one-sex mixing populations were obtained by Nold (1980), Hethcote and Yorke (1984), Hyman and Stanley (1988, 1989), Jacquez et al. (1988, 1989), Blythe and Castillo-Chavez (1989), Castillo-Chavez and Blythe (1989), Gupta et al. (1989), and Anderson et al. (1989). A representation theorem describing all solutions as random perturbations of random (proportionate) mixing, based on the work of Blythe and Castillo-Chavez (op. cits.), was obtained by Busenberg and Castillo-Chavez (1989, 1991). Models that follow pairs of individuals (two-sex models) can be found (in a demographic context) in the works fo Kendall (1948), Keyfitz (1972), Parlett (1972), and J.H. Pollard (1973). Formulations of the standard two-sex mixing pair-formation framework are found in the work of Fredrickson (1971) and McFarland (1972). Application of the Fredrickson-McFarland framework to epidemiological models has been carried out by Dietz (1988), Dietz and Hadeler (1988), Castillo-Chavez (1989), Waldstatter (1989), Hadeler (1989a, b, 1991), and Castillo-Chavez et al. (1991). In this section we provide an alternative approach to the process of pair formation. This axiomatic framework was introduced in Castillo-Chavez et al. (1990), where some special solutions were found. We use the set of mixing probabilities $\{p_{ij}(t)$ and $q_{ji}(t) : i = 1, ...L$ and $j = 1, ..., N\}$ to describe the mixing/pair formation in a heterosexually active population through the following set of properites or axioms:

Definition 1. $(p_{ij}(t), q_{ji}(t))$ is called a mixing/pair-formation matrix if and only if it satisfies the following properties (at all times):

(A1) $0 \leq p_{ij} \leq 1, \quad 0 \leq q_{ji} \leq 1,$

(A2) $\sum_{j=i}^{N} p_{ij} = 1 = \sum_{i=1}^{L} q_{ji}$, whenever $c_i T_i^m \neq 0 \neq b_j T_j^f$.

(A3) $c_i T_i^m p_{ij} = b_j T_j^f q_{ji}, \quad i = 1, ..., L; \;\; j = 1, ..., N.$

(A4) $p_{ij} \equiv q_{ji} \equiv 0$ by definition if $c_i b_j T_i^m T_j^f = 0$ for some $i, 0 \leq i \leq L$ or for some $j, 0 \leq j \leq N$.

Note that (A3) can be viewed as a conservation of partnerships law or a group reversibility property, while (A4) asserts that the mixing of "non-existing" or non-sexually active subpopulations cannot be arbitrarily defined. For the gonorrhea model, and most deterministic models for STD's, subpopulations that are sexually active do not become extinct and do remain sexually active for all time. We now proceed to characterize a useful solution, namely Ross's solution.

We note that (A2) and (A3) imply the relation

$$\sum_{i=1}^{L} c_i T_i^m = \sum_{j=1}^{N} b_j T_j^f \tag{7}$$

which states that the total rate of axquisition of of female partners must equal the total rate of acquisition of male partners. In fact, summing (A3) over j and i and using (A2) we get

$$c_i T_i^m = \sum_{j=1}^{N} b_j T_j^f q_{ji},$$

$$\sum_{i=1}^{L} c_i T_i^m = \sum_{i=1}^{L} \sum_{j=1}^{N} b_j T_j^f q_{ji}.$$

Changing the order of summmation we obtain (7) since

$$\sum_{j=1}^{N} \sum_{i=1}^{L} b_j T_j^f q_{ji} = \sum_{j=1}^{N} b_j T_j^f.$$

Definition 2. A two-sex mixing/pair-formation function is called separable if and only if

$$p_{ij} = \tilde{p}_i p_j \text{ and } q_{ji} = \tilde{q}_j q_i, \; i = 1, ..., L; \; j = 1, ..., N$$

This definition leads us to the following useful characterization of two-sex separable mixing functions.

Theorem 1 *The only separable solution is the Ross solution given by* $(p_{ij}, q_{ji}) = (\overline{p}_j, \overline{q}_i)$ *where*

$$\overline{p}_j = \frac{b_j T_j^f}{\sum_{i=1}^L c_i T_i^m}, \quad \overline{q}_i = \frac{c_i T_i^m}{\sum_{j=1}^N b_j T_j^f}; \quad j = 1, ..., N \text{ and } i = 1, ..., L.$$

Proof. Suppose that (p_{ij}, q_{ji}) is a separable mixing function satisfying (A1)–(A4). By (A2), whenever $c_i b_j T_i^m T_j^f \neq 0$, we have for all j and i

$$1 = \tilde{q}_j \sum_{i=1}^L q_i = \tilde{q}_j \frac{1}{k}, \quad k \text{ a constant}$$

$$1 = \tilde{p}_i \sum_{j=1}^N q_j = \tilde{p}_i \frac{1}{\ell}, \quad \ell \text{ a constant}$$

which implies $\tilde{q}_j = k$ and $\tilde{p}_i = \ell$, for all i, j, hence,

$$q_{ji} = \tilde{q}_j q_i = k q_i \equiv \overline{q}_i \tag{8}$$

$$p_{ij} = \tilde{p}_i p_j = \ell p_j \equiv \overline{p}_j. \tag{9}$$

If (8) and (9) are substituted into (A3) then

$$c_i T_i^m \ell p_j = b_j T_j^f k q_i \quad \text{or} \quad c_i T_i^m \overline{p}_j = b_j T_j^f \overline{q}_i. \tag{10}$$

Summing over i, we get

$$\overline{p}_j \sum_{i=1}^L c_i T_i^m = b_j T_j^f \sum_{i=1}^L \overline{q}_i = b_j T_j^f,$$

since from (A2) and (8) $\sum_{i-1}^L \overline{q}_i = 1$. Thus

$$\overline{p}_j = \frac{b_j T_j^f}{\sum_{i=1}^L c_i T_i^m} \quad j = 1, ..., N. \tag{11}$$

Summing (10) over j and using (A2), we have

$$c_i T_i^m \sum_{j=1}^N \overline{p}_j = \overline{q}_i \sum_{j=1}^N b_j T_j^f \quad \text{or} \quad c_i T_i^m = \overline{q}_i \sum_{j=1}^N b_j T_j^f$$

Thus

$$\overline{q}_i = \frac{c_i T_i^m}{\sum_{j=1}^n b_j T_j^f}, \quad i = 1, ..., L. \tag{12}$$

Conversely, using (7), it is easy to see that $(\overline{p}_j, \overline{q}_i)$ satisfies (A1)- (A3), and we note that it vacuously satisfies (A4). □

Remark 1. Note that from (A3) it follows that, if $q_{ji} \neq 0$, then

$$\frac{p_{ij}}{q_{ji}} = \frac{b_j T_j^f}{c_i T_i^m} = \frac{\overline{p}_j}{\overline{q}_i}, \tag{13}$$

and hence using (A4), we see that $p_{ij} = 0$ if and only if $\overline{p}_j = 0$. Thus, the support of any two-sex mixing function is equal to the support of $(\overline{p}_j, \overline{q}_i)$.

We now use Equations (11), (12) and (13), to generate more solutions to axioms (A1)-(A4). We begin by introducing some new terms. Let

$(\phi_{ij}^m) \equiv$ The males' structural covariance matrix $(0 \leq \phi_{ij}^m)$ denoting the degree of preference or affinity (i.e., the deviation from random mixing) that group i-males have for group j-females, $j = 1, ..., N$ $i = 1, ..., L$.
$\ell_i^m \equiv \sum_{k=1}^N \overline{p}_k \phi_{ik}^m \equiv$ The weighted average preference of group i males , $i = 1, ..., L$.

$$R_i^m \equiv 1 - \ell_i^m, \; i = 1, ..., L. \tag{14}$$

We require that $R_j^m \geq 0$, and that

$$\sum_{i-1}^{L} \ell_i^m \overline{p}_i = \sum_{i=1}^{L} \sum_{k=1}^{N} \overline{p}_k \phi_{ik}^m \overline{p}_i < 1. \tag{15}$$

Similarly, let

$(\phi_{ji}^f) \equiv$ The females' structure covariance matrix $(0 \leq \phi_{ji}^f)$ denoting the degree of preference or affinity (i.e., the deviation from random mixing) that group j-females have for group i-males, $j = 1, ..., N, i, ..., L$.
$\ell_j^f \equiv \sum_{k=1}^L \overline{q}_k \phi_{jk}^f \equiv$ The weighted average preference of group j-females, $j = 1, ..., N$.

$$R_j^f \equiv 1 - \ell_j^f, \; j = 1, ..., N \tag{16}$$

Again, we require that $R_j^f \geq 0$, and that

$$\sum_{j=1}^{N} \ell_j^f \overline{q}_j = \sum_{j=1}^{N} \sum_{k=1}^{L} \overline{q}_k \phi_{jk}^f \overline{q}_j < 1. \tag{17}$$

With these assumptions and definitions, and with the additional condition (22) which is given below, we observe that a solution to axioms (A1) - (A4) is given (formally) by the following multiplicative perturbations to the separable mixing solution $(\overline{p}_j, \overline{q}_i)$

$$p_{ij} = \overline{p}_j \left[\frac{R_j^f R_i^m}{\sum_{k=1}^N \overline{p}_k R_k^f} + \phi_{ij}^m \right], \quad i = 1, ..., L; \; j = 1, ..., N, \tag{18}$$

$$q_{ji} = \overline{q}_i \left[\frac{R_i^m R_j^f}{\sum_{k=1}^L \overline{q}_k R_k^m} + \phi_{ji}^f \right]. \tag{19}$$

We now show that (p_{ij}, q_{ji}), $i = 1, ..., L$, $j = 1, ..., N$ given by (18) and (19) is a two-sex mixing matrix. The fact that axiom (A4) holds follows immediately from (18) and (19). In order to show that (A1) and (A2) hold, note that

$$\sum_{j=1}^{N} p_{ij} = R_i^m \left[\frac{\sum_{j=1}^{N} \overline{p}_j R_j^f}{\sum_{k=1}^{N} \overline{p}_k R_k^f}\right] + \left[\sum_{j=1}^{N} \overline{p}_j \phi_{ij}^m\right]$$

$$= R_i^m + \sum_{j=1}^{N} \overline{q}_j \phi_{ij}^m = R_i^m + (1 - R_i^m) = 1,$$

and, similarly

$$\sum_{i=1}^{L} q_{ji} = 1,$$

thus (A1) and (A2) are satisfied.

Note that axiom (A3) is satisfied if

$$c_i T_i^m \overline{p}_j \left[\frac{R_j^f R_i^m}{\sum_{k=1}^{N} \overline{p}_k R_k^f} + \phi_{ij}^m\right] = b_j T_j^f \overline{q}_i \left[\frac{R_i^m R_j^f}{\sum_{k=1}^{L} \overline{q}_k R_k^m} + \phi_{ij}^f\right]. \tag{20}$$

By observing that $c_i T_i^m \overline{p}_j = b_j T_j^f \overline{q}_i$, due to the fact that $(\overline{p}_j, \overline{q}_i)$, is a two-sex mixing function, we see that (20) holds outside the common support of $(\overline{p}_j, \overline{q}_i)$ and (p_{ij}, q_{ji}), if and only if

$$\left[\frac{R_j^f R_i^m}{\sum_{k=1}^{N} \overline{p}_k R_k^f} + \phi_{ij}^m\right] = \left[\frac{R_i^m R_j^f}{\sum_{k=1}^{L} \overline{q}_k R_k^m} + \phi_{ji}^f\right]. \tag{21}$$

Further, (21) holds if and only if

$$\phi_{ij}^m - \phi_{ji}^f = R_i^m R_j^f \left[\frac{1}{\sum_{k=1}^{L} \overline{q}_k R_k^m} - \frac{1}{\sum_{k=1}^{N} \overline{p}_k R_k^f}\right]$$

$$= R_i^m R_j^f \left[\frac{\sum_{k=1}^{N} \overline{p}_k R_k^f - \sum_{k=1}^{L} \overline{q}_k R_k^m}{\left(\sum_{k=1}^{L} \overline{q}_k R_k^m\right)\left(\sum_{k=1}^{N} \overline{p}_k R_k^f\right)}\right]$$

$$= R_i^m R_j^f \left[\frac{\sum_{k=1}^{L} \overline{q}_k \ell_k^m - \sum_{k=1}^{N} \overline{p}_k \ell_k^f}{\left(\sum_{k=1}^{L} \overline{q}_k R_k^m\right)\left(\sum_{k=1}^{N} \overline{p}_k R_k^f\right)}\right]$$

or equivalently, if and only if

$$\phi_{ij}^m = \phi_{ji}^f + R_i^m R_j^f \left[\frac{\sum_{k=1}^{N} \overline{p}_k R_k^f - \sum_{k=1}^{L} \overline{q}_k R_k^m}{\left(\sum_{k=1}^{L} \overline{q}_k R_k^m\right)\left(\sum_{k=1}^{N} \overline{p}_k R_k^f\right)}\right]. \tag{22}$$

In order to show that every solution of axioms (A1)-(A4) is given by Equations (18)-(19) we proceed as follows. Using property (A4) we observe that $p_{ij}/\overline{p}_j$ and $q_{ji}/\overline{q}_i$ are well defined on the support Δ of $(\overline{p}_j, \overline{q}_i)$, and therefore from (13)

$$\frac{p_{ij}}{\overline{p}_j} = \frac{q_{ji}}{\overline{q}_i} \geq 0 \text{ on } \Delta.$$

Properties (A1) and (A2) imply that there exist $\epsilon > 0$ and a set of positive integers $Q \subset Z_+^2$ such that $p_{ij}/\overline{p}_j = q_{ji}/\overline{q}_i > \varepsilon$. Thus we can define

$Q \equiv \{(i,j) : \frac{p_{ij}}{p_j} > \varepsilon\}$, and a set related to Q defined as follows $\overline{Q} \equiv \{i : (i,j) \in Q \text{ for some } j\}$. We now define the following functions

$$R_i^m \equiv \varepsilon\chi_{\overline{Q}}(i)\sum_{k=1}^{L}\chi_{\overline{Q}}(k)\overline{q}_k,$$

$$R_j^f \equiv \varepsilon\chi_{\overline{Q}}(j)\sum_{k=1}^{N}\chi_{\overline{Q}}(k)\overline{p}_k,$$

where χ denotes the characteristic (or indicator function) of a set. Note that we can think of $\overline{Q}$ as a "connectivity set" which specifies all male groups which have contacts with the jth female group.

We now note that

$$\sum_{i=1}^{L} R_i^m\overline{q}_i = \varepsilon\left(\sum_{k=1}^{L}\chi_{\overline{Q}}(k)\overline{q}_k\right)^2, \tag{23}$$

and

$$\sum_{j=1}^{N} R_j^f\overline{p}_j = \varepsilon\left(\sum_{k=1}^{N}\chi_{\overline{Q}}(k)\overline{q}_k\right)^2. \tag{24}$$

Hence

$$\frac{R_j^f R_i^m}{\sum_{k=1}^{N} R_k^f\overline{p}_k} = \varepsilon\chi_{\overline{Q}}(i)\chi_{\overline{Q}}(j)\frac{\sum_{k=1}^{L}\chi_{\overline{Q}}(k)\overline{q}_k}{\sum_{k=1}^{N}\chi_{\overline{Q}}(k)\overline{p}_k}, \tag{25}$$

and

$$\frac{R_j^f R_i^m}{\sum_{k=1}^{L} R_k^m\overline{q}_k} = \varepsilon\chi_{\overline{Q}}(i)\chi_{\overline{Q}}(j)\frac{\sum_{k=1}^{N}\chi_{\overline{Q}}(k)\overline{q}_k}{\sum_{k=1}^{L}\chi_{\overline{Q}}(k)\overline{q}_k}. \tag{26}$$

Now let

$$\phi_{ij}^m \equiv \frac{p_{ij}}{\overline{p}_j} - \varepsilon\chi_{\overline{Q}}(i)\chi_{\overline{Q}}(j)\frac{\sum_{k=1}^{L}\chi_{\overline{Q}}(k)\overline{q}_k}{\sum_{k=1}^{N}\chi_{\overline{Q}}(k)\overline{p}_k},$$

and

$$\phi_{ji}^f \equiv \frac{q_{ji}}{\overline{q}_i} - \varepsilon\chi_{\overline{Q}}(i)\chi_{\overline{Q}}(j)\frac{\sum_{k=1}^{N}\chi_{\overline{Q}}(k)\overline{q}_k}{\sum_{k=1}^{L}\chi_{\overline{q}}(k)\overline{q}_k}.$$

From the last two expressions we see that

$$\sum_{j=1}^{N}\phi_{ij}^m\overline{p}_j = 1 - \varepsilon\chi_{\overline{Q}}(i)\sum_{k=1}^{L}\chi_{\overline{Q}}(k)\overline{q}_k = l_i^m,$$

and

$$\sum_{i=1}^{L} \phi^f_{ji}\overline{q}_i = 1 - \varepsilon\chi_{\overline{Q}}(j)\sum_{k=1}^{N}\chi_{\overline{Q}}(k)\overline{p}_k = l^f_j .$$

Further, since

$$\phi^m_{ij} - \phi^f_{ji} = \varepsilon\chi_{\overline{Q}}(i)\chi_{\overline{Q}}(j)\left[\frac{\sum_{k=1}^{N}\chi_{\overline{Q}}(k)\overline{p}_k}{\sum_{k=1}^{L}\chi_{\overline{Q}}(k)\overline{q}_k} - \frac{\sum_{k=1}^{L}\chi_{\overline{Q}}(k)\overline{q}_K}{\sum_{k=1}^{N}\chi_{\overline{Q}}(k)\overline{p}_k}\right].$$

We see by using (23)-(26), that Equation (22) is automatically satisfied. From the definition of ϕ^m_{ij} and from (25) we obtain (18), and (19) is obtained similarly using (26).

Hence we have established the following results:

Theorem 2 *Let $\{\phi^m_{ij}\}$ and $\{\phi^f_{ji}\}$ be two nonegative matrices. Let $\ell^m_i \equiv \sum_{k=1}^{N}\overline{p}_k\,\phi^m_{ik}$ and $\ell^f_j \equiv \sum_{k=1}^{L}\overline{q}_k\phi^f_{jk}$ where $\{(\overline{p}_j,\overline{q}_i)\ j = 1,...,N$ and $i = 1,...,L\ \}$ denotes the set composed of the Ross solutions. We also let $R^m_i \equiv 1-\ell^m_i, i = 1,...,L$ and $R^f_j \equiv 1-\ell^f_j, j = 1,...,N$, and assume that ϕ^m_{ij} and ϕ^f_{ji} are chosen in such a way that (22) holds and R^m_i and R^f_j remain nonnegative for all time $t \geq 0$. We further assume that*

$$\sum_{i=1}^{L}\ell^m_i\overline{p}_i = \sum_{i=1}^{L}\sum_{k=1}^{N}\overline{p}_k\phi^m_{ik}\overline{p}_i < 1,$$

and

$$\sum_{j=1}^{N}\ell^f_j\overline{q}_j = \sum_{j=1}^{N}\sum_{k=1}^{L}\overline{q}_k\phi^f_{jk}\overline{q}_j < 1.$$

Then equations (18) and (19) give a solution of axioms (A1)-(A4). Conversely, any solutions to axioms (A1)-(A4) is given by Equations (18) and (19) with $\{\phi^m_{ij}\}$ and $\{\phi^f_{ji}\}$ satisfying the above conditions.

Remark 2. ϕ^m_{ij} and ϕ^f_{ji} can always be chosen in such a way that R^m_i and R^f_j remain nonnegative for all time (for example, let them be in the interval [0,1]). However, there is no recipe for specifying necessary conditions for guaranteeing condition (7) because it is intimately connected to the time-dependent values values of T^f_i and T^m_j, and hence to the behavior of the dynamical system. Consequently, the admissible dynamical systems must be structured so that (7) is satisfied for all time (including $t = 0$).

An important question is whether it is possible to have a separable solution in one one of the two sexes and not in the other. This is settled in our next results which serves to ellucidate the meaning of the preference matrices ϕ^m_{ij} and ϕ^f_{ij}. These matrices, of course, reflect the actual proportions of pairings that occur rather than the personal preferences of the individuals in these pairs. Consequenly, the balance law (22), which is imposed by the symmetry of pairings, fixes the structure of one of these matrices once the other is given. In the case where one of the sexes has a separable solution, the condition imposed on the other by (22) is quite strong, as is seen by the following theorem.

Theorem 3 *If either* $\phi_{ij}^m = \alpha,\ 0 \le \alpha < 1,\ \forall\, i,\ j$ or if $\phi_{ji}^f = \beta,\ 0 \le \beta < 1,\ \forall\, i,\ j,$ *then* $p_{ij} = \overline{p}_j$ *and* $q_{ij} = \overline{q}_i$, *that is* (18) *and* (19) *reduce to the unique separable Ross solution.*

Proof. Suppose that $\phi_{ji}^f = \alpha,\ 0 \le \alpha < 1$, for all $i,\ j$. Then $\ell_j^f = \alpha$, and $R_j^f = 1-\alpha$ for all j. Thus

$$q_{ji} = \overline{q}_i \left[(1-\alpha) \frac{R_i^m}{\sum_{k=1}^L \overline{q}_k R_k^m} + \alpha \right]. \tag{27}$$

But from (22)

$$\phi_{ij}^m = \alpha + R_i^m (1-\alpha) \left[\frac{(1-\alpha) - \sum_{k=1}^L \overline{q}_k R_k^m}{(1-\alpha) \sum_{k=1}^L \overline{q}_k R_k^m} \right], \tag{28}$$

which implies that

$$\ell_i^m = \sum_{j=1}^N \phi_{ij}^m \overline{p}_j = 1 - R_i^m = \alpha + R_i^m \left[\frac{(1-\alpha) - \sum_{k=1}^L \overline{q}_k R_k^m}{\sum_{k=1}^L \overline{q}_k R_k^m} \right].$$

Thus

$$1 = \alpha + R_i^m \frac{(1-\alpha)}{\sum_{k=1}^L \overline{q}_k R_k^m}$$

which implies that $R_i^m = \delta$, an arbitrary constant. By the definition of R_i^m we have $0 < \delta < 1$. Now, using (28) we get $\phi_{ij}^m = 1 - \delta = \beta$, with $0 < \beta < 1$. From (23) we obtain

$$q_{ji} = \overline{q}_i \left(\frac{(1-\alpha)\delta}{\delta} + \alpha \right) = \overline{q}_i.$$

Similarly, starting with $\phi_{ij}^m = \beta$, and using the above argument we have $p_{ij} = \overline{p}_j$, and the proof is completed. □

Remark 3. As in the one-sex framework, the only separable solution is proportionate mixing. Theorem 3 shows that solutions cannot be separable in one sex and not the other. Solutions where one sex chooses while the other does not are applicable to models for vector-transmitted diseases in which the vector exhibits strong host preference, while the host is just a "moving" target. Clearly, the balance condition (22) imposed by the pairing hypothesis imposes an automatic "preference" restriction on the host even though the preferential seeking is performed by the vector only.

Remark 4. Several other one-sex special solutions have been discussed in the literature. These include "preferred" mixing, like-with-like mixing, etc. (see Nold

1980, Hethcote and Yorke 1984, Blythe and Castillo-Chavez 1989, Castillo-Chavez and Blythe 1989, Jacquez et al. 1988, Hyman and Stanley 1989, Gupta et al. 1989, Blythe et al. 1989, etc.), and the several examples of their general solution given by Busenberg and Castillo-Chavez (1989, 1991). Blythe and Castillo-Chavez (1991a) have established explicitly that all these solutions are special cases of the general solution of Busenberg and Castillo-Chavez (1989, 1991).

A derivation of this general solution which explains the steps on the basis of demographic reasoning through the budgeting of rates is found in Blythe *et al* (1991a).

Remark 5. The gonorrhea model found in this section, but for one-sex populations, was introduced (along with some generalizations) by Castillo-Chavez and Blythe (1990) as a simple device to easily test mixing patterns. A thorough numerical analysis of these mixing matrices (one-sex framework) is found in Blythe and Castillo-Chavez (1990b). A discussion of methods for estimating the mixing matrices (one-sex framework) from data can be found in Blythe et al. (1991), and Pugliese (1990).

4 Two-sex age-structured models

We formulate two-sex models of the SI type with age-structured models. One follows individuals while the other follows pairs. Extensions to models for other diseases such as AIDS or gonorrhea that require a different epidemiological and compartmental structure can be easily formulated following the approach found in Busenberg and Castillo-Chavez (1989, 1991) and Castillo-Chavez et al. (1991). To formulate these models, we need a description of mixing functions that incorporate age (risk can be easily incorporated, see the above references). Pairing is defined through the mixing functions:

$p(a,a',t)$ = proportion of partnerships of males of age a with females of age a' at time t,

$q(a',a,t)$ = proportion of partnerships of females of age a with males of age a at time t,

and we let

$C(a,t)$ = expected or average number of partners of a male of age a at time t per unit time,

$D(a',t)$ = expected or average number of partners of a female of age a' at time t per unit time.

The following natural conditons characterize these mixing functions:

(B1) $p,q \geq 0$.

(B2) $\int_0^\infty p(a,a',t)da' = \int_0^\infty q(a',a,t)da = 1$,

(B3) $p(a,a',t)C(a,t)T^m(a,t) = q(a',a,t)D(a',t)T^f(a',t)$,

(B4) $C(a,t)T^m(a,t)D(a',t)T^f(a',t) = 0 \rightarrow p(a,a',,t) = q(a',a,t) = 0$,

Conditions (B1) and (B2) are due to p and q being proportions. Condition (B3) simply states that the rate of pair formation of males of age a with females of age a' equals the rate of pair formation of females of age a' with males of age a (all per unit time and age). Condition (B4) says that there is no mixing in the age and activity levels where there are no active individuals; i.e., on the set $\mathcal{L}(t) = \{(a,a',t) : C(a,t)T^m(a,t)D(a',t)T^f(a',t) = 0\}$. This last condition is usually vacuously satisfied in most applications. The need to state it derives from the proof of the Representation Theorem (Theorem 2).

The pair (p,q) is called a two-sex mixing function if and only if it satisfies axioms (B1-B4). Further, a two-sex mixing function is called separable if and only if

$$p(a,a',t) = p_1(a,t)p_2(a',t) \text{ and } q(a,a',t) = q_1(a,t)q_2(a',t).$$

If we let

$$h_p(a,t) = C(a,t)T^m(a,t) \tag{29}$$

and

$$h_q(a',t) = D(a',t)T^f(a',t) \tag{30}$$

then, omitting t to simplify the notation, we establish the following result:

Theorem 4 *The only two-sex separable (Ross) mixing function satisfying conditions* (B1-B4) *is given by* $(\overline{p},\overline{q})$, *where*

$$\overline{p}(a') = \frac{h_q(a')}{\int_0^\infty h_p(u)du}, \tag{31}$$

$$\overline{q}(a) = \frac{h_p(a)}{\int_0^\infty h_q(u)du}. \tag{32}$$

The proof is found in Castillo-Chavez *et al.* (1991).

We now let $m(a,t)$ denote the density of (uninfected) males of age a who are not in pairs at time t, and let $f(a',t)$ denote the density of (uninfected) females of age a' who are not in pairs at time t. We assume that D and C (as defined above) and μ_m and μ_f are functions of age (the mortality rates for males and females), σ denotes the constant rate of separation, and we let that $w(a,a',t)$ denote the age-specific density of heterosexual (uninfected) pairs (where a denotes the age of the male and a' the age of the female). Using the two-sex mixing functions p and q, we arrive at the following demographic model for heterosexual (uninfected) populations:

$$\begin{aligned} \frac{\partial m}{\partial t} + \frac{\partial m}{\partial a} &= -C(a)m(a,t)\int_0^\infty p(a,a',t)da' \\ &\quad - \mu_m(a)m(a,t) + \int_0^\infty [\mu_f(a') + \sigma]w(a,a',t)da', \end{aligned} \tag{33}$$

$$\begin{aligned}\frac{\partial f}{\partial t}+\frac{\partial f}{\partial a'} &= -D(a')f(a',t)\int_0^\infty q(a',a,t)da \\ &- \mu_f(a')f(a',t)+\int_0^\infty [\mu_m(a)+\sigma]w(a,a',t)da,\end{aligned} \tag{34}$$

$$\begin{aligned}\frac{\partial w}{\partial t}+\frac{\partial w}{\partial a}+\frac{\partial w}{\partial a'} &= D(a')f(a',t)q(a,a',t) \\ &- [\mu_f(a')+\mu_m(a)+\sigma]w(a,a',t).\end{aligned} \tag{35}$$

To complete this model we need to specify the initial and boundary conditions. To this effect we let λ_m and λ_f denote the female age- and sex-specific fertility rates, and let m_0, f_0, and w_0 denote the intial age densities. Hence, the initial and boundary conditions are given by

$$m(0,t)=\int_0^\infty \lambda_m(a')w(a,a',t)da, \tag{36}$$

$$f(0,t)=\int_0^\infty \lambda_f(a')w(a,a',t)da', \tag{37}$$

$$w(0,0,t)=0 \tag{38}$$

$$f(a,0)=f_0(a),\ m(a,0)=m_0(a),\ w(a,a',0)=w_0(a,a'). \tag{39}$$

A preliminary analysis of this demographic model in found in Castillo-Chavez *et al.* (1991). If we let $\sigma \to \infty$ then (formally) the above system approaches ther classical McKendrick/Von Foerster model (see loc. cit.) This demographic model, in conjunction with the McKendrick/Von Foerster model, will be used to formulate epidemiological models through the usual creation of the appropriate epidemiological compartments (see Hoppensteadt 1974, Dietz 1988, Dietz and Hadeler 1988, Castillo-Chavez 1989).

We begin by letting $T^m(a,t)$ and $T^f(a',t)$ denote, respectively, the male and female densities of single infected individuals. Hence, the heterosexual pairs are denoted by: $w_{mf}(a,a',t)$, $w_{Mf}(a,a',t)$, $w_{mF}(a,a',t)$, and $w_{MF}(a,a',t)$. If we use the notation with the appropriate indexing (that is f m, F, or M) in order to denote susceptible females and males and infective females and males, respectively. We then arrive at the following epidemiological model that follows pairs:

$$\begin{aligned}\frac{\partial m(a,t)}{\partial t}+\frac{\partial m(a,t)}{\partial a} &= -C_{mf}(a,t)m(a,t)\int_0^\infty p_{mf}(a,a',t)da' \\ &- C_{mF}(a,t)m(a,t)\int_0^\infty p_{mF}(a,a',t)da' - \mu_m(a)m(a,t) \\ &+ \int_0^\infty [\mu_f(a')+\sigma(a',a)]w_{mf}(a,a',t)da' + \int_0^\infty [\mu_F(a') \\ &+ \sigma(a',a)]w_{mF}(a,a',t)da',\end{aligned} \tag{40}$$

$$\frac{\partial f(a',t)}{\partial t}+\frac{\partial f(a',t)}{\partial a'} = -D_{fm}(a',t)f(a',t)\int_0^\infty q_{fm}(a',a,t)da$$

$$- D_{fM}(a',t)f(a',t)\int_0^\infty q_{fM}(a',a,t)da - \mu_f(a')f(a',t) +$$

$$\int_0^\infty [\mu_m(a)+\sigma(a',a)]w_{mf}(a,a',t)da$$

$$+\int_0^\infty [\mu_M(a)+\sigma(a',a)]w_{Mf}(a,a',t)da, \tag{41}$$

$$\frac{\partial M(a,t)}{\partial t}+\frac{\partial M(a,t)}{\partial a} = -C_{Mf}(a,t)M(a,t)\int_0^\infty p_{Mf}(a,a',t)da'$$

$$- C_{MF}(a,t)M(a,t)\int_0^\infty p_{MF}(a,a',t)da' - \mu_M(a)M(a,t)$$

$$+\int_0^\infty [\mu_f(a')+\sigma(a',a)]w_{Mf}(a,a',t)da'$$

$$+\int_0^\infty [\mu_F(a')+\sigma(a',a)]w_{MF}(a,a',t)da', \tag{42}$$

$$\frac{\partial F(a',t)}{\partial t}+\frac{\partial F(a',t)}{\partial a'} = -D_{FM}(a',t)F(a',t)\int_0^\infty q_{Fm}(a,a',t)da$$

$$- D_{FM}(a',t)F(a',t)\int_0^\infty q_{Fm}(a,a',t)da - \mu_F(a')F(a',t)$$

$$+\int_0^\infty [\mu_m(a)+\sigma(a',a)]w_{mF}a,a',t)da$$

$$+\int_0^\infty [\mu_M(a)+\sigma(a',a)]w_{MF}(a,a',t)da, \tag{43}$$

$$\frac{\partial w_{fm}(a,a',t)}{\partial a}+\frac{\partial w_{fm}(a,a',t)}{\partial a'}+\frac{\partial w_{fm}(a,a',t)}{\partial t} = D_{fm}(a')f(a',t)q_{fm}(a',a,t)$$
$$-(\sigma(a',a)+\mu_m(a)+\mu_f(a'))w_{fm}(a,a',t), \tag{44}$$

$$\frac{\partial w_{Fm}(a,a',t)}{\partial a}+\frac{\partial w_{Fm}(a,a',t)}{\partial a'}+\frac{\partial w_{Fm}(a,a',t)}{\partial t} = D_{Fm}(a')f(a',t)q_{Fm}(a',a,t)$$
$$-(\sigma(a',a)+\mu_m(a)+\mu_F(a'))w_{Fm}(a,a',t), \tag{45}$$

$$\frac{\partial w_{fM}(a,a',t)}{\partial a}+\frac{\partial w_{fM}}{\partial a'}+\frac{\partial w_{fM}}{\partial t} = D_{fM}(a',)f(a',t)q_{fM}(a',a,t)$$
$$-(\sigma(a',a)+\mu_M(a)+\mu_f(a'))w_{fM}, \tag{46}$$

$$\frac{\partial w_{FM}(a,a',t)}{\partial a}+\frac{\partial w_{FM}(a,a',t)}{\partial a'}+\frac{\partial w_{FM}(a,a',t)}{\partial t} = D_{FM}(a')f(a',t)q_{FM}(a',a,t)$$
$$-(\sigma(a',a)+\mu_M(a)+\mu_F(a'))w_{FM}(a,a't), \tag{47}$$

with appropriate initial and boundary conditions (see Castillo-Chavez 1989). It is important to note that we have used "restricted" mixing functions, that is, mixing functions that deal exclusively with certain "pairs"(namely, mf, fM, Mf, and MF), and hence the mixing axioms (B1)-(B4) have to be re-interpreted in this context (see the above references).

An SI model that does not follow pairs but individuals is therefore given by the following set of equations:

$$\frac{\partial m(a,t)}{\partial t}+\frac{\partial m(a,t)}{\partial a}=-C_m(a,t)m(a,t)\int_0^\infty \beta_{Fm}(a,a')p_{(m+M)(f+F)}(a,a',t)\frac{F(a',t)}{F(a',t)+f(a',t)}da'\mu_m(a)m(a,t), \tag{48}$$

$$\frac{\partial f(a',t)}{\partial t}+\frac{\partial f(a',t)}{\partial a'}=-D_f(a',t)f(a',t)\int_0^\infty \beta_{Mf}(a,a')q_{(f+F)(m+M)}(a,a',t)\frac{M(a,t)}{M(a,t)+m(a,t)}da-\mu_m(a)f(a',t), \tag{49}$$

$$\frac{\partial M(a,t)}{\partial t}+\frac{\partial M(a,t)}{\partial a}=C_m(a,t)m(a,t)\int_0^\infty \beta_{Fm}(a,a')p_{(m+M)(f+F)}(a,a',t)\frac{F(a',t)}{F(a',t)+f(a',t)}da'-\mu_M(a)M(a,t), \tag{50}$$

$$\frac{\partial F(a',t)}{\partial t}+\frac{\partial F(a't)}{\partial a'}=D_f(a',t)f(a',t)\int_0^\infty \beta_{Mf}(a,a')q_{(f+F)(m+M)}(a,a't)\frac{M(a,t)}{M(a,t)+m(a,t)}da-\mu_F(a')F(a',t), \tag{51}$$

where $\beta_{Fm}(a',a)$ and $\beta_{Mf}(a,a')$ represent the appropriate transmission coefficients. For a detailed derivation of a related model for one-sex populations see Busenberg and Castillo-Chavez (1989, 1991).

5 Conclusion

In this paper we have given an axiomatic definition and found a representation theorem for the general solution of the two-sex mixing problem. This representation theorem is based on multiplicative perturbations of the Ross solutions which are the only separable solutions of this problem. We have shown that there are no solutions that allow for one-sex preferential sexual systems with proportional or random mixing in the other sex. These results generalize the corresponding theorems for the one-sex mixing problem that we previously obtained (Busenberg and Castillo-Chavez 1989, 1991.) We have also formulated a model of the SIS type for a discrete number of groups. We outline generalizations to age-structed populations through the introduction of two epidemiological models that incorporate this mixing framework at the level of individual interactions or at the level of pair dynamics. We point out that although models of this type have been formulated before (see Dietz 1988, Dietz and Hadeler 1988, Castillo-Chavez 1989), here they have been formulated explicity under a unified framework.

Finally, we note that S. P. Blythe (1991) has shown that our original solution (Busenberg and Castillo-Chavez, 1989, 1991) provides a representation theorem for the n-sex problem. Nevertheless, the separation of the mixing into two mixing matrices (p and q) provides useful results, such as the imposibility of single sex preferential solutions (see Theorem 3) that are not immediate from our original formulation. This extra information arises from the breaking up of the group reversability property (Axiom A3) through the use of the connectivity properties of the groups involved (for example, individuals of the same sex do not mix, and all pairings involve one member from each sex group).

Acknowledgment

This research has been partially supported by the Center for Applied Mathematics at Cornell University and NSF grant DMS-8703631 to Stavros Busenberg, and by NSF grant, NIAID Grant R01 A129178-01, and Hatch project grant NYC 151-409, USDA to Carlos Castillo-Chavez. Over the last two years, Carlos Castillo-Chavez has held continuous discussions of these ideas with S. P. Blythe; many of the ideas discussed in this manuscript have originated throughout these discussions. We thank him for his valuable comments.

References

1. Anderson, R. M., Blythe, S. P., Gupta, S., and Konings, E. (1989): The transmission dynamics of the Human Immunodeficiency Virus Type I in the male homosexual community in the United Kindom: the influence of changes in sexual behavior. Phil. Trans. Roy. Soc. London. B **325**, 145-198
2. Blythe, S. P. (1991): Heterogeneous sexual mixing in populations with arbitrarily connected multiple groups. To appear in Math. Pop. Studies
3. Blythe, S. P. and Castillo-Chavez, C. (1989): Like-with-like preference and sexual mixing models. Math. Biosci. **96**, 221-238
4. Blythe, S.P. and Castillo-Chavez, C. (1991a): The one-sex mixing problems: a choice of solutions? To appear in J. Math. Biology
5. Blythe, S.P. and Castillo-Chavez, C.(1990b): Like-with-like mixing and sexually transmitted disease epidemics in one-sex populations. Biometrics Unit Technical Report # BU-1078-M, Cornell University
6. Blythe, S. P., Castillo-Chavez, C., and Casella, G. (1991): Empirical methods for the estimation of the mixing probabilities for socially structured populations from a single survey sample. To appear in Math. Pop. Studies
7. Blythe, S.P., Castillo-Chavez, C., J.S. Palmer, Cheng, M. (1991): Towards a unified theory of mixing and pair formation. To appear in Math. Biosci.
8. Busenberg, S., and Castillo-Chavez, C. (1989): Interaction, pair formation and force of infection terms in sexually transmitted diseases. In (C. Castillo-Chavez, ed.) Mathematical and Statistical Approaches to AIDS Epidemiology. Lecture Notes in Biomathematics **83**, Springer-Verlag, Berlin-Heidelberg-New York, 289-300

9. Busenberg, S. and Castillo-Chavez C. (1991): On the role of preference in the solution of the mixing problem, and its application to risk- and age-structured epidemic models. To appear in IMA J. of Math Applic. to Med. and Biol.
10. Castillo-Chavez, C. (1989): Review of recent models of HIV/AIDS transmission. In (S. A. Levin, T. G. Hallam, and L. J. Gross, eds.,) Applied Mathematical Ecology, Biomathematics **18**, Springer-Verlag, Berlin-Heidelberg-New York, 253-262
11. Castillo-Chavez, C. and Blythe, S. P. (1989): Mixing framework for social/sexual behavior. In (Castillo-Chavez, ed.) Mathematical and Statistical Approaches to AIDS Epidemiology. Lecture Notes in Biomathematics **83**, Springer-Verlag, Berlin-Heidelberg-New York, 275-288
12. Castillo-Chavez, C. and Blythe, S. P. (1990): A "test-bed" procedure for evaluating one-sex mixing frameworks. Manuscript
13. Castillo-Chavez, C., Busenberg, S., Gerow, K. (1991): Pair formation in structured populations. To appear in Differential Equations with Applications to Biology, Physics and Engineering, (J. Golstein, F. Kappel, W. Schappacher, Eds.), Marcel Dekker, New York
14. Cooke, K. L. and Yorke, J. A. (1973): Some equations modelling growth processes and gonorrhea epidemics. Math. Biosci., **58**, 93-109
15. Dietz, K. (1988): On the transmission dynamics of HIV. Math Biosci. **90**, 397-414
16. Dietz, K. and Hadeler, K. P. (1988): Epidemiological models for sexually transmitted diseases. J. Math. Biol. **26**, 1-25
17. Fredrickson, A. G. (1971): A mathematical theory of age structure in sexual populations: Random mating and monogamous marriage models. Math. Biosci. **20**, 117-143
18. Gupta, S., Anderson, R.M., May, R.M. (1989): Network of sexual contacts: implications for the pattern fo spread of HIV. AIDS ; **3**: 1-11
19. Hadeler, K. P. (1989a): Pair formation in age-structured populations. Acta Applicandae Mathematicae **14**, 91-102
20. Hadeler, K. P. (1989b): Modeling AIDS in structured populations. Manuscript
21. Hadeler, K. P. (1990): Homogeneous delay equations and models for pair formation. Manuscript
22. Hethcote, H. W. and Yorke, J. A. (1984): Gonorrhea transmission dynamics and control. Lecture Notes in Biomathematics **56**, Springer-Verlag, Berlin-Heidelberg-New York
23. Hoppensteadt, F. (1974): An age dependent epidemic model. J. Franklin Instit. 297, 325-333
24. Hyman, J. M. and Stanley, E. A. (1988): Using mathematical models to understand the AIDS epidemic. Math Biosci. **90**, 415-473
25. Hyman, J. M. and Stanley, E. A. (1989): The effect of social mixing patterns on the spread of AIDS. In Mathematical approaches to problems in resource management and epidemiology, (C. Castillo-Chavez, S. A. Levin, and C. Shoemaker, eds.) Lecture Notes in Biomathematics **81**, Springer-Verlag, Berlin-Heidelberg-New York, 190-219
26. Jacquez, J. A., Simon, C. P., Koopman, J., Sattenspiel, L., Perry, T. (1988): Modelling and analyzing HIV transmission: the effect of contact patterns. Math Biosci. **92**, 119-199
27. Jacquez, J. A., Simon, C. P., and Koopman, J. (1989):Structured mixing: heterogeneous mixing by the definition of mixing groups. Mathematical and Statistical Approaches to AIDS epidemiology (C. Castillo-Chavez, ed.) Lecture Notes in Biomathematics **83**, Springer-Verlag, Berlin-Heidelberg-New York, 301-315

28. Kendall, D. G. (1949): Stochastic processes and population growth. Roy. Statist. Soc., Ser B2, 230-264
29. Keyfitz, N. (1949): The mathematics of sex and marriage. Proceedings of the Sixth Berkeley Symposium on Mathematical Statistics and Probability. Vol. IV: Biology and Health, 89-108
30. McFarland, D. D. (1972): Comparison of alternative marriage models. In Population Dynamics, (Greville, T. N. E., ed.), Academic Press, New York, London: 89-106
31. Nold, A. (1980): Heterogeneity in disease-transmission modeling. Math. Biosci. **52**, 227-240
32. Parlett, B. (1972): Can there be a marriage function? In Population Dynamics (Greville, T. N. E., ed.) Academic Press, New York, London: 107-135
33. Pollard, J. H. (1973): Mathematical Models for Growth of Human Populations, Chapter 7: The two sex problem. Cambridge University Press
34. Pugliese, A. (1989): Contact matrices for multipopulation epidemic models: how to build a consistent matrix close to data? Technical report UTM 338, Dipartimento di Matematica, Università degli Studi di Trento
35. Ross, R. (1911): The Prevention of Malaria (2nd edition, with Addendum). John Murray, London
36. Ross, R. and Hudson, H. P. (1916): An application of the theory of probabilities to the study of a priori pathometry.- Part I. Proc. R. Soc. Lond. A **93**, 212-225
37. Waldstätter, R. (1989): Pair formation in sexually transmitted diseases. In Mathematical and Statistical Approaches to AIDS Epidemiology, (C. Castillo-Chavez, ed.). Lecture Notes in Biomathematics, **83**, Springer-Verlag, Berlin-Heidelberg-New York, 260-274

Modelling the Effects of Screening in HIV Transmission Dynamics

Ying-Hen Hsieh

Department of Applied Mathematics, National Chung-Hsing University, Taichung, Taiwan, ROC

Abstract

In this paper, we study the effect of screening on the transmission dynamics of HIV virus in a male homosexual population using a simple mathematical model along the lines of those proposed by Anderson, et al. (1986) and others. Analytical results will be given on threshold for a successful screening program which will prevent an epidemic for the simple case when infectivity is assumed to be constant throughout the incubation period. For the model with variable infectivity, we will use numerical simulation and phase plane analysis to explore the possible effects (positive and negative) of random screening as a control policy for AIDS. We also give a formulation of a heterogeneous population model with proportionate mixing and discuss two random screening schemes for the model.

1 Introduction

Since the discovery of the human immunodeficiency virus (HIV), the etiological agent for AIDS, in 1983 (see Barre-Linoussi et al. (1983)- and Gallo et al. (1983)), the basic research on the transmission dynamics of the HIV virus in human populations has become a major research topic in mathematical biology.

AIDS is different from other sexually transmitted diseases (STD) in many respects. Its method of transmission is much more varied than most STD's since HIV virus can be transmitted through blood transfusion, sharing of needle, births, etc, as well as sexual contacts. The individuals who are seropositive could be infected with HIV virus for a long period of times, possibly lifelong, without developing symptoms of AIDS while stay infectious with varying degree of infectivity.

Mathematical models have been developed to study various factors of transmission of STD's. More recently, AIDS models have been proposed by May (1986), Anderson et al. (1986, 1988), May and Anderson (1987), Jacquez et al.

(1988), Castillo-Chavez et al. (1989a, 1989b, 1989c), to name a few. Most models are developed with the purpose to study certain specific characteristics of the transmission dynamics of the AIDS epidemic, such as the long infectious period and incubation period of the disease (Castillo-Chavez et al. 1989a, 1989b), the effect of heterogeneity in sexual activity (Jacquez et al. 1988) or its demographic consequences (Anderson et al. 1988 May et al. 1988). For a review of these models, see Castillo-Chavez (1988) .

The purpose of this work is to study the effects of screening and quarantine as a possible control measure for the spread of HIV infections. The model we will use is a simple epidemiological model of a single population—the group of sexually active male homosexuals. Following the models developed by Anderson et al. (1986) and Castillo-Chavez et al. (1989a), we divide the population into four basic groups—the susceptibles, the first infectious group, asymptomatic carriers, and the second infectious group. The motivation for such a population structure is the growing evidence that the amount of free virus in the body of an infected person is relatively high immediately after infection, then remains low for several years before becoming high again as some symptoms of full-blown AIDS begin to appear (see Francis et al. 1984, Salahuddin et al. 1984 and Lange et al. 1986). The most well-known study on variable infectivity is by Hyman and Stanley (1988) in which a curve plotting probability of infection per contact versus time since infection is given. We also make the assumption that an infection always lead to AIDS, eventually, although it might take years for AIDS symptoms to develop. Therefore, we do not include a compartment for the full-blown AIDS patients in our model since all infected individual will eventually move into this group which is inactive sexually and hence has no bearing on the transmission dynamics of HIV virus in the general population. We also assume no emigration of infectives and that sexual contact is the only mean of HIV transmission since, among male homosexuals, it is the most predominant method of transmitting the disease.

In the subsequent sections, we will present our model with random screening of fixed number of individuals in the population each year and a removal from the pool of sexually active homosexual individuals. (In Gail et al. 1989, it is assumed that those tested positive will result in a decrease in their probability of transmission.) The present methods of testing consist mainly of a serodiagnostic test for antibody to HIV in the circulation system of an individual in order to determine if a person has contracted HIV virus. There are two different tests. One test used on all persons tested is inexpensive, but with high false-positive rate. The second test, which is usually used to double check those tested positive in the first test, is highly reliable but also much more expensive.

In Sect. 2, we will give a threshold for screening in a simple mathematical model assuming constant infectivity. We will show that there is a *threshold* for a successful screening program which serves as a bifurcation parameter for endemic equilibrium and epidemic-free equilibrium. In Sect. 3 we will consider the problem with variable infectivity. We assume that the infected individuals are only being screened out of the population at the two highly infectivity stages when the amount of free virus in one's body is high. The reason for such assumption is twofold : (1) It would lessen the probability of false-positive tests (see Meyer and Pauker 1987) if only the less expensive test is used in a comprehensive screening program, (2) we wish to keep the number of people screened out at a minimum, for economical and social considerations. We assume the seropositive individuals at the carrier stage pose little threat to the general population due to their low infectivity, hence only individuals from the two highly infectious groups are removed. We also assume that all individuals tested out positive will be successfully removed from the active population although, in reality, such assumption would require government-forced removal and civil right issues in many instances.

The central issue of infection control by screening is not without controversy in its social and moral ramifications (see, e.g. Barry et al. 1986). The opponents of such program argue that in addition to the immense expenses (for both the testing and the follow-up quarantine program), there are also ethical questions of confidentiality and management, since some patients might lose their jobs, health care insurance, schooling etc., while others might refuse testing or refuse any quarantine measure when tested positive. Furthermore, it was argued that such measures cannot in practice remove all infected individuals from the sexually active community, but the false sense of security would result in increased sexual contacts by the yet-unidentified infectives (Gerberding and Anderson 1987). This last point is extremely crucial since a recent study by the Hudson Institute reveals that of the heterosexual couples that use condom, one out of five were infected by partner when one of the couple has contracted HIV virus elsewhere, hence the campaign to advocate use of condom might actually be responsible for an increase in HIV infection. In our model, we assume the contacts rate (the mean number of sex partners per individual per year) remain constant for all time thus excluding the possibility of a decrease in contact rate when epidemic spreads or an increase in contact rate when a control measure such as screening is implemented.

Our analysis in Sect. 3 will be in two parts. First, we will highlight the main results obtained in Hsieh (1990) in Sect. 3.1 and 3.2. Section 3.3 consists of numerical examples we use to illustrate the above results. Section 4 will be devoted to phase plane analysis of the effect of screening in the early stages, i.e. whether the doubling time of the HIV transmission can be prolonged significantly with a screening and removal program implemented at the outset of the epidemic. We introduce a heterogeneous population model with screening in Section 5 and discuss some recent results on heterogeneous mixing related to our model with screening. Finally we will give conclusions based on the above results and general observations which we can make concerning the effectiveness of a screening policy. Some mathematical details are given in an appendix.

We should remark, at the point, that the results obtained in this article is by no means a prediction on the consequences of a screening program, since mathematical models in general only serve to suggest the possible relationship between various factors in consideration and their relative importance rather than provide a description of real-life situations—least of all a simple and clearcut answer to practical problems. It is our belief that this study would be helpful in designing a screening program which involves many issues of social, economical, and ethical nature.

2 The model

The population to be considered is the group of male homosexuals. Assuming that sexual contact is the only mean of infection for AIDS, we consider four subgroups : S(susceptibles), X_1(first infectious group), X_2(asymptomatic carriers), and X_3(second infectious group), The model is as follows :

$$S'(t) = \Lambda - cS\frac{\beta_1 X_1 + \beta_3 X_3}{N} - \mu S, \tag{2.1}$$

$$X_1'(t) = cS\frac{\beta_1 X_1 + \beta_3 X_3}{N} - (\nu_1 + \mu)X_1 - \sigma(N)X_1, \tag{2.2}$$

$$X_2'(t) = \nu_1 X_1 - (\nu_2 + \mu)X_2, \tag{2.3}$$

$$X_3'(t) = \nu_2 X_2 - (\nu_3 + \mu)X_3 - \sigma(N)X_3. \tag{2.4}$$

Here Λ is the recruitment rate into the population, c is the contact rate—the mean number of sex partners for each individual per year, β_i is the probability of infection when in contact with a member of X_i, and μ is the natural mortality rate. The term $1/\nu_i$ denotes the average amount of time in years an individual spends as a member of X_i and $N = S + X_1 + X_2 + X_3$ is the total population.

$\sigma(N)$ is the fraction of infectives being screened out and removed from the population. In some epidemiological models, it is assumed to be independent of the actual size of the total population (see Cromer 1988). However, since we assume that screening took place in the population randomly, the number of infectives screened will depend on the size of the total population. In our model, we let $\sigma(N) = \sigma/N$, where σ is the fixed number of individuals screened randomly each year, so the total number of removal from X_i is $\sigma X_i/N$—the number screened times the probability that an individual screened is in X_i. This model was first proposed by Hsieh (1990).

It should be noted that, in practice, the number of individuals screened out from the two highly infectious stages might not have the same proportionality to the total number of persons at each stage, σ/N. The reason for this departure is:

(1) The first infectious stage is relatively short (2 weeks–6 months compared to 1–3 years for second infectious stage, see Hyman and Stanley 1988) with little presence of antibodies to HIV while the screening is done only once a year.

(2) The seropositive individuals in the low infectious carrier stage can be tested regularly and removed as soon as number of free virus in the body becomes dangerously high. Thus, with a good case-tracing program, significantly greater number of persons will be screened out at the second infectious stage than the first.

In Gail et al. (1989), a model of voluntary confidential screening was proposed where it concludes that, in an isolated high-risk population, it requires less than

100 tests to prevent one infection. There are also numerous other possibilities to be considered in a screening strategy. For example, one can choose to implement the more accurate but expensive test on a small number of individuals over the less accurate but inexpensive test on a large number of persons. We can also apply the testing randomly on all individuals never tested instead of over the whole population since, given that the individual was not infected in the last test, the conditional probability of infection since tested is smaller than the probability of infection in general. However, such considerations call for a far more complicated model which is beyond the scope of this article.

2.1 Threshold for screening when infectivity is constant

For the rest of this section, we will show that a threshold for screening exists in a simplified version of (2.1)–(2.4). The simplest mathematical model for HIV transmission is to assume the infectivity of infectives remains constant throughout the incubation period. The resulting equations are as follow :

$$S'(t) = \Lambda - c\beta S\frac{X}{X+S} - \mu S, \tag{2.5}$$

$$X'(t) = c\beta S\frac{X}{X+S} - (\nu+\mu)X. \tag{2.6}$$

Here S denotes the number of susceptibles, X is the number of infectives, Λ is the constant recruitment rate into the population, c is contact rate—the mean number of sex partners for each individual per year, β is the constant probability of infection when in contact with an infective, μ is the natural mortality rate, and ν is the constant removal rate out of the infectious group X.

It is easy to see the system has an epidemic-free equilibrium $(\Lambda/\mu, 0)$ and an endemic equilibrium $(\bar{S}, \bar{X})$ if $c\beta > \nu + \mu$:

$$\bar{S} = \frac{\Lambda}{c\beta - \nu}, \tag{2.7}$$

$$\bar{X} = \Lambda\frac{c\beta - \nu - \mu}{(c\beta - \nu)(\nu + \mu)}. \tag{2.8}$$

The stability of the equilibriums can be determined by the ***basic reproductive number*** R_0 which is the mean number of secondary infections caused by single infective in a population of susceptibles. This number is important in our analysis since in many epidemiological models, $R_0 > 1$ implies asymptotic stability of the endemic equilibrium, i.e. the epidemic will spread in the population; while $R_0 \leq 1$ means the disease will die out. (See Hethcote 1976 or Anderson et al. 1986). In our present model,

$$R_0 = \frac{c\beta}{\nu + \mu}. \tag{2.9}$$

It is easy to show that when $R_0 \leq 1$, $X(t) \to 0$ as $t \to \infty$ for all value of initial infective population $X(0) \geq 0$. If the initial number of infectives is greater than or equal to one and the basic reproductive number is greater than 1, then the number of infectives will always be greater than 1 and the epidemic-free equilibriums is unstable. Since $(\bar{S}, \bar{X})$ is the only other equilibrium when $c\beta > \nu + \mu$ and is locally asymptotically stable by a simple local stability analysis, $(S(t), X(t)) \to (\bar{S}, \bar{X})$ as $t \to \infty$ when $R_0 > 1$.

Hence R_0 is a global bifurcation parameter since the global asymptotic stable equilibrium switches between the epidemic-free equilibrium and the endemic equilibrium as the value of R_0 crossed 1.

Suppose $R_0 > 1$ (or $c\beta > \nu + \mu$), then the epidemic would spread without any control measure. Now we put screening in our model (2.5)–(2.6) and get

$$S'(t) = \Lambda - c\beta S\frac{X}{X+S} - \mu S, \tag{2.5a}$$

$$X'(t) = c\beta S\frac{X}{X+S} - (\nu+\mu)X - \sigma\frac{X}{X+S}. \tag{2.6a}$$

where σ is the number of individuals screened randomly each year.

First we shall consider the stability of the epidemic-free equilibrium $(\Lambda/\mu, 0)$. Let the basis reproductive number at $(\Lambda/\mu, 0)$ be defined by

$$R_0 = \frac{1}{\nu+\mu}(c\beta - \frac{\sigma}{\Lambda/\mu}) \tag{2.10}$$

we then have the following stability result for the epidemic-free equilibrium :

Theorem 1 *If $R_0 > 1$, then $(\Lambda/\mu, 0)$ is an unstable equilibrium for the system* (2.5a)-(2.6a). *On the other hand, if $R_0 \leq 1$, then $(\Lambda/\mu, 0)$ is locally asymptotically stable.*

The proof of Theorem 1 is given in the Appendix.

Since $R_0 > 1$ is equivalent to $\sigma < \Lambda(c\beta - \nu - \mu)/\mu$, the above theorem asserts that when the number of individuals screened randomly each year is too small, the epidemic-free equilibrium is unstable and the population will suffer an epidemic. However, since the stability result is only local for the case $\sigma \geq \Lambda(c\beta - \nu - \mu)/\mu$, we need to look further at the stability of the endemic equilibrium.

If we let $\sigma^* = \Lambda(c\beta - \nu - \mu)/\mu$, the following theorem for the system (2.5a)–(2.6a) is proven in the Appendix :

Theorem 2 *If $\sigma \geq \sigma^*$, the closed rectangle in R^2 with vertices (0,0), $(\Lambda/\mu, 0)$, $(\Lambda/\mu, X^*)$, $(0, X^*)$ is a positive invariant set of the system* (2.5a)-(2.6a) *for X^* sufficiently large.*

We will denote the closed rectangle by $\Re$.

The system (2.5a)–(2.6a) has a unique (positive) endemic equilibrium in $\Re$:

$$\bar{S}_c = \frac{K + \sqrt{K2 + 4\sigma\Lambda c\beta(\nu - c\beta)}}{2c\beta(c\beta - \nu)}, \tag{2.11}$$

$$\bar{X}_c = \frac{(c\beta - \nu - \mu)\bar{S}c - \sigma}{\nu + \mu}, \tag{2.12}$$

where $K = c\beta\sigma + \mu\sigma + \Lambda c\beta$.

$\bar{S}_c$ is positive since $c\beta > \nu + \mu > \nu$ and the term in the square root in (2.10) is a convex function of σ which maps $0, \infty)$ into $(0, \infty)$.

Since $S'(t) < 0$ when $S \geq \Lambda/\mu$, it follows $\bar{S}_c < \Lambda/\mu$ for all $\sigma \geq 0$. Together with the formula for $\bar{X}_c$ in (2.12), we have

$$\bar{X}_c < \frac{(c\beta - \nu - \mu)\Lambda/\mu - \sigma}{\nu + \mu} = \frac{\sigma^* - \sigma}{\nu + \mu}. \tag{2.13}$$

But for an endemic equilibrium with positive $\bar{X}_c$, (2.13) implies $\sigma^* > \sigma$, thus we obtain the following lemma by contradiction :

Lemma. *When $\sigma \geq \sigma^*$, the endemic equilibrium in* (2.11)-(2.12) *does not exist since $\bar{X}_c \leq 0$.*

When $\sigma \geq \sigma^*$, the only equilibrium in the compact positive invariant set $\Re$ is the epidemic-free equilibrium $(\Lambda/\mu, 0)$ which is locally asymptotically stable and all trajectories outside $\Re$ enter it eventually ($S'(t) < 0$ if $S \geq \Lambda/\mu$ and $X'(t) < 0$ if $S < \lambda/\mu, X > X^*$), hence the epidemic-free equilibrium must also be globally asymptotically stable. This together with Theorem 1 yield the main result of this section :

Theorem 3 *Suppose $c\beta > \nu + \mu$, the population in* (2.5)-(2.6) *will approach the endemic equilibrium in* (2.7)-(2.8) *without screening. If the screening term σ in* (2.5a)-(2.6a) *satisfies $\sigma \geq \Lambda(c\beta - \nu - \mu)/\mu$, the population will approach the epidemic-free equilibrium $(\Lambda/\mu, 0)$. However, if $\sigma < \Lambda(c\beta - \nu - \mu)/\mu$, the population will still tend toward the endemic equilibrium in* (2.11)-(2.12).

Theorem 3 asserts that $\sigma^* = \Lambda(c\beta - \nu - \mu)/\mu$ is a global bifurcation parameter for the system (2.5a)–(2.6a) which bifurcates between the endemic and epidemic-free equilibrium as σ cross σ^*. In terms of epidemiology, it is a threshold for screening below which the control policy of screening and removal will not alter the course of an epidemic.

3.1 Model without screening

In Sections 3.1 and 3.2, we will highlight some of the results given in Hsieh 1990—. First let us consider the dynamics of our model in (2.1)–(2.4) without screening and removal. That is, we let $\sigma(N) = 0$ so that (2.1)–(2.4) becomes

$$S'(t) = \Lambda - cS\frac{\beta_1 X_1 + \beta_3 X_3}{N} - \mu S, \tag{3.1}$$

$$X_1'(t) = cS\frac{\beta_1 X_1 + \beta_3 X_3}{N} - (\nu_1 + \mu)X_1, \tag{3.2}$$

$$X_2'(t) = \nu_1 X_1 - (\nu_2 + \mu)X_2, \tag{3.3}$$

$$X_3'(t) = \nu_2 X_2 - (\nu_3 + \mu)X_3. \tag{3.4}$$

This model was studied by Castillo-Chavez (personal communication). One can show that the system described by (3.1)-(3.4) has endemic equilibrium $(S^*, X_1^*, X_2^*, X_3^*)$ where, for $0 < S^* < \Lambda/\mu$,

$$\begin{aligned} X_1^* &= \frac{\Lambda - \mu S^*}{\nu_1 + \mu}, \\ X_2^* &= \frac{(\Lambda - \mu S^*)\nu_1}{(\nu_1 + \mu)(\nu_2 + \mu)}, \\ X_3^* &= \frac{(\Lambda - \mu S^*)\nu_1\nu_2}{(\nu_1 + \mu)(\nu_2 + \mu)(\nu_3 + \mu)}. \end{aligned} \tag{3.5}$$

When $S \geq \Lambda/\mu$, $S'(t) < 0$ thus the equilibriums in (3.5) are the only positive equilibriums. To determine the stability of these equilibriums, let us consider the basic reproductive number of HIV infection, R_0, which is the mean number of secondary infections caused by an infected individual in a population of susceptibles.

The mean number of secondary infections caused by an infected individual is

$$R_0 = c\left[\frac{\beta_1}{\nu_1 + \mu} + \frac{\beta_3\nu_1\nu_2}{(\nu_1 + \mu)(\nu_2 + \mu)(\nu_3 + \mu)}\right]. \tag{3.6}$$

In practice, the value of R_0 is usually larger than 4, given the available data on HIV infections (see Anderson et al. 1986— or Hyman and Stanley 1988—). Thus endemic equilibriums in (3.5) seem to be globally attracting without screening.

3.2 Model with screening and removal

We now go back to the system (2.1)-(2.4) with $\sigma(N) = \sigma/N$ where σ is the (fixed) number of individuals screened randomly each year. Hence $\sigma(N)X_i = \sigma \cdot X_i/N$ is the number of infectious persons screened out in a given year. The model then becomes :

$$S'(t) = \Lambda - cS\frac{\beta_1 X_1 + \beta_3 X_3}{N} - \mu S, \tag{3.1a}$$

$$X_1'(t) = cS\frac{\beta_1 X_1 + \beta_3 X_3}{N} - (\nu_1 + \mu + \frac{\sigma}{N})X_1, \tag{3.2a}$$

$$X_2'(t) = \nu_1 X_1 - (\nu_2 + \mu)X_2, \tag{3.3a}$$

$$X_3'(t) = \nu_2 X_2 - (\nu_3 + \mu + \frac{\sigma}{N})X_3. \tag{3.4a}$$

Similar to the case in Section 3.1, the only endemic equilibriums $(\hat{S}, \hat{X}_1, \hat{X}_2, \hat{X}_3)$ occur at

$$\hat{X}_1 = \frac{\Lambda - \mu\hat{S}}{\nu_1 + \mu + \sigma/\hat{N}},$$
$$\hat{X}_2 = \frac{(\Lambda - \mu\hat{S})\nu_1}{(\nu_1 + \mu + \sigma/\hat{N})(\nu_2 + \mu)}, \tag{3.7}$$
$$\hat{X}_3 = \frac{(\Lambda - \mu\hat{S})\nu_1\nu_2}{(\nu_1 + \mu + \sigma/\hat{N})(\nu_2 + \mu)(\nu_3 + \mu + \sigma/\hat{N})},$$

with $0 < \hat{S} < \Lambda/\mu$.

Here $\hat{N} = \hat{S} + \hat{X}_1 + \hat{X}_2 + \hat{X}_3$ is the unique positive root of the cubic equation

$$\begin{aligned}&(\nu_1 + \mu)(\nu_2 + \mu)(\nu_3 + \mu)N^3 + \{(\nu_2 + \mu)\sigma(\nu_1 + \nu_3 + 2\mu) \\ &- \hat{S}(\nu_1 + \mu)(\nu_2 + \mu)(\nu_3 + \mu) - (\Lambda - \mu\hat{S})[(\nu_2 + \mu)(\nu_3 + \mu) \\ &+ \nu_1(\nu_3 + \mu) + \nu_1\nu_2]\}N^2 + \{\sigma^2(\nu_2 + \mu) - \sigma[(\Lambda - \mu\hat{S})(\nu_1 + \nu_2 + \mu) \\ &+ \hat{S}(\nu_2 + \mu)(\nu_1 + \nu_3 + 2\mu)]\}N - \hat{S}(\nu_2 + \mu)\sigma^2 = 0\end{aligned} \tag{3.8}$$

The basic reproductive number R_0 at the endemic equilibrium in (3.7) is

$$R_0 = c\left[\frac{\beta_1}{\nu_1 + \mu + \sigma/\hat{N}} + \frac{\beta_3\nu_1\nu_2}{(\nu_1 + \mu + \sigma/\hat{N})(\nu_2 + \mu)(\nu_3 + \mu + \sigma/\hat{N})}\right] \tag{3.9}$$

Since $1 \geq \sigma/\hat{N} > 0$ when $\sigma > 0$, it is obvious that the basic reproductive number is decreased with screening. In the following section, we will turn to numerical simulation for some intuitive understanding.

3.3 Some numerical examples

All simulations used the following values for the parameters : $\beta_1 = \beta_3 = 0.03$, $\mu = 0.02$, $c = 36$, $\nu_1 = 2$, $\nu_2 = 1/5$, $\nu_3 = 1/3$. All parameters were take from existing data to reflect practical situations as closely as possible. We also let $S(0) = 100,000$, $\Lambda = 1,500$ and use 4th order Runge-Kutta method to obtain Fig. 1. We let $X_1(0) = 1$, $X_2(0) = X_3(0) = 0$ and $\sigma = 0$. That is, we assume one infective was introduced into a population of 100,000 susceptibles initially. With no screening, the population will approach endemic equilibrium in (3.5) with the number of susceptibles S^* at approximately 4,600 and the total number of (active) infectious person at (approximately) 10,600. Hence we have an epidemic with more than 2/3 of the active population infected. If we add a screening term $\sigma = 10,000$, i.e. 10,000 individuals were screened at random each year, then we have Fig. 2 where the number of susceptibles is about 16,000 and the total number of infectious persons is around 5,800. Thus, with random screening of 10,000 individuals imposed at the onset of the epidemic, the epidemic cannot be avoided, but less than 30% of the population are infected. Hence even a partial screening and removal program can decrease the number of (active) infected population.

However, in the case of Fig. 2, the number of infected individuals screened out of the community, X_4, given by the additional equation

$$X_4'(t) = \frac{\sigma}{N}(X_1(t) + X_3(t)) - \mu X_4(t) \tag{3.13}$$
$$- \sigma\nu_3 \left[\frac{X_3(t)}{N(t)} + \frac{\nu_1}{\nu_1+\mu} \cdot \frac{\nu_2}{\nu_2+\mu} \cdot \frac{X_1\left(t - \frac{1}{\nu_1+\mu} - \frac{1}{\nu_2+\mu}\right)}{N\left(t - \frac{1}{\nu_1+\mu} - \frac{1}{\nu_2+\mu}\right)} \right],$$

is 24,344, more than the total active population! Therefore a screening policy which is not sufficiently comprehensive could decrease the number of active infectives, but more than double the actual infected population.

Under our assumption that sexual contact is the only mean of HIV transmission, and assuming no emigration of infectives into the population, if the whole population is screened simultaneously at any time during a year, all infectives will be screened out and the spread of HIV virus will stop. Further simulations show that using the same set of data as in Fig. 1, the total infected population will go to zero if we use a screening term of σ greater than or equal to 65,000. A theoretical study on a threshold for screening in the case of variable infectivity will appear in a subsequent article being prepared by the author.

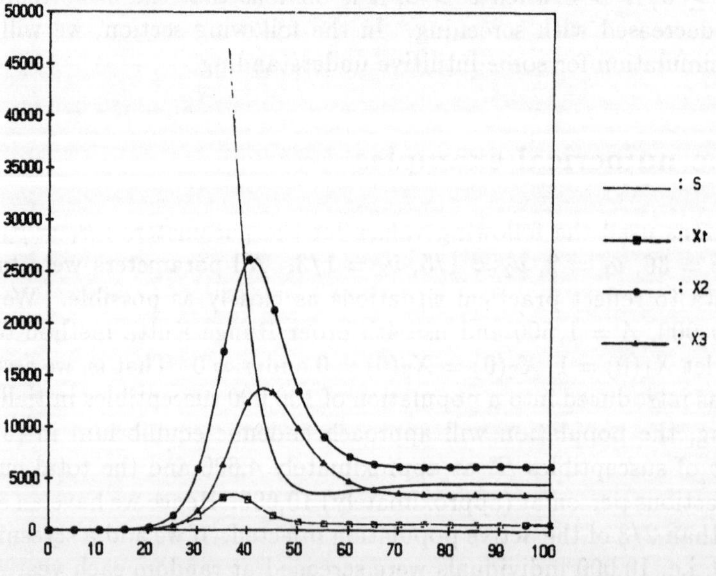

Fig. 1. The active population without screening ($\sigma = 0$). $\beta_1 = \beta_3 = 0.003$, $\mu = 0.002$, $c = 36$, $\nu_1 = 2$, $\nu_2 = 0.2$, $\nu_3 = 1/3$, $\Lambda = 1,500$ and $S(0) = 100,000$.

4 Screening at early stages of the epidemic

In much of the early studies on mathematical models of HIV transmission dynamics (May 1986, Anderson et al. 1986), the analysis focuses on the early stages of the epidemic. Attempts were made to find explicit formula for initial growth of the epidemic as well as the time it takes for the number of infections to double. The doubling time t_d of the epidemic is important since it allows us to know how fast the disease is spreading at the onset of the epidemic in order for one to understand the gravity of the situation and to implement control measures.

Anderson et al. (1986) showed that, in their simple mathematical model of HIV transmission, the number of infectives is given by an exponential function of time at the early stages, while Hyman and Stanley (1988) have found that the number of diagnosed AIDS cases in United States is well approximated by a cubic polynomial. The discrepancy between the two studies is insignificant, since in the neighborhood of $t = 0$, exponential and cubic functions approximate each other.

At the early stages of the epidemic when $S \simeq N$ and $X_1 + X_2 + X_3$ small compare to S, we do not distinguish the three stages of incubation period in (2.1)–(2.4), thus arriving at :

$$S'(t) \simeq \Lambda - c\beta X - \mu S, \tag{4.1}$$
$$X'(t) \simeq (c\beta - \nu - \mu)X. \tag{4.2}$$

From which we have the number of infectives and susceptibles :

$$\begin{aligned} X(t) &\simeq X(0)e^{(c\beta-\nu-\mu)t}, \\ S(t) &= e^{-\mu t}\left[S(0) + \int_0^t e^{\mu\tau}(\Lambda - c\beta X(0)e^{(c\beta-\nu-\mu)\tau})\,d\tau\right]. \end{aligned} \tag{4.3}$$

Thus once again, we have exponential growth with the doubling time

$$t_d = \ln 2/(c\beta - \nu - \mu). \tag{4.4}$$

Now if we take screening into consideration, (3.1a)–(3.4a) gives us (with $S \simeq N$),

$$S'(t) \simeq \Lambda - c\beta X - \mu S, \tag{4.5}$$
$$X'(t) \simeq (c\beta - \nu - \mu)X - \frac{\sigma}{S}X. \tag{4.6}$$

Analytic solution of the system (4.5)–(4.6) is not possible, but we can get a basic understanding of its dynamical behavior by a simple phase plane analysis given in Figs. 3 and 4.

For $\sigma < \sigma^* = \Lambda(c\beta - \nu - \mu)/\mu$, the threshold for screening in Sect. 2, the system has a epidemic-free equilibrium at $X = 0$, $S = \Lambda/\mu$ which is clearly unstable since $X(t)$ increases for $S > \sigma/(c\beta - \nu - \mu)$, the system also has an endemic equilibrium at

$$X = \frac{\sigma^* - \sigma}{\sigma^* c\beta}, \qquad S = \sigma/(c\beta - \nu - \mu). \tag{4.7}$$

We can see from the please diagram Fig. 3 that the endemic equilibrium is an attractor as all solutions spiral in toward it. When $\sigma \geq \sigma^*$ as in Fig. 4, the epidemic-free equilibrium at $X = 0$, $S = \Lambda/\mu$ is the only equilibrium and all solutions are attracted toward it.

Although this result would seem to duplicate the results in Sect. 2, the readers must be reminded that the phase plane analysis done in this section are for the early stages of an epidemic, when the infected persons are small in number and mainly of the first infectious group (or of the noninfectious carrier group). Hence a phase plane analysis for the stability of system when time is large is not really meaningful, what is meaningful to us is that (i) $X(t)$ is decreasing unless S is large, i.e. $S > \Lambda/\mu$. (ii) $X(t)$ and $S(t)$ are never both increasing in Fig. 4, and are both increasing in Fig. 3 only in the shaded region $\Re^*$ where $S > \sigma/(c\beta - \nu - \mu)$ and $X < (\Lambda - \mu S)/c\beta$. Thus exponential initial growth of the number of HIV infections is possible only for certain values of initial population when screening is implemented.

The above observation is illustrated with Figs. 5 and 6, where all data are as in Fig. 2, except $S(0) = 10,000$ in Fig. 5, and $S(0) = 50,000$ in Fig. 6. We designate the population in Fig. 5 as population A and the one in Fig. 6 as population B.

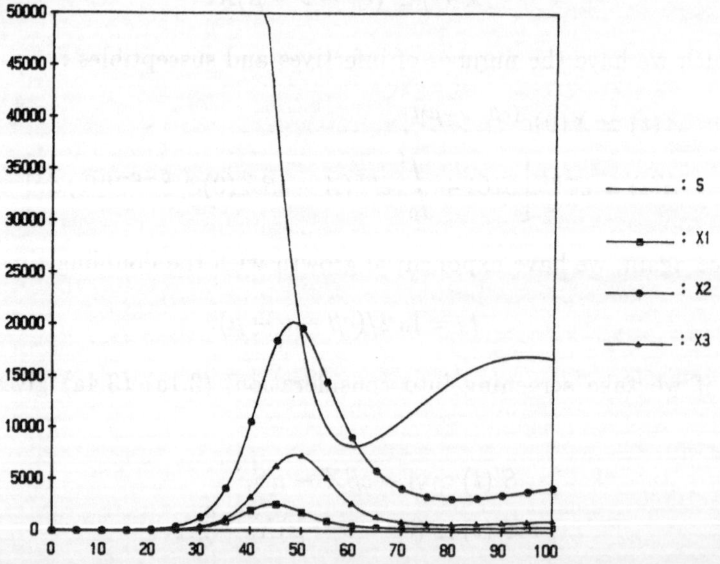

Fig. 2. The active population without screening ($\sigma = 10,000$) and all other parameters same as in Fig. 1.

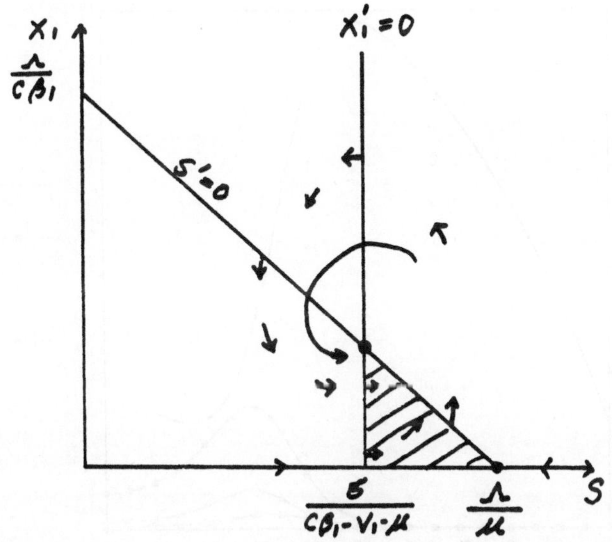

Fig. 3. Phase diagram for $\sigma < \sigma^*$ at early stages of the epidemic when $S \simeq N$. The shaded region is $\Re^*$.

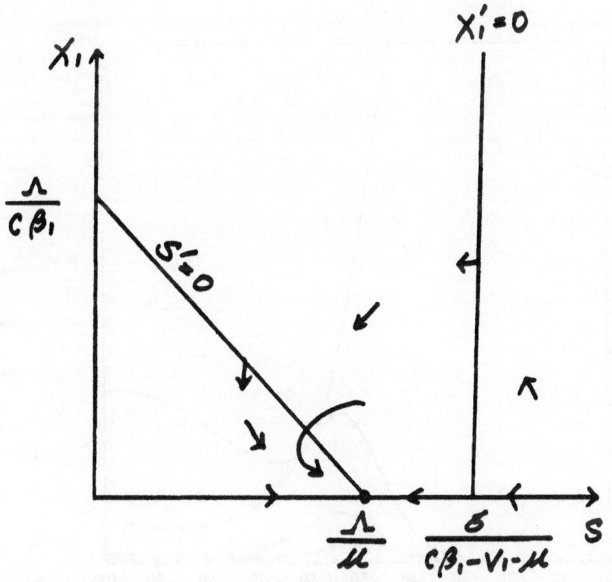

Fig. 4. Phase diagram for $\sigma < \sigma^*$ at early stages of the epidemic when $S \simeq N$.

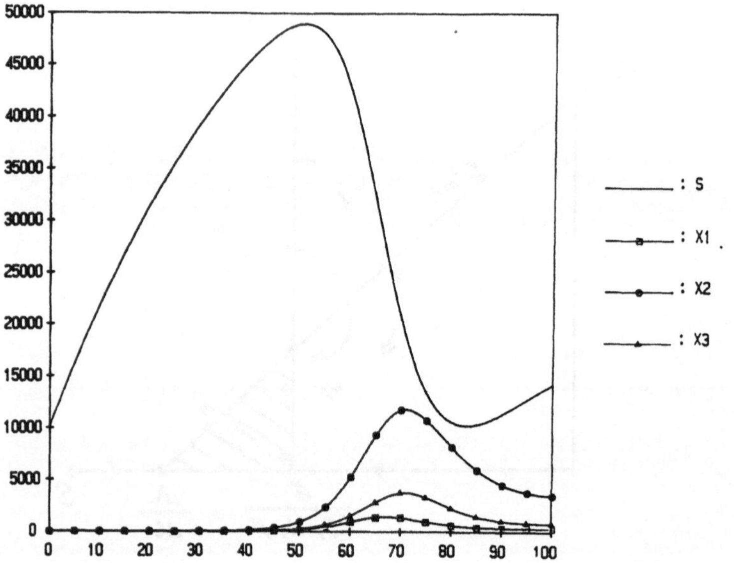

Fig. 5. The active population without screening. All parameters same as in Fig. 2 except $S(0) = 10,000$.

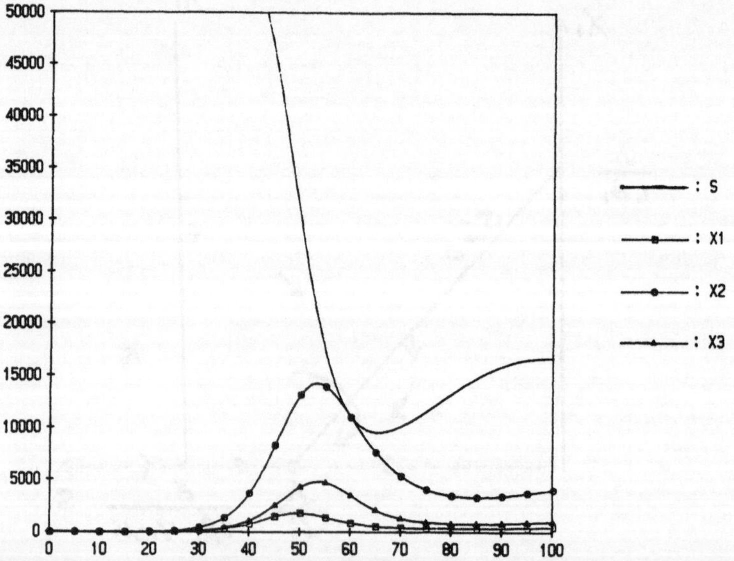

Fig. 6. The active population without screening. All parameters same as in Fig. 2 except $S(0) = 50,000$.

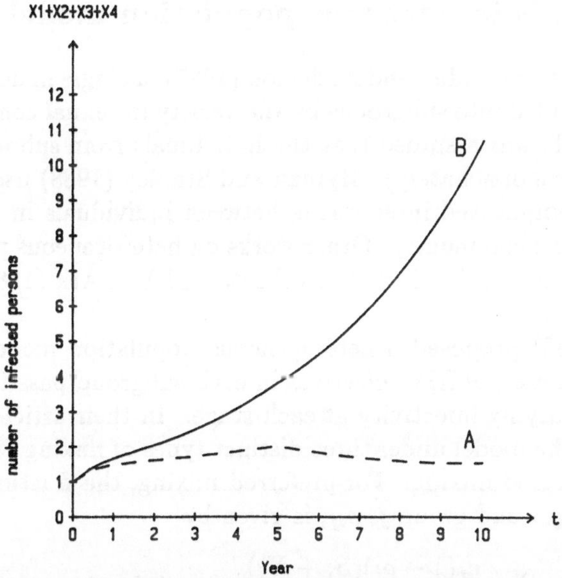

Fig. 7. The total infected population $X_1 + X_2 + X_3 + X_4$ in the simulated populations for Fig. 5 and Fig. 6 for the first 10 years. Dashed curve: population A (Fig. 5), solid curve : population B (Fig. 6).

Due to the difference in initial values, the two populations converge to distinct endemic equilibriums. However the large deviation in initial value only result in a (relative) small difference in the numbers for the endemic equilibrium—with the difference less than 25% in each of the four groups S, X_1, X_2, and X_3. But in the first 10 years, the total number of infectives in the population, $X_1 + X_2 + X_3 + X_4$, increases exponentially in population B but only linearly in population A (see Fig. 7). The reason being the initial data for population B is contained in the region $\Re^*$ in Fig. 3 where both X and S are increasing in time. Hence, with the same screening program, a population could either exhibit minor initial growth in AIDS victims as population A in Fig. 7, or increases 10-fold in 10 years as in population B. It is also worthwhile to note that the initial doubling time is about 12 years in population A but only 2 years in population B. Thus a carefully designed screening program can slow down the full thrust of the epidemic significantly.

It should be noted that theoretically, given an initial population, one can always choose σ large enough ($\sigma > \sigma^*$) so that the initial growth will not be exponential, but in practice it could mean a very comprehensive screening program which might not be desirable.

5 Screening in a heterogeneous population model

In Anderson et al. (1986) and May and Anderson (1987), a single male homosexual population was divided into subgroups by the variety in sexual contacts each individual practiced. It was assumed that the individuals from subpopulations mixed randomly (or proportionately). Hyman and Stanley (1988) use a similar model to study the complicated interactions between individuals in a sexually active and IV-drug-using community. Other works on heterogeneous population model with proportionate mixing include Hethcote and Van Ark (1987), Isham (1988), Kaplan (1989a).

Jacquez et al. (1988) proposed a heterogeneous population model of n interacting subgroups where the HIV infectives in each subgroup pass through m infective stages with varying infectivity at each stage. In their article, Jacquez et al. (1988) examine the model under three distinct types of mixing : restricted, proportional, and preferred mixing. For preferred mixing, the fraction of partnerships between group i and group j, p_{ij}, is given by

$$p_{ij} = \rho_i \delta_{ij} + (1 - \rho_i) \frac{c_j(1 - \rho_j)(S_j + X_j)}{\sum_{k=1}^{n} c_k(1 - \rho_k)(S_k + X_k)}, \quad i, j = 1, \ldots, n. \tag{5.1}$$

Here S_j and X_j are the numbers of susceptibles and infectives in the j-th group, respectively. δ_{ij} is the Kronecker delta, c_j is the average number of partners per year for a person in group j, and ρ_i is the fraction of group i's contacts reserved for contacts within the group. Jacquez et al. (1988) noted that "$\rho_i = 0$ for all i" is proportional mixing and "$\rho_i = 1$ for all i" is restricted mixing, hence when $0 \leq \rho_i \leq 1$ for all i, the resulting matrix $\mathbf{P} = (p_{ij})$ is a general *mixing matrix* which covers all three types of mixing mentioned above. Among other results, Jacquez et al. (1988) gave a threshold below which the epidemic-free equilibrium is the only equilibrium and above which a unique endemic equilibrium occurs. More recently, Lin (1990a) showed that the stability modulus of the Jacobian matrix for the system proposed by Jacquez et al. (1988) at the epidemic-free equilibrium is indeed a threshold. Lin (1990a) also proved : (1). the local asymptotic stability of the endemic equilibrium when it exists, (2). the equivalence between his threshold condition and the one proposed by Jacquez et al. (1988), thus extending the results on the model of Jacquez et al. (1988).

Recently, Kaplan and Lee (1990) obtained a worst-case bound for heterogeneous population models with nonrandom heterogeneous mixing and showed numerically that the worst case number of infected persons in the endemic equilibrium is within 10% of the number of infected persons that would result from random mixing. Thus, for the purpose of public health decision-making process, a simple heterogeneous model with proportionate mixing would be sufficient. (For additional discussion on the usefulness of simple random mixing model, also see Kaplan 1989b.) Finally, although the infectivity of an infective varies in time during the long incubation period, denoted by $\beta(t)$, we can quite appropriately use a constant infectivity parameter β if one define it to be given by

$$\beta = \frac{1}{T_I}\int_0^{\infty} \beta(t)e^{-\mu t} Pr\{A > t\}\, dt, \tag{5.2}$$

where T_I is the mean incubation time, μ is the natural mortality rate, and A is a random variable denoting the incubation time of an infected person. (See Kaplan and Lee 1990)

Hence a heterogeneous population model with screening is formulated as follows. Let S_i, X_i be the number of susceptibles and infectives in group i, respectively; and let p_{ij} be defined as in (5.1) with $\rho_i = 0$ for all i, i.e.,

$$p_{ij} = \frac{c_j(S_j + X_j)}{\sum\limits_{k=1} nc_k(S_k + X_k)}. \tag{5.3}$$

Moreover, let μ, c_i be defined as above and β as in (5.2). We also assume Λ_i is the constant recruitment rate into group i, γ_i is the mean transfer rate to AIDS victim for infected persons in group i, and σ_i is the fixed number of persons screened randomly each year out of group i. The model then becomes :

$$\frac{dS_i}{dt} = \Lambda_i - c_i S_i \beta \sum_{j=1} \frac{np_{ij}X_j}{S_j + X_j} - \mu S_i, \tag{5.4}$$

$$\frac{dX_i}{dt} = c_i S_i \beta \sum_{j=1} \frac{np_{ij}X_j}{S_j + X_j} - (\mu + \gamma_i)X_i - \frac{\bar{\sigma}_i(t)X_i}{S_i + X_i}, \; i = 1, \ldots, n. \tag{5.5}$$

Here $\bar{\sigma}_i(t)$ is the *adjusted* number of persons screened in group i and is given by

$$\bar{\sigma}_i(t) = \min\{\sigma_i, S_i(t) + X_i(t)\}. \tag{5.6}$$

The adjustment on σ_i is necessary due the possibility that the number of persons $S_i + X_i$ in one particular subgroup may well drop below σ during some time interval. (For single population model, such adjustment is unnecessary since once the total population drop below the number of persons screened, we can screen the whole population effectively.) This extra consideration occurs due to the practical reason that the policymaker may wish to set up a screening program by subgroups. In case one wishes to uncomplicate the system in (5.4)–(5.5) a bit, we can assume a comprehensive screening program for all persons in the population regardless of subgroup, then (5.5) can be simplified to :

$$\frac{dX_i}{dt} = c_i S_i \beta \sum_{j=1} \frac{np_{ij}X_j}{S_j + X_j} - (\mu + \gamma_i)X_i - \frac{\sigma X_i}{\sum\limits_{k=1} n(S_k + X_k)}, \; i = 1, \ldots, n, \tag{5.5a}$$

where σ is the fixed number of persons screened randomly out of the whole population. One can see either (5.5) or (5.5a) presents problem of its own in terms of analysis.

When $\sigma = 0$, our model is described by (5.4) together with

$$\frac{dX_i}{dt} = c_i S_i \beta \sum_{j=1} \frac{np_{ij}X_j}{S_j + X_j} - (\mu + \gamma_i)X_i, \quad i = 1, \ldots, n, \tag{5.7}$$

which is just a special case of the model proposed by Jacquez et al. (1988). The stability results for the case $\sigma = 0$ can be obtained from the work of Castillo-Chavez et al. 1989c— and Lin 1990a,1990b— on more general model when there are different infectivity rates for distinct subgroups, i.e., β_{ij} is the probability of infection for a susceptible in group i when in contact with an infective in group j. (This assumption is especially valid when sexual contact is not the only mean of HIV transmission in the population.)

The open question to be studied is : When a positive endemic equilibrium exists for the system (5.4) and (5.7), is there a threshold value for screening, σ_i^* or σ^*, so that either the system described by (5.4) and (5.5), or (5.4) with (5.5a), will have epidemic-free equilibrium as the globally asymptotically stable equilibrium if the threshold value for screening is attained. It seems reasonable to conjecture that such threshold value do exists, since under our assumption all new recruits into the population are uninfected so that if the screening is comprehensive enough, all infectives can be removed out of the population successfully. However, to find an explicit form for the threshold like we have for the single population model in Section 2 is quite difficult. The Jacobian matrix for the system is extremely complicated due to the extra screening term in (5.5) and (5.5a). My guess is that basic reproductive number probably is still the best path to success.

6 Conclusion

From the analysis and numerical simulations given in the previous sections, we can reach the following conclusion on the effect of screening in our model of HIV transmission dynamics :

(i) Under the assumption of constant infectivity, an explicit threshold for screening can be obtained so that random screening is successful if and only if the number of individuals screened is above or equal to the threshold.

(ii) For the case of variable infectivity, it is shown numerically that screening can prevent the spread of epidemic if the program is comprehensive enough, but if not sufficiently large-scaled, it could increase the total infected population.

(iii) A more significant effect of screening appears to be the lengthening of doubling time of the disease at the early stages if a sufficiently comprehensive screening and removal program is implemented at the early stages of the epidemic.

The results of this study caution us against being over-optimistic in the effect of a screening and removal program as control policy for the AIDS epidemic. However, if implemented judiciously, screening could be used as a temporary and partial solution to the fight against AIDS, especially in regions where HIV infection is at its very early stage. A selective screening program (for high risk groups, pregnant women, surgeons, etc.) could effectively delay the spread of epidemic if implemented early enough, thus buying more time for medical research to find a vaccine. Other than that, the high economic and social cost of a large-scale screening policy could make it both unrealistic as well as ineffective, not to mention the adverse consequences a false sense of security could bring!

7 Appendix

Theorem 1 *If $R_0 > 1$, then $(\Lambda/\mu, 0)$ is an unstable equilibrium for the system* (2.5a)–(2.6a). *On the other hand, if $R_0 \leq 1$, then $(\Lambda/\mu, 0)$ is locally asymptotically stable.*

Proof. The Jacobian matrix for the right-hand side of (2.5a)–(2.6a) is

$$J = \begin{bmatrix} -\mu - c\beta\frac{X^2}{(X+S)^2} & -c\beta\frac{S^2}{(X+S)^2} \\ \sigma\frac{X}{(X+S)^2} + c\beta\frac{X^2}{(X+S)^2} & c\beta\frac{S^2}{(X+S)^2} - (\nu+\mu) - \frac{\sigma S}{(X+S)^2} \end{bmatrix}. \tag{A1}$$

At the epidemic-free equilibrium $S = \Lambda/\mu$, $X = 0$, the characteristic equation for J is

$$det(\lambda I - J) = \lambda 2 + a_1\lambda + a_2 = 0, \tag{A2}$$

where

$$a_1 = 2\mu + \nu - c\beta + \frac{\sigma\mu}{\Lambda}, \qquad a_2 = \mu(\nu + \mu + \frac{\sigma\mu}{\Lambda} - c\beta). \tag{A3}$$

When $R_0 > 1$, $c\beta > \nu + \mu + \sigma\mu/\Lambda$ which implies $a_2 < 0$ and $(\Lambda/\mu, 0)$ is unstable. When $R_0 < 1$, $c\beta < \nu + \mu + \sigma\mu/\Lambda$ which yields $a_1 > 0$, $a_2 > 0$ which satisfy the Routh-Hurwitz condition for (locally) asymptotic stability. For the case $R_0 = 1$, $c\beta = \nu + \mu + \sigma\mu/\Lambda$ and we only have locally stability ($a_1 > 0$, $a_2 = 0$). Therefore we go back to the equations (2.5a)–(2.6a) to show asymptotic stability. For positive initial data $X(0) > 0$ and $S(0) > 0$, $S'(t) < 0$ if $S \geq \Lambda/\mu$, therefore we can assume without loss of generality $S(t) < \Lambda/\mu$ for t sufficiently large.

Using the fact $c\beta = \nu + \mu + \sigma\mu/\Lambda$ when $R_0 = 1$, (2.6a) becomes

$$\begin{aligned} X'(t) &= c\beta S\frac{X}{X+S} - (c\beta - \frac{\sigma\mu}{\Lambda})X - \frac{\sigma X}{X+S} \\ &= -c\beta\frac{X^2}{X+S} - \sigma X(\frac{1}{X+S} - \frac{\mu}{\Lambda}). \end{aligned} \tag{A4}$$

With $\frac{\mu}{\Lambda} < \frac{1}{S}$ and $\frac{1}{X+S} - \frac{\mu}{\Lambda} > \frac{1}{X+S} - \frac{1}{S}$, we have

$$\begin{aligned} X'(t) &< -c\beta\frac{X^2}{X+S} - \sigma X(\frac{1}{X+S} - \frac{1}{S}) \\ &= \frac{X^2}{X+S}(\frac{\sigma}{S} - c\beta). \end{aligned} \tag{A5}$$

Since $\frac{\Lambda}{\mu} = \frac{\sigma}{c\beta - \nu - \mu} > \frac{\sigma}{c\beta}$, we can pick sufficiently small neighborhood of $(\Lambda/\mu, 0)$ so that $S > \frac{\sigma}{c\beta}$ and hence $X(t) \to 0$ as $t \to \infty$. □

Theorem 2 *If* $\sigma \geq \sigma^*$, *the closed rectangle in* $\mathbb{R}^2$ *with vertices (0,0),* $(\Lambda/\mu, 0)$, $(\Lambda/\mu, X^*)$, $(0, X^*)$ *is a positive invariant set of the system* (2.5a)-(2.6a) *for* X^* *sufficiently large.*

Proof. When $S \geq \Lambda/\mu$, $S'(t) < 0$. Since $\sigma \geq \Lambda(c\beta - \nu - \mu)/\mu$,

$$X'(t) \leq c\beta S \frac{X}{X+S} - (\nu + \mu)X - \frac{\Lambda}{\mu}(c\beta - \nu - \mu)\frac{X}{X+S} \tag{A6}$$

$$= \frac{X}{X+S}\left[(c\beta - \nu - \mu)S - (\nu + \mu)X + \frac{\Lambda}{\mu}(\nu + \mu - c\beta)\right].$$

Since we already supposed that $c\beta > \nu + \mu$, and $S < \Lambda/\mu$ in the closed rectangle. If we pick $X^* = \frac{\sigma^*}{\nu + \mu}$, it follows that $X'(t) < 0$, if $X \geq X^*$ and $S \leq \Lambda/\mu$. □

Acknowledgment

This research was supported partially by a grant from the National Science Council of ROC (NSC78-0208-M005-03) for which the author is grateful. The work has benefited greatly through discussions with Carlos Castillo-Chavez and Herb Hethcote. The author also wishes to thank Karl Hadeler and Fabio Milner for their comments on threshold for screening.

References

1. Anderson, R.M., Medley, G.F., May, R.M., Johnson, A.M., (1986): A preliminary study of the transmission dynamics of the human immunodeficiency virus (HIV), the causative agent of AIDS. IMA. J. Math. Med. Biol. **3**, 229–263
2. Anderson, R.M., May, R.M., McLean, A.R., (1988): Possible demographic consequences of AIDS in developing countries. Nature, **332** (6161), 228–234
3. Barre-Sinoussi, F. et al., (1983): Isolation of a T-lymphotropic retrovirus from a patient at risk for acquired immune deficiency (AIDS). Science. **220**, 868–871
4. Barry, M.J., Cleary, P.D., Fineberg, V.H., (1986): Screening for HIV infection : risks, benefits, and the burden of proof. Law, Medicine, and Health Care. **14**, 259–267
5. Castillo-Chavez, C., (1988): Review of recent models of HIV/AIDS transmission. In Applied Mathematical Ecology, Levins, S.A., Hallam, T.G., Gross,L.J., eds., Lecture Notes in Biomathematics **18**, Springer-Verlag, Berlin-Heidelberg-New York
6. Castillo-Chavez, C., Cooke, K., Huang, W., Levin, S.A., (1989a): The role of long incubation period in the dynamics of acquired immunodeficiency syndrome (AIDS). Part I. Single populations models. J. Math. Biol. **27**, 373–398

7. Castillo-Chavez, C., Cooke, K., Huang, W., Levin, S.A., (1989b): The role of long period of infectiousness in the dynamics of acquired immunodeficiency syndrome (AIDS). In Mathematical Approaches to Ecological and Environmental Problem Solving. Castillo-Chavez, C., Levin, S.A., Shoemaker, C., eds., Lecture Notes in Biomathematics, **83**, Springer-Verlag, Berlin-Heidelberg-New York
8. Castillo-Chavez, C. K. Cooke, W. Huang, and S. A. Levin (1989c): Results on the dynamics for models for sexual transmission of the human immunodeficiency virus. Appl. Math. Lett. **2**. 327-331
9. Cromer, T.L., (1988): Seasonal control for an endemic disease with seasonal fluctuations. Theoret. Pop. Biol. **33**, 115-125
10. Francis, D.P., Feorino, P.M., Broderson, J.R., et al., (1984): Infection of chimpanzees with lymphadenopathy-associated virus. Lancet, December 1, 1276–77
11. Gail, M.H., Preston, D., Piantadosi, S., (1989): Disease prevention models of voluntary confidential screening for human immunodeficiency virus (HIV). Stat. in Medicine, **8**, 59-81
12. Gallo, R.C., et al., (1983): Frequent detection and isolation of cytopathic retroviruses (HTLV-III) from patients with AIDS and at risk for AIDS. Science, **224**, 500–503
13. Gerberding, J.L., Henderson, D.K., (1987): Design of rational infection control policies for Human Immunodeficiency Virus infection. J. Infect. Dis., **156** (6), 861–4
14. Hethcote, H.W., (1976): Qualitative analysis for communicable disease models. Math. Biosci., **28**, 335–356
15. Hethcote, H. W. and J. W. Van Ark, (1987): Epidemiological models for heterogeneous populations: proportionate mixing, parameter estimation, and immunization programs. Math Biosci **84**, 85-118
16. Hsieh, Y.H., (1990): An AIDS model with screening. Proceedings of 7th ICMCM. Math. Comput. Modelling, to appear
17. Hyman, J.M., Stanley, E.A., (1988): Using mathematical models to understand the AIDS epidemic. Math. Biosci., **90**, 415–473
18. Isham, V. (1988): Mathematical modeling of the transmisson dynamics of HIV infections and AIDS. A review. J. Roy. Stat. Soc. **A151**, 5-30
19. Jacquez, J.A., C.P. Simon, J. Koopman, L. Sattenspiel, and T. Perry. (1988): Modeling and analyzing HIV transmission: the effect of contact patterns. Math. Biosci, **92**, 119-199
20. Kaplan, E. H. (1989a): What are the risks of risky sex? Modeling the AIDS epidemic. Operations Research **37**, 198-209
21. Kaplan, E. H. (1989b): Can bad models suggest good policies? Sexual mixing and the AIDS epidemic. J. Sex. Resear. **26**, 301-314
22. Kaplan, E. H. and Y. S. Lee (1990): How bad can it get? Bounding worst case endemic heterogeneous mixing models of HIV/AIDS. Math Biosci **99**, 157-180
23. Lange, J.M. et al., (1986): Persistent HIV antigenaemia and decline of HIV core antibodies associated with transition to AIDS. Br. Med. J., **293**, 1459–62
24. Lin, X. (1990): Qualitative analysis on a HIV transmission model. Math Biosci (to appear)
25. Lin, X. (1990): On the uniqueness of endemic equilibrium of an HIV/AIDS transmission model for a heterogeneous population (preprint)
26. May, R.M., (1986): Notes on AIDS models. Unpublished manuscript
27. May, R.M., Anderson, R.M., (1987): Transmission dynamics of HIV infection. Nature, **326**, 137–142

28. May, R.M., Anderson, R.M., McLean, A.R., (1988): Possible demographic consequences of HIV/AIDS : I, assuming HIV infection always leads to AIDS. Math. Biosci., **90**, 475–505
29. Meyer, K.D., Pauker, S.G., (1987): Screening for HIV : Can we afford the false-positive rate? N. Engl. J. Med., **317**, 238–41
30. Salahuddin, S.Z., et al., (1984): HTLV-III in symptom-free zero negative persons. Lancet, December **22–29**, 1418–20

An S→E→I Epidemic Model with Varying Population Size

Andrea Pugliese

Dipartimento di Matematica, Università di Trento, Povo (TN), Italy

Abstract

An $S \to E \to I$ epidemic model with a general shape of density–dependent mortality and incidence rate is studied analytically and numerically.

The combined effect of a latent period and of varying population size can produce oscillations in this ODE model. When fertility of exposed individuals is the same as that of susceptibles, there is a clear threshold. On the contrary, when both exposed and infectives do not contribute to birth rate, there may exist multiple endemic states also below the threshold.

When the contact rate is independent of population size, the global behaviour is established: all trajectories converge to an equilibrium.

1 Introduction

In a previous paper [12] I analyzed a simple epidemic model, with vertical transmission, an arbitrary density–dependent demography, and a density–dependent contact rate; it was assumed that the disease increased the death rate and/or decreased the birth rate; therefore population size varied with disease prevalence.

The main result of the analysis was that the qualitative conclusions obtained for the constant population model were still valid. There exists a threshold under which the disease-free equilibrium is globally stable; above the threshold there exists also an endemic equilibrium, which is stable and attracts all initial points with $I_0 > 0$. The threshold relates to the disease transmissibility at the demographic equilibrium population size.

A similar model has been studied by Brauer [6], with somewhat different assumptions; he also allows for a generic distribution of the infectious period (before disease-related death occurs); he finds that, while an exponential distribution (the one used in an ODE formulation) always yields (local) stability of the endemic equilibrium, other distributions may render the equilibrium unstable.

Models with varying population size have also been considered by Anderson and May [3], Busenberg *et al.* [7], Busenberg and van den Driessche [8], Andreasen [4], Kretzschmar [10]. They all assumed that birth and death rates did not depend on population size, and therefore, without epidemics, there would be exponential population growth. Their results are therefore not directly comparable to [12].

The most similar model has been studied by Anderson *et al.* [2] as a model for fox rabies. They assumed that population growth (in absence of epidemics) is logistic, and a standard mass action law term for the incidence rate. The main difference with the model studied in [12] is that they allow for a latent period following infection, i.e. a period in which the animal is not contagious. They found numerically conditions under which the endemic equilibrium is unstable, and there exist stable limit cycles.

The purpose of this paper is to introduce a latent period in the model I considered before, in order to compare the results of [12] with those of Anderson *et al.* [2].

2 Formulation of the model

The population (N) is divided between susceptibles (S), exposed (E) and infectives (I).

As for the demography, I assume non-disease related mortality to be a non-decreasing function, $m(N)$ of total population size; infectives suffer also an additional mortality μ. The fertility of susceptibles is a, that of exposed is $a(1-\delta_1)$, that of infectives is $a(1-\delta_2)$, $0 \leq \delta_1 \leq \delta_2 \leq 1$. I assume that all newborns are susceptibles, not allowing therefore for vertical transmission; this is done because in [12] vertical transmission does not change the results, but only make the expressions more complicated. Again for the sake of simplicity, I assume that only mortality (and not also fertility) is density–dependent.

Since it is difficult to study the resulting system for all combinations of δ_1 and δ_2, I will explore in detail only two possible choices, which yield a rather different picture. In the first case, I will assume that exposed individuals have no disease symptoms at all; therefore their fertility is equal to that of susceptible individuals, namely $\delta_1 = 0$. In the second case, I will assume, following Anderson *et al.* [2] and Brauer [5] [6], that both exposed and infected individuals do not contribute at all to the birth rate, namely $\delta_1 = \delta_2 = 1$. I suspect that this second choice is not very realistic; I study it for ease of comparison with other results.

As for the rate of new infections, the probability for a susceptible of getting infected is equal to $\sigma(N)I$ (note that the contact rate per individual $c(N)$ is the equal to $\sigma(N)N$); in case of environmentally transmitted diseases one normally assumes $\sigma(N) \equiv \beta$; for sexually transmitted diseases $\sigma(N) = \beta/N$. Since these are considered to be the extremes, in general one may assume $\sigma(N)$ to be a non-increasing function, while $N\sigma(N)$ is a non-decreasing function (see [9] for the same assumptions). The case $\sigma(N) = \beta/N$ is easier to study, since it leads to a reduction of the dimension of the problem [11], and will be discussed at the end. Here I assume that $c(N)$ is strictly increasing.

The equations resulting from these assumptions are

$$\begin{cases} \frac{dS}{dt} = (a - m(N))S + a(1-\delta_1)E + a(1-\delta_2)I - \sigma(N)SI \\ \frac{dE}{dt} = \sigma(N)SI - m(N)E - \nu E \\ \frac{dI}{dt} = \nu E - (m(N) + \mu)I \end{cases} \tag{1}$$

where $N = S + E + I$.

The exact assumptions that will be made are:

(H) $a > 0, \quad \nu > 0, \quad \mu \geq 0, \quad 0 \leq \delta_1 \leq \delta_2 \leq 1, \quad \mu + a\delta_2 > 0$
$\sigma(N)$ is a non-increasing function, while $c(N) \equiv N\sigma(N)$ is a strictly increasing C^1 function on $(0, \infty)$ such that $\lim_{N\to 0^+} c(N) = 0$.
$m(N)$ is C^1 and non-decreasing on $[0, \infty)$. There exists $(N_1, N_2) \subset (0, \infty)$ such that $m(N)$ is strictly increasing on (N_1, N_2), with $m(N_1) < a < m(N_2)$.

I assumed c and m to be C^1, in order to be able to perform the linearization at equilibria. In the examples, I will also use functions that are only piecewise C^1; it is still possible to perform the linearization, as long as m and c are differentiable at the equilibria.

From the assumptions made on m, it follows that there exists a unique $N^* > 0$ such that $m(N^*) = a$. This is the demographic equilibrium, in absence of the disease.

Summing the three equations in (1), one obtains

$$\frac{dN}{dt} = (a - m(N))N - a\delta_1 E - a\delta_2 I - \mu I \tag{2}$$

One can clearly use, instead of system (1), a system obtained using (2) and any two equations of system (1), remembering that $N = S + E + I$.

Existence and uniqueness of solutions of (1) for nonnegative initial data follows from standard results; the fact that $c(N)$ is not assumed to be Lipschitz at $N = 0$ can be handled as in [12]. It is easy to see that the set $\mathcal{A} = \{(S, E, I) : 0 \leq S, E, I, \quad S + E + I \leq N^*\}$ is positively invariant.

An equilibrium for the system, using as variables (N, E, I), is found as $(N^*, 0, 0)$. The linearization matrix there is

$$A = \begin{pmatrix} -m'(N^*)N^* & -a\delta_1 & -a\delta_2 - \mu \\ 0 & -(m(N^*) + \nu) & \sigma(N^*)N^* \\ 0 & \nu & -(m(N^*) + \mu) \end{pmatrix} \tag{3}$$

whose eigenvalues are $\lambda_1 = -m'(N^*)N^* < 0$ and the eigenvalues of

$$B = \begin{pmatrix} -(m(N^*) + \nu) & \sigma(N^*)N^* \\ \nu & -(m(N^*) + \mu) \end{pmatrix} \tag{4}$$

We have tr $B < 0$, while

$$\det B = (a + \nu)(a + \mu) - \nu\sigma(N^*)N^*$$

If we define

$$R_0 = \frac{\sigma(N^*)N^*}{m(N^*)+\mu} \cdot \frac{\nu}{a+\nu} \tag{5}$$

we have thus the following

Proposition 1 *If $R_0 < 1$, the equilibrium $(N^*, 0, 0)$ of system (1) is locally asymptotically stable. If $R_0 > 1, (N^*, 0, 0)$ is unstable.*

3 Existence of endemic equilibria

I now look for the existence of an equilibrium $(\bar{S}, \bar{E}, \bar{I})$ with $\bar{I}, \bar{E}, \bar{S} > 0$. From (1c) we must have

$$\bar{E} = \frac{\mu + m(\bar{N})}{\nu} \bar{I} \tag{6}$$

then, using (6) in (1b)

$$\sigma(\bar{N})\bar{S} = (m(\bar{N}) + \nu)\frac{\mu + m(\bar{N})}{\nu} \tag{7}$$

Finally from (2)

$$\bar{I}/\bar{N} = \frac{(a - m(\bar{N}))\nu}{\nu(\mu + a\delta_2) + a\delta_1(m(\bar{N}) + \mu)} \tag{8}$$

From (6) and (8) we have

$$\bar{S} = \bar{N}\left(1 - \frac{(a - m(\bar{N}))(m(\bar{N}) + \mu + \nu)}{\nu(\mu + a\delta_2) + a\delta_1(m(\bar{N}) + \mu)}\right) \tag{9}$$

Finally (7) yields

$$\sigma(\bar{N})\bar{N} = \frac{(m(\bar{N}) + \mu)(m(\bar{N}) + \nu)\left[\nu(\mu + a\delta_2) + a\delta_1(m(\bar{N}) + \mu)\right]}{\nu\left[\nu(\mu + a\delta_2) + a\delta_1(m(\bar{N}) + \mu) - (a - m(\bar{N}))(m(\bar{N}) + \mu + \nu)\right]} \tag{10}$$

Since we need $\bar{I} > 0$, (8) yields

$$a - m(\bar{N}) > 0 \tag{11}$$

or $\bar{N} < N^*$. Since we need $\bar{S} > 0$, (9) implies

$$(a - m(\bar{N}))(m(\bar{N}) + \mu + \nu) < \nu(\mu + a\delta_2) + a\delta_1(m(\bar{N}) + \mu) \tag{12}$$

As discussed above, I will explore (10) only in two extreme cases: $\delta_1 = 0; \quad \delta_2 = \delta \geq 0$; or $\delta_1 = \delta_2 = 1$.

In the former case, (10) reduces to

$$\sigma(\bar{N})\bar{N} = F(m(\bar{N})) \tag{13}$$

where

$$F(x) = \frac{(x+\nu)(x+\mu)(\mu+a\delta)}{G(x)} \tag{14}$$

$$G(x) = \nu(\mu+a\delta) - (a-x)(x+\mu+\nu)$$

We have

$$\begin{aligned} F'(x) &= \frac{(\mu+a\delta)}{G(x)^2}[(2x+\mu+\nu)(\nu(\mu+a\delta)-(a-x)(x+\mu+\nu)) \\ &\quad + (x+\mu)(x+\nu)(-2x+a-\mu-\nu)] \\ &= \frac{(\mu+a\delta)}{G(x)^2}a[-(2x+\mu+\nu)(x+\mu+\nu(1-\delta)) + (x+\mu)(x+\nu)] \\ &= -\frac{(\mu+a\delta)}{G(x)^2}a\left[(x+\nu)\nu(1-\delta) + (x+\mu)(x+\mu+\nu(1-\delta)\right] \end{aligned}$$

Therefore $F'(x) < 0$ for $x > 0$.

It is easy to see that the solutions of $G(x) = 0$ are either one positive and one negative; or both negative. Let $\hat{x}$ be the larger one.

If $m(0) > \hat{x}$, we have

$$F(m(0)) > 0 = \lim_{N\to 0^+} \sigma(N)N.$$

If $m(0) \leq \hat{x}$, there exists $\hat{N} < N^*$ such that $m(\hat{N}) = \hat{x}$. $F(m(N))$ is negative for $0 \leq N < \hat{N}$ and

$$\lim_{N\to \hat{N}^+} F(m(N)) = +\infty > \sigma(\hat{N})\hat{N}.$$

In either case, there exists a solution of (13) satisfying (11) and (12) if and only if

$$\sigma(N^*)N^* > F(m(N^*)) = \frac{(a+\nu)(a+\mu)}{\nu} \tag{15}$$

Remembering the meaning of R_0, we can state

Proposition 2 *Let $\delta_1 = 0$. If $R_0 > 1$, there exists a unique positive equilibrium of (1). If $R_0 \leq 1$, there are no positive equilibria.*

If $\delta_1 = \delta_2 = 1$, (10) reduces to

$$\sigma(\bar{N})\bar{N} = F(m(\bar{N})) \tag{16}$$

where

$$F(x) = \frac{\mu\nu + a(x+\mu+\nu)}{\nu} \tag{17}$$

is increasing. In this case, it is easy to construct instances in which (16) has multiple solutions (see Fig. 1), Let

$$\sigma(N) = \beta; \quad m(N) = \begin{cases} b_0 & \text{for } N \leq N_0 \\ b_0 + \kappa(N-N_0) & \text{for } N > N_0 \end{cases}$$

Then, if $0 \leq b_0 < a$, $a\kappa > \beta\nu > 0$, $\frac{(a+\mu)(a+\nu)}{\beta\nu} - \frac{a-b_0}{\kappa} > N_0 > \frac{\mu(a+\nu)+a(b_0+\nu)}{\beta\nu}$, there exist two solutions of (16) with $N \in (0, N^*)$.

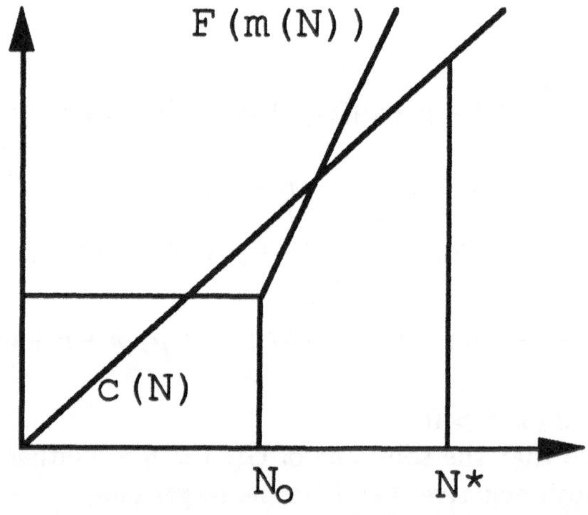

Fig. 1. Possible plot of $c(N)$ and $F(m(N))$ when $\sigma(N)$ and $m(N)$ are defined as above. The two intersections (both with $N<N^*$) correspond to the two endemic equilibria of the system.

Note that multiple solutions cannot occur when $\sigma(N) \equiv \beta$ and $m(N) = b_0 + \kappa N$, the case considered in [2]; in this case, in fact, both sides of (16) are straight lines, and, if they do not coincide, they have at most one point in common.

One can study this as a bifurcation in the strength of infection, parametrized by a real variable β. Let

$$H(\beta, N) = \beta c(N) - F(m(N));$$

for a given value of β a value of $N \in (0, N^*)$ such that $H(\beta, N) = 0$ is an endemic equilibrium for (1). It is clear that, for all $N > 0$, $\beta(N) = \frac{F(m(N))}{c(N)}$ is positive and solves $H(\beta(N), N) = 0$. We therefore have a function $\beta(N)$ defined on $(0, +\infty)$, such that $\lim_{N\to 0^+} \beta(N) = +\infty$. If $\beta'(N) < 0$ for all N, we can then invert the function and obtain that for each $\beta \in (\frac{F(m(N^*))}{c(N^*)}, +\infty)$ we have a unique solution in $(\hat{N}, N^*)$ of $H(\beta, N) = 0$. Otherwise, if there are $N < N^*$ such that $\beta'(N) > 0$, there can be no global inversion (see Fig. 5). Summarizing we can state

Proposition 3 *Let $\delta_1 = \delta_2 = 1$. If $R_0 > 1$, there exists at least one positive equilibrium of* (1). *If $R_0 \leq 1$, there may or may not be positive equilibria.*

4 Local stability of the endemic equilibria

We now study the (local) stability of the endemic equilibria.

First consider the case $\delta_1 = 0$; $\delta_2 = \delta \geq 0$. We now use as variables (N, S, I); the linearization matrix at $(\bar{N}, \bar{S}, \bar{I})$ is

$$\begin{pmatrix} a - m(\bar{N}) - m'(\bar{N})\bar{N} & 0 & -(\mu + a\delta) \\ a - m'(\bar{N})\bar{S} - \sigma'(\bar{N})\bar{S}\bar{I} & -m(\bar{N}) - \sigma(\bar{N})\bar{I} & -a\delta - \sigma(\bar{N})\bar{S}, \\ \nu - m'(\bar{N})\bar{I} & -\nu & -(m(\bar{N}) + \mu + \nu) \end{pmatrix}$$

The characteristic polinomial of A is, after a change of sign,

$$z^3 + a_1 z^2 + a_2 z + a_3.$$

After some computations we find

$$\begin{aligned} a_1 &= m'(\bar{N})\bar{N} + B \\ a_2 &= m'(\bar{N})\bar{N}B + C \\ a_3 &= m'(\bar{N})\bar{N}C + \gamma D \end{aligned}$$

with

$$\begin{aligned} B &= \frac{a(a-b)\left[b + \mu + \nu(1-\delta)\right]}{\nu(\mu + a\delta) - (a-b)(b+\mu+\nu)} + 2b + \mu + \nu > 0 \\ C &= \frac{a(a-b)\left[\nu(1-\delta)(2b+\mu+\nu) + (b+\mu)^2\right]}{\nu(\mu + a\delta) - (a-b)(b+\mu+\nu)} > 0 \\ D &= (a-b)(b+\mu)(b+\nu) > 0 \end{aligned}$$

where

$$\gamma = \frac{\bar{N}c'(\bar{N})}{c(\bar{N})},$$

$b = m(\bar{N})$, $c(N) = N\sigma(N)$. From the assumptions that $c(N)$ is increasing, while $\sigma(N)$ is non-increasing, we see that for all N, $0 \leq \gamma \leq 1$.

From these expressions, it is clear that a_1, a_2, and a_3 are positive. The (local) stability of $(\bar{N}, \bar{S}, \bar{I})$ depends only on the sign of $a_1 a_2 - a_3$. Setting, for ease of notation, $\alpha = m'(\bar{N})\bar{N}$, we have

$$a_1 a_2 - a_3 = \alpha^2 B + \alpha B^2 + BC - \gamma(a-b)(b+\mu)(b+\nu) \tag{18}$$

This expression is not definite in sign; therefore, according to the parameters values, $(\bar{N}, \bar{S}, \bar{I})$ may be either stable or unstable.

How does $a_1 a_2 - a_3$ depend on the parameters? It is clear that, as α increases, so does $a_1 a_2 - a_3$ (a stronger density–dependence of the mortality renders stable the endemic equilibrium); while, as γ increases, $a_1 a_2 - a_3$ decreases (the shape of $\sigma(N)$ therefore influences the stability); at the limit of $\gamma = 0$, $(\bar{N}, \bar{S}, \bar{I})$ is always stable. The dependence on the other parameters is less clear, and has been explored through numerical evaluations of (18).

In Fig. 2 I show how the minimum α necessary to render (18) positive depends on $D = \frac{1}{\mu}$, the average length of the infectious period before death and $L = \frac{1}{\nu}$, the average length of the latent period, for fixed values of a, δ, γ and $b = m(\bar{N})$. Note that $c(\bar{N})$ varies with D and L, since by (13) we have $c(\bar{N}) = F(m(\bar{N}))$. It appears that when L and D differ , the endemic equilibrium tends to be always stable, while when L and D are similar, a relevant density–dependence is necessary for stability.

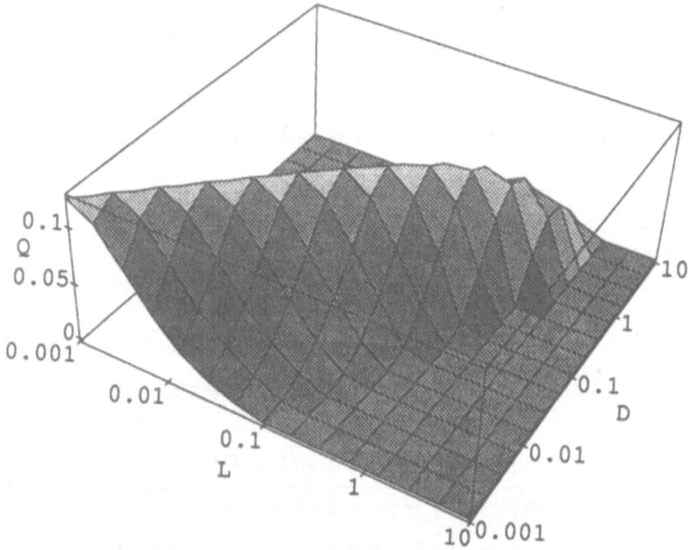

Fig. 2. The minimum $Q=\frac{\alpha}{\delta}\geq 0$ such that (18) is greater or equal to 0, vs. $L=\frac{1}{\nu}$ and $D=\frac{1}{\mu}$. The other parameter values are $a=0.15$, $b=0.1$, $\delta=0.9$, $\gamma=1$.

In Fig. 3 I again show the stability region, assuming however that $c(N) = \beta N$, and $m(N) = b_0 + \frac{a-b_0}{K} N$ (logistic demography with carrying capacity K). For fixed values of a, b_0, δ, β and μ, we show how the stability of the endemic equilibrium varies with ν and K. It appears, as shown also by Anderson *et al.*, that a large K tends to destabilize the endemic equilibrium, although this does not occur for all values of ν. For a fixed value of K one can note again that the endemic equilibrium tends to be unstable when $L = \frac{1}{\nu} \approx D = \frac{1}{\mu}$, while it becomes stable when L becomes either larger or smaller. We already know [12] that in the limit $L = 0$, the endemic equilibrium is always stable.

When $\delta_1 = \delta_2 = 1$ (the case considered by Anderson et al.), the characteristic polynomial of the linearization matrix at the endemic equilibrium is (after a change of sign):

$$z^3 + a_1 z^2 + a_2 z + a_3$$

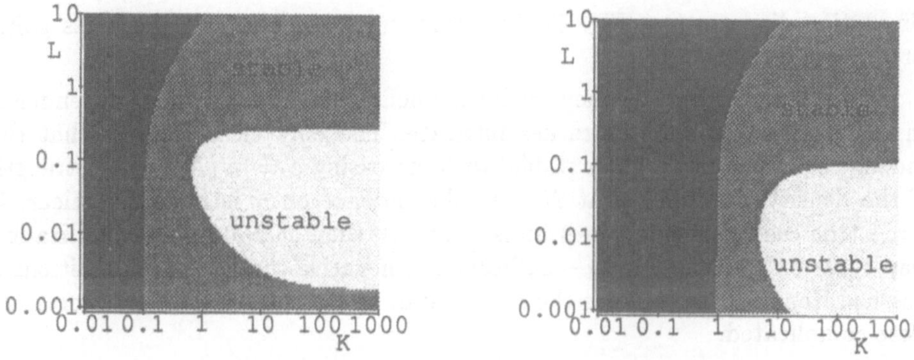

Fig. 3. **The region of stability of the endemic equilibrium of system (1). We assume $\sigma(N)\equiv\beta$, and $m(N)=b_0+(a-b_0)N/K$. On the x-axis K; on the y-axis $L=1/\nu$. In the black region of (K,L) values, there is no endemic equilibrium; in the darker grey part the endemic equilibrium exists and is stable; in the lighter part the endemic equilibrium is unstable. Parameter values are $\beta=80$, $a=1$, $b_0=0.5$, $\delta=0.9$. On the left part of the figure we have $\mu=7$, i.e. $D=0.143$; on the right part $\mu=73$, i.e. $D=0.0137$.**

with

$$a_1 = \alpha + B$$
$$a_2 = \alpha B - C(1-\gamma)$$
$$a_3 = -\alpha C + \gamma D$$

where

$$B = 2b + \mu + \nu > 0$$
$$C = \frac{a(b+\mu)(b+\nu)(a-b)}{a(b+\mu+\nu)+\mu\nu} > 0$$
$$D = (a-b)(b+\mu)(b+\nu) > 0$$
$$\alpha = m'(\bar{N})\bar{N}$$
$$b = m(\bar{N})$$
$$\gamma = \frac{c'(\bar{N})\bar{N}}{c(\bar{N})}$$

Finally we have

$$a_1a_2 - a_3 = \alpha^2 B + \alpha\left(B^2 + \gamma C\right) - (BC(1-\gamma) + \gamma D)$$

Note first that $a_3 > 0$ is equivalent to $\frac{d}{dN}\frac{F(m(N))}{c(N)}|_{N=\bar{N}} > 0$ (remember that by definition $F(m(\bar{N})) = c(\bar{N})$). Therefore the parts going backwards of the bifurcation diagram (see Fig. 5) are always unstable, while in the parts going forward we have $a_3 > 0$.

As for a_2 and $a_1a_2 - a_3$, they are both indefinite in sign, and their dependence on the parameters appears rather intricate. The only clear thing is that the endemic equilibrium becomes stable, with increasing $\alpha = m'(\bar{N})\bar{N}$, the strength of the density–dependence at $\bar{N}$. The dependence on γ is instead unclear. If $\gamma = 1$ (the case considered in Anderson *et al.*), then $a_2 > 0$ and destabilization may occur only through $a_1a_2 - a_3$ becoming negative (under some conditions a classical Hopf bifurcation); on the other hand, if $\gamma < 1$, the situation is certainly more complicated.

In order to see graphically the possible cases, I assume, analogously to Fig. 3, a logistic demography with carrying capacity K. In Fig. 4 I show the regions of the plane (K,L) where zero, one or two endemic equilibrium exist (L is the length of the latent period).

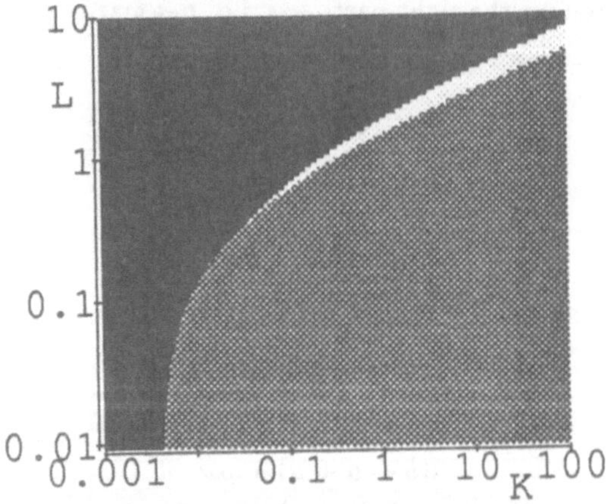

Fig. 4. The region of existence of endemic equilibria of system (1). We assume $\sigma(N) = \beta N^{1/4}$, and $m(N) = b_0 + (a - b_0)N/K$. On the x-axis K; on the y-axis $L = 1/\nu$. In the black region of (K,L) values, there is no endemic equilibrium; in the darker grey part there is one endemic equilibrium; in the lighter part there are two endemic equilibria. Parameter values are $\beta=10$, $a=2$, $b_0=0.25$, $\mu=0.5$.

The parameters for Figs. 4 and 5 have been chosen so as to give rise to multiple endemic equilibria with a logistic demography; if mortality had been chosen as in Fig. 1, multiple equilibria would arise more easily.

In Fig. 5, for two values of L, I show the bifurcation diagram of the equilibria with varying K. As seen above, when there are two endemic equilibria, the larger one is always unstable ($a_3 < 0$). The stability of the smaller one depends on the parameter values. In the left part of Fig. 5, at the bend of the branch of endemic equilibria, the real positive eigenvalue (remember that on the backward part $a_3 < 0$) of the linearization matrix at the equilibrium crosses into the negative half plane; for a while the equilibrium is locally stable; further on the endemic equilibrium loses stability with a couple of eigenvalues crossing the imaginary axis. In the right part of Fig. 5, at the bend of the branch a second real eigenvalue crosses into the positive half plane; the lower endemic equilibrium is always unstable.

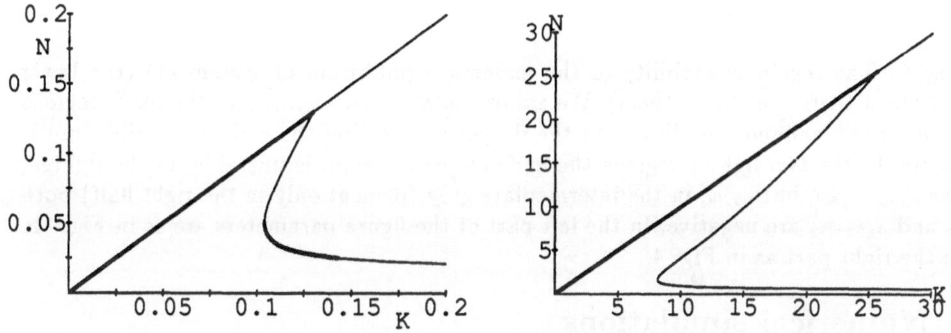

Fig. 5. The bifurcation diagram of the equilibria of system (1) with varying K, the carrying capacity of a logistic demography. The straight line corresponds to the equilibrium without infectives ($N=K$); the line branching off is the line of endemic equilibria. The thick part of the lines corresponds to a locally stable equilibrium; the thin part to an unstable equilibrium. In the left part of the graph L (the length of the latent period) is 0.7; in the right part of the graph L is 4. Other parameter values as in Fig. 4.

In Fig. 6 I show more extensively how the stability of the endemic equilibrium (the smaller one when there are two of them) depends on K and L. In the right part of the figure parameter values are as in Fig. 4. In the left part of the figure parameter values are as in Fig. 3, in order to allow a comparison. Clearly the stability region is about the same as in Fig. 3 for small values of L; for large values of L, this case gives rise to a much larger region of instability.

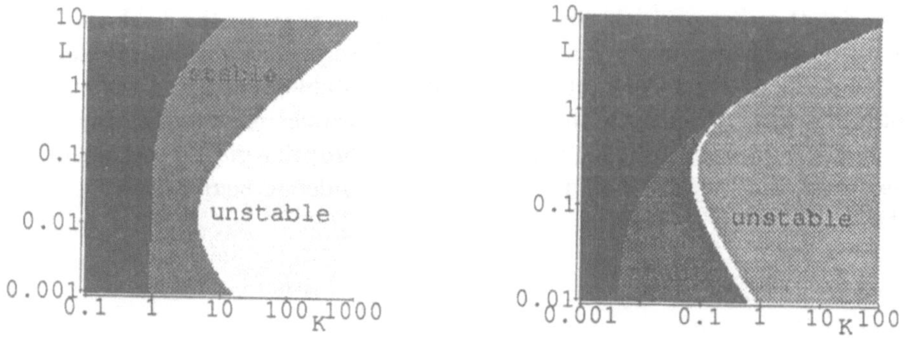

Fig. 6. The region of stability of the endemic equilibrium of system (1) (the lower one when there are two of them). We assume $m(N)=b_0+(a-b_0)N/K$. In the black regions there are no endemic equilibria. In the darker grey region the endemic equilibrium is stable. In the two lighter regions the endemic equilibrium is unstable; in the lightest one $a_1a_2-a_3<0$, but $a_2>0$; in the intermediate grey (present only in the right half) both a_2 and $a_1a_2-a_3$ are negative. In the left part of the figure parameters are as in Fig. 3b. In the right part as in Fig. 4.

5 Numerical simulations

In order to have more information about the behaviour of the solutions of (1), numerical solutions of system (1) have been obtained with an adaptive Runge–Kutta algorithm.

In Fig. 7 I show some results in the case $\delta_1 = 0$. The two trajectories shown (starting from the same initial value) correspond to the case $\gamma = 1$ and $\gamma = 1/2$; β has been adjusted so as to give rise to the same endemic equilibrium. When $\gamma = 1$ the equilibrium is unstable, and the trajectory apparently converges to a limit cycle. When $\gamma = 1/2$ the equilibrium is stable and the trajectory converges to the equilibrium in weakly damped oscillations. Note that in this example $m(N)$ has been chosen as in Fig. 1; at the endemic equilibrium there is no density–dependence.

In Fig. 8 we have $\delta_1 = \delta_2 = 1$. $m(N)$ is as in Fig. 1; other parameters have been chosen so as to give rise to two equilibria, both unstable as in Fig. 5b. The two trajectories shown start from very close to the two endemic equilibria; apparently both approach the same limit cycle, one from the inside, the other one from the outside. Note that, although the system is tridimensional, the trajectories appear to be adequately described by the projections on the plane (N, I).

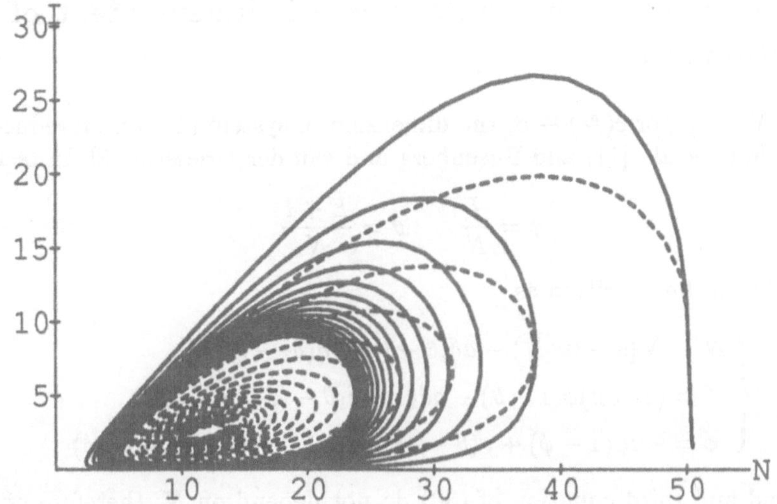

Fig. 7. Numerical solutions of system (1) with $\delta_1=0$. We used $m(N)=0.1$ for $N<50$; $m(N)=0.1(N-49)$ for $N\geq 50$; $c(N)=\beta N^{\gamma}$. The solid line is obtained with $\gamma=1$, $\beta=0.025$. The dashed line with $\gamma=1/2$, $\beta=0.085940$. Other parameter values are $a=0.15$, $\mu=0.1$, $\nu=1$, $\delta_2=0.9$.

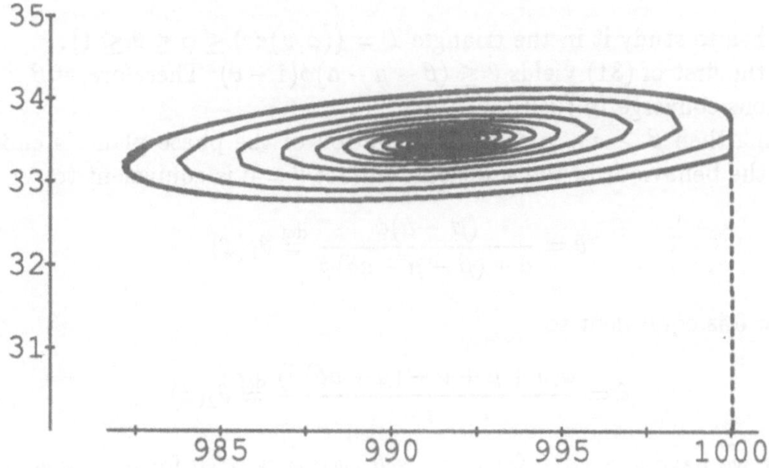

Fig. 8. Numerical solutions of system (1) with $\delta_1=\delta_2=1$. We used $m(N)=0.1$ for $N<1000$; $m(N)=0.1(N-999)$ for $N\geq 1000$; $c(N)=0.047N^{1/2}$. Other parameter values are $a=0.15$, $\mu=1$, $\nu=0.5$, $\delta_2=0.9$. The solid line is obtained starting from $(N,S,I)=(991.6, 884.4, 33.5)$. The dashed line starting from $(1000.2, 937.8, 19.3)$. Not shown is the trajectory starting from $(1000.2, 937.9, 19.2)$ which apparently converges to the disease free equilibrium $(1000.5, 1000.5, 0)$

6 The case when the contact rate is independent of population size

When $\sigma(N) = \frac{\beta}{N}$, or $c(N) \equiv \beta$, the dimension of system (1) can be reduced, as noted by May et al. [11] and Busenberg and van der Driessche [8]. In fact, let

$$\phi = \frac{I}{N} \qquad \theta = \frac{E+I}{N}.$$

System (1) can be rewritten as

$$\begin{cases} \dot{N} = N\left[a - m(N) - a\delta_1\theta - (\mu + a(\delta_2 - \delta_1))\phi\right] \\ \dot{\theta} = (\beta-\mu)\phi(1-\theta) - a\theta(1 - \delta_1\theta - (\delta_2-\delta_1)\phi) \\ \dot{\phi} = -\mu\phi(1-\phi) + \nu(\theta-\phi) - a\phi(1-\delta_1\theta - (\delta_2-\delta_1)\phi) \end{cases} \tag{30}$$

The second and third equation of (30) do not depend on N; therefore one can study a bidimensional system in the variables θ and ϕ.

Again I study (30) in the two cases: $\delta_1 = 0$, and $\delta_1 = \delta_2 = 1$.

When $\delta_1 = 0$, the system to be studied is:

$$\begin{cases} \dot{\theta} = (\beta-\mu)\phi(1-\theta) - a\theta(1-\delta\phi) \\ \dot{\phi} = -\mu\phi(1-\phi) + \nu(\theta-\phi) - a\phi(1-\delta\phi) \end{cases} \tag{31}$$

and one has to study it in the triangle $T = \{(\phi,\theta) : 0 \le \phi \le \theta \le 1\}$.

In T the first of (31) yields $\dot{\theta} \le (\beta - \mu - a)\phi(1-\theta)$. Therefore, if $\beta < a + \mu$, all solutions converge to $(0,0)$.

Assume then $\beta \ge a + \mu$. The examination of the phase plane is enough to find out the behaviour of the solutions. In fact $\dot{\theta} = 0$ is equivalent to

$$\theta = \frac{(\beta-\mu)\phi}{a + (\beta - \mu - a\delta)\phi} \stackrel{\text{def}}{=} \theta_1(\phi)$$

while $\dot{\phi} = 0$ is equivalent to

$$\theta = \frac{\phi(a+\mu+\nu-(\mu+a\delta)\phi)}{\nu} \stackrel{\text{def}}{=} \theta_2(\phi)$$

Moreover we have $\dot{\theta} > [<]\, 0$ for $\theta < [>]\, \theta_1(\phi)$; $\dot{\phi} > [<]\, 0$ for $\theta > [<] \theta_2(\phi)$. See Fig. 9.

As for the relative position of $\theta_1(\phi)$ and $\theta_2(\phi)$, we first note that

$$\theta_1(0) = \theta_2(0) = 0 \qquad \theta_1(1) = \frac{\beta-\mu}{\beta-\mu+a(1-\delta)} < 1 < \frac{\nu + a(1-\delta)}{\nu} = \theta_2(1)$$

We first look for endemic equilibria, i.e. $\phi > 0$ such that $\theta_1(\phi) = \theta_2(\phi)$. Such ϕ would satisfy $Q(\phi) = 0$ where

$$\begin{aligned} Q(\phi) = &-(\beta-\mu-a\delta)(\mu+a\delta)\phi^2 + [(\beta-\mu-a\delta)(a+\mu+\nu) \\ &- a(\mu+a\delta)]\phi + a(a+\mu+\nu) - \nu(\beta-\mu). \end{aligned}$$

Since $Q(1) = a(1-\delta)(\beta-\mu+\nu+a(1-\delta)) > 0$, while $\lim_{\phi\to\infty} Q(\phi) = -\infty$, there exists at least one solution of $Q(\phi) = 0$ in $(1,\infty)$. Since Q is a second-degree polynomial, there exists at most one solution of $Q(\phi) = 0$ in $(0,1)$.

Therefore, if $Q(0) < 0$, that is equivalent to $\beta\nu > (a+\mu)(a+\nu)$, or $\theta_1'(0) > \theta_2'(0)$, there exists one solution in $(0,1)$ of $\theta_1(\phi) = \theta_2(\phi)$, i.e. an endemic equilibrium for (31). Note that the equilibrium is necessarily in T, since $\theta_1(\phi) < 1$, while $\theta_2(\phi) > \phi$ for all $0 < \phi < 1$.

If $\theta_1'(0) \le \theta_2'(0)$, or $Q(0) \ge 0$, it follows that there is no endemic equilibrium.

Global convergence to the endemic equilibrium in the first case, to $(0,0)$ in the second case is then easily established by considering the phase plane (see Fig. 9).

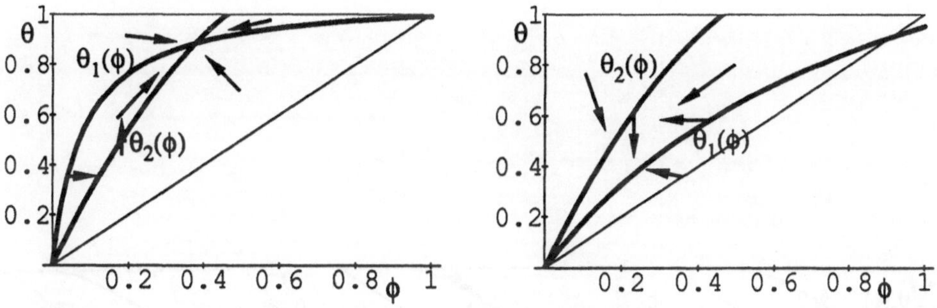

Fig. 9. The phase plane of system (31). In a) $\beta\nu>(a+\mu)(a+\nu)$; in b) $\beta\nu\le(a+\mu)(a+\nu)$

The behaviour of N then follows from the first of (30). If $\beta\nu \le (a+\mu)(a+\nu)$, N tends to N^*. If $\beta\nu > (a+\mu)(a+\nu)$, let $\bar\phi$ be the solution in $(0,1)$ of $Q(\phi) = 0$. If $a \le m(0) + (\mu + a\delta)\bar\phi$, N tends to 0; otherwise it tends to the solution $\bar N$ of $m(N) = a - (\mu + a\delta)\bar\phi$.

Consider now system (30) when $\delta_1 = \delta_2 = 1$. The resulting system is

$$\begin{cases} \dot\theta = (1-\theta)\left[(\beta-\mu)\phi - a\theta\right] \\ \dot\phi = -\mu\phi(1-\phi) + \nu(\theta-\phi) - a\phi(1-\theta) \end{cases} \tag{32}$$

As above, we find the isoclines. $\dot\theta = 0$ is equivalent to

$$\theta = 1 \quad \text{or} \quad \theta = \frac{\beta-\mu}{a}\phi \stackrel{\text{def}}{=} \theta_1(\phi)$$

while $\dot\phi = 0$ is equivalent to

$$\theta = \frac{\phi(a+\nu+\mu(1-\phi))}{a\phi+\nu} \stackrel{\text{def}}{=} \theta_2(\phi)$$

We have $\theta_2(0)=0$, $\theta_2(1)=1$, and θ_2 concave. Moreover, if $\nu \geq \mu$, θ_2 is increasing on $[0,1]$; if $\mu > \nu$, $\theta_2(\nu/\mu) = 1$, $\theta_2(\phi) < 1$ for $0 \leq \phi < \nu/\mu$, $\theta_2(\phi) > 1$ for $\nu/\mu < \phi < 1$.

It follows that if $\theta_1'(0) \geq \theta_2'(0)$, or $\beta\nu \geq (a+\mu)(a+\nu)$, $\theta_1(\phi) > \theta_2(\phi)$ for all $0 < \phi \leq 1$, and the phase plane is as in Fig. 10a. From all initial points except $(0,0)$ the solutions converge to $(1,1)$ if $\nu \geq \mu$; to $(\nu/\mu, 1)$ if $\nu < \mu$.

If $(a+\mu)(a+\nu) > \beta\nu > \mu\nu + a\max\{\mu,\nu\}$, there exists an endemic equilibrium which is a saddle point (Fig. 10b). Depending on the starting point, a different equilibrium is approached: there exists a separatrix line, starting from which solutions converge to the endemic equilibrium; starting below that line solutions converge to $(0,0)$; starting above to $(1,1)$ if $\nu \geq \mu$, to $(\nu/\mu, 1)$ if $\nu < \mu$.

Finally, if $\beta\nu \leq \mu\nu + a\max\{\mu,\nu\}$, in T $\theta_1(\phi)$ is always below $\theta_2(\phi)$, and we have global convergence to $(0,0)$ except from the line $\theta = 1$ (Fig. 10c).

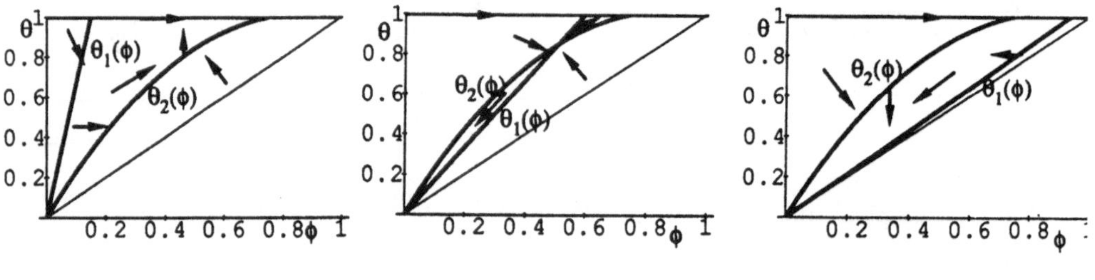

Fig. 10. The phase plane of system (32). In a) $\beta\nu \geq (a+\mu)(a+\nu)$; in b) $(a+\mu)(a+\nu) > \beta\nu > \mu\nu + a\max\{\mu,\nu\}$; in c) $\beta\nu \leq \mu\nu + a\max\{\mu,\nu\}$

The behaviour of N follows from the first of (30). When (ϕ,θ) converge to $(0,0)$, N tends to N^*. When (ϕ,θ) converge to $(1,1)$ or $(\nu/\mu, 1)$, N tends to 0. In the nongeneric case that N converges to the endemic equilibrium $(\bar{\phi}, \bar{\theta})$, N tends to 0, if $a(1-\bar{\theta}) \leq m(0) + \mu\bar{\phi}$; otherwise it tends to the solution $\bar{N}$ of $m(N) = a(1-\bar{\theta}) - \mu\bar{\phi}$.

7 Discussion

In this paper I have analyzed an ODE model for a disease without recovery, with a latent period and with variable population size: demographic parameters depend on epidemic state.

It appears that this model, as already noted by Anderson *et al.* [2] can produce either convergence to an equilibrium (with or without infectives), or oscillations. This in contrast both with models without latent period [12], and with constant population size (see for instance [1]). Although neither the latent period, nor variable population size, can produce by themselves oscillations in an ODE model (but see [6]), their combination can.

Another aspect resulting from this work is the qualitative difference between the case where fertility of exposed (but not infective) individuals is equal to that of susceptibles to the case where it is reduced. In the first case we have a clear threshold, below which the disease–free equilibrium is stable, and no endemic equilibrium exists; and above which the disease–free equilibrium is unstable, and a unique endemic equilibrium exists. Building a bifurcation diagram, the branch of endemic equilibria inherits the (local) stability from the disease–free equilibrium; destabilization may occur only with two conjugate eigenvalues crossing the imaginary axis. On the other hand, when fertility of exposed and infective individuals is zero (as assumed in [2] and [6]) the threshold phenomenon is weaker: below the threshold the disease–free equilibrium is stable; above it is unstable, and there exists an endemic equilibrium. However, the endemic equilibrium needs not be unique, and it may exist also below the threshold. Multiple equilibria do not arise if mortality and the contact rate per individual are linear with population size (as in [2]), but may arise as soon as one of these assumptions is relaxed.

Summarizing, the model with $\delta_1 = 0$, which appears to be more respondent to the biology, gives rise to a picture similar to that of most epidemic models, with also the possibility of oscillations. The dynamics of the model with $\delta_1 = \delta_2 = 1$ appears to be more complicated for certain parameter values, and is probably not completely captured by the few simulations presented in Fig. 8; further analysis of the qualitative behaviour would probably be interesting. Note, however, that when the latent period is not very large, the two models do not differ sensibly (see Fig. 3b and 4a).

The different behaviour of the two cases shows clearly when contact rate is independent of population size (Section 6). When $\delta_1 = 0$, one again has a threshold, with global convergence to the disease free equilibrium below the threshold, to the endemic equilibrium above. When $\delta_1 = \delta_2$, the endemic equilibrium exists only in an intermediate range of values and is always unstable; from almost all initial values the population converges either to the disease free equilibrium, or to zero population.

References

1. Anderson, R.M., Directly transmitted viral and bacterial infections of man. In Population Dynamics of Infectious Diseases, Anderson, R.M., ed., 1-37, Chapman and Hall, London
2. Anderson, R.M., Jackson, H.C., May, R.M., and Smith, A.M., (1981): Population dynamics of fox rabies in Europe. Nature **289**, 765-771
3. Anderson, R.M., and May, R.M., (1978): Regulation and stability of host-parasite population interactions. J. Anim. Ecol. **47**, 219-247
4. Andreasen, V., (1989): Disease regulation of age–structured host populations. Theor. Pop. Biol. **36**, 214-239.
5. Brauer, F., (1989): Epidemic models in populations of varying size. In Mathematical Approaches to Ecological and Environmental Problem Solving, C. Castillo-Chavez, S.A. Levin, and C. Shoemaker, eds., Lecture Notes in Biomathematics, **83** 109-123, Springer-Verlag, Berlin-Heidelberg-New York
6. Brauer, F., (1990): Models for the spread of universally fatal diseases. J. Math. Biol. **28**, 451-462.
7. Busenberg, S., Cooke, K.L., and Pozio, M.A., (1983): Analysis of a model of a vertically transmitted disease. J. Math. Biol. **17**, 305-329
8. Busenberg S. and van den Driessche, P., (1990): Analysis of a disease transmission model in a population of varying size. J. Math. Biol. **28**, 257-270.
9. Castillo-Chavez, C., Cooke, K.L., Huang, W., and Levin, S.A., (1989): The role of long incubation periods in the dynamics of acquired immunodeficiency syndrome (AIDS)., I. J. Math. Biol., **27**, 373-398
10. Kretzschmar M., (1989): Persistent solutions in a model for parasitic infections. J. Math. Biol. **27**, 549-573
11. May, R.M., Anderson, R.M., and McLean, A.R., (1988): Possible demographic consequences of HIV/AIDS epidemics. Math. Biosci. **90**, 475-505
12. Pugliese, A., (1990): Population models for diseases with no recovery. J. Math. Biol. **28**, 65-82

Stability Change of the Endemic Equilibrium in Age-Structured Models for the Spread of S—I—R Type Infectious Diseases

Horst R. Thieme

Department of Mathematics, Arizona State University, Tempe, AZ 85287-1804, USA

Abstract

Age-structured endemic models of $S \rightarrow I \rightarrow R$ type can exhibit a stability change for the endemic equilibrium if the rate of a susceptible individual to be infected by an infective individual does not depend of the age of the susceptible individual, but is highly concentrated in a particular age class of the infectives.

1 Introduction

The biennial outbreaks of measles have fascinated mathematical modelers of infectious diseases for many years and different mechanisms have been suggested to explain their occurrence. See Schenzle (1984), Hethcote, Stech and van den Driessche (1981), Aron (1989), and Hethcote and Levin (1989) for reviews. Schenzle (1984) convincingly argues that the outbreak patterns are presumably caused by the interaction of age-structure and seasonal forcing due to the school system. We remark that, in this paper, by *age* we always understand *chronological age*, i.e. the elapse of time since birth, in contrast to so-called *class age* which gives the elapsed time since entering a specific demographic or epidemiologic class of the population.

Schenzle's (1984) work, however, which mainly consists of numerical simulations, does not completely clarify the mathematical mechanisms which cause the biennial oscillations. Several authors — Andreasen (1989a,b, preprint a), Busenberg, Cooke and Iannelli (1988), Busenberg, Iannelli and Thieme (to appear, preprint), Greenhalgh (1987, 1988a,b), Gripenberg (1983), Inaba (1990) — have therefore studied the question whether the introduction of age-structure alone can be responsible for undamped oscillations in endemic models which have a stable endemic equilibrium without age-structure.

Before we elaborate on this subject, we like to mention that there are at least four other reasons for incorporating age-structure into epidemic models:

assessing the demographic impact of infectious diseases with significant fatalities, see, e.g., Bernoulli (1760), McLean (1986), and May, Anderson and McLean (1988, 1989);

the estimation of parameters from age-specific data, see, e.g., Dietz (1975), Anderson and May (1982, 1985), Dietz and Schenzle (1985a), Aron (1989);

the protection of age-specific risk-groups (like in the case of rubella), see, e.g., Knox (1980), Dietz (1981), Anderson and May 1983), Katzmann, Dietz (1984) and Hethcote (1988, 1989);

the design of vaccination programs, see, e.g., Dietz (1981), Anderson and May (1982, 1983, 1985), Anderson, Grenfell and May (1984), Katzmann and Dietz (1984), Schenzle (1984, 1985), Hethcote (1988, 1989), Aron (1989).

(The references are anything else but complete, and the reader is referred to the literature cited in the just-mentioned papers.)

Though the importance of chronological age for epidemic modeling was recognized (Kermack and McKendrick, 1932; Discussion), class age (elapse of time since infection) has been incorporated much earlier than chronological age into epidemic models (McKendrick (1926), Kermack and McKendrick (1927, 1932, 1933). Bernoulli's (1760) paper on the increase of life expectation by inoculation against smallpox is no counter-example, because it does not really deal with an epidemic model as has been pointed out by Dietz (1988). Hoppensteadt (1974, 1975) considers epidemic models with both class and chronological age and settles the well-posedness of the model equations. Dietz (1975) uses an age-structured epidemic model (with age-independent parameters) to relate basic reproductive number, life expectation, and mean age at infection. For the further development of age-structured epidemic modeling the reader is asked to consult the literature mentioned above and below; see also the references in Andreasen (1989b) and Castillo-Chavez (1989).

From the many interesting problems in the age-dependent epidemiology of infectious diseases, this note only deals with the question whether the introduction of age-structure can make an otherwise stable endemic equilibrium unstable. In other works: Are age-dependent rates of infection and/or of (natural) mortality possible sources of undamped oscillations of the disease prevalence. It is well-known that $S \to I \to S$, $S \to I \to R(\to S)$ and $S \to E \to I \to R(\to S)$ models — with exponentially distributed periods of latency, infection and immunity, and with bilinear incidence rate, but without age-structure and seasonal forcing — exhibit a unique **locally asymptotically stable** endemic equilibrium if the basic reproductive number exceeds 1. For $S \to I \to S$ and $S \to I \to R(\to S)$ type models the endemic equilibrium is not only locally but globally asymptotically stable. For the $S \to E \to I \to R(\to S)$ model, this is conjectured also, but has not yet been proved (as far as the author knows). See Hethcote (1976), e.g., for $S \to I \to S$ and $S \to I \to R$ type models and Liu, Hethcote and Levin (1987) for $S \to E \to I \to R(\to S)$ models. Recall that S, E, I, R stand for *susceptible*, *exposed* i.e. in the latency period, *infective*, and *removed*.

For which of these models can the introduction of age-structure destroy the stability of the endemic equilibrium?

We emphasize that we restrict this question to such models where the total population has time-independent size and age distribution and the disease does not affect the overall demographics. Under these circumstances undamped oscillations of the disease prevalence — if they occur at all — cannot be driven either by the disease or the demographics alone, but must originate from an interplay between epidemiologic and demographic factors. A model where a disease with significant fatalities spreads in an age-structured population with time-dependent size (also in the absence of the disease) has been analyzed by Andreasen (1989b). (See there for further references.)

For a very general class of age-structured $S \to I \to S$ models one can show that the endemic equilibrium (if it exists) is globally asymptotically stable — see Busenberg, Cooke and Iannelli (1988), Busenberg, Iannelli and Thieme (to appear, preprint).

This note — as before Andreasen (1989a, preprint a), Greenhalgh (1987, 1988a,b), Gripenberg (1983) and Inaba (1990) — addresses the introduction of age-structure into an $S \to I \to R$ endemic model. Gripenberg's (1983) model also includes class age and so covers $S \to E \to I(\to S)$ models as well. For a more precise explanation of the effects we introduce the rate

$$k(a, \tilde{a})$$

at which an average susceptible individual with age a is infected by an average infective individual with age $\tilde{a}$. As in the model without age-structure, a number R_0 can be defined such that the disease vanishes from the population if $R_0 < 1$ and endemic equilibria exist if $R_0 > 1$ — see Inaba (1990). Whether this implies that the disease persists if $R_0 > 1$ is not known in general, though it is presumably true. In general R_0 is given as the spectral radius of a linear operator. An explicit expression for R_0 can be given under the assumption of *proportionate mixing*

$$k(a, \tilde{a}) = k_1(a)k_2(\tilde{a}). \tag{1.1}$$

In this case the endemic equilibrium is unique provided it exists. See Dietz and Schenzle (1985b). Whether multiple endemic equilibria are possible, if proportionate mixing is not assumed, does not seem to be known at the moment, but we can conclude from Inaba's (1990) global stability result for $R_0 < 1$ that multiple equilibria do not originate by backwards bifurcation from the disease-free equilibrium as they can do in multiple group models for HIV/AIDS transmission dynamics — see Huang (thesis). Inaba (1990) gives a condition for uniqueness which does not require proportionate mixing but restricts k in other respects.

Inaba (1990), Proposition 5.8 (iii) also shows — without assuming proportionate mixing — that an endemic equilibrium is *locally asymptotically stable* if the prevalence of the disease at this equilibrium is sufficiently low. Locally asymptotic stability of an endemic equilibrium means, roughly speaking, that the disease dynamics, once they get close to the endemic equilibrium, stay close and finally converge to it. It does not necessarily mean that, for arbitrary initial conditions, the disease dynamics will ever get close to the endemic equilibrium.

Inaba (1990), Lemmata 5.1 and 5.2, also lays the functional analytic basis for associating the locally asymptotic stability of an endemic equilibrium with the location (in the complex plane) of the eigenvalues of a certain linear operator. Under the assumption of proportionate mixing this can be translated to the location of the roots of a *characteristic equation* — compare Gripenberg (1983): If all roots of the characteristic equation lie strictly in the left complex half plane, the endemic equilibrium is locally asymptotically stable; if at least one root strictly lies in the right complex half plane, the endemic equilibrium is unstable.

Instability of an endemic equilibrium is an important source of undamped oscillations — though this conclusion is only intuitive, not rigorous. Rigorously, instability only means that one can prescribe a neighborhood of the endemic equilibrium such that, arbitrarily close to the equilibrium, disease dynamics can be found which have to leave that neighborhood after some time (which does not exclude that they may return later).

Even under the assumption of proportionate mixing, no general characterization has been given of the conditions under which the endemic equilibrium is locally asymptotically stable. Such a characterization does not only have to include the functions k_1, k_2 in (1.1) but also the probability

$$P(a)$$

of an average individual to survive up to age a. Even if k_1, k_2 are constants the answer is not known for general P. The special cases which have been studied so far — Andreasen (1989a, preprint a), Greenhalgh (1987, 1988a,b), Inaba (1990) — and the results for age-structured $S \to I \to S$ models — Busenberg, Cooke and Iannelli (1988), Busenberg, Iannelli and Thieme (to appear, preprint) — might suggest the conjecture that the endemic equilibrium is always locally asymptotically stable. Enderle (thesis), however, has numerically obtained periodic solutions — with a period of about eight years — for a discrete-time Reed-Frost type model with asymmetric contact rate between three age-classes. Further there is numerical and analytical indication that the introduction of age-dependent mortality into a two-strain influenza type model (with partial cross-immunity) makes the endemic equilibrium unstable (Castillo-Chavez, Hethcote, Andreasen, Levin and Liu, 1989; Andreasen, 1989a, preprint b).

It is the purpose of this note to give an analytic proof that the introduction of age-structure into a (continuous time) model of standard $S \to I \to R$ type can destabilize the endemic equilibrium. This is even the case under the assumption of proportionate mixing with the infection rate being asymmetric in so far as $k_1 \neq k_2$. More specifically we prove that the endemic equilibrium can loose its stability provided that $k_1(a)$ is constant — i.e. the rate of being infected does not depend on the age of the susceptible individual — and that $k_2(a)P(a)$ is sufficiently concentrated in one particular age class. This condition is not realistic, but it shows that a careful further investigation is needed and that no easy answer can be expected.

2 The model

We consider an age-structured population whose age-density at time t is given by $n(t,a)$ such that

$$N(t) = \int_0^\infty n(t,a)da \tag{2.1}$$

gives the population size at time t. We describe the population development by the *McKendrick* equations

$$\left(\frac{\partial}{\partial t} + \frac{\partial}{\partial a}\right) n(t,a) = -\mu(a)n(t,a) \tag{2.2}$$

$$n(t,0) = \int_0^\infty \beta(a)n(t,a)da. \tag{2.3}$$

$\mu(a)$ and $\beta(a)$ are the age-dependent per capita mortality and birth rates. As everyone is born at age 0, $n(t,0)$ in (2.3) is the population birth rate at time t.

In this age-structured population we now consider the spread of an infectious disease which does not affect the mortality and reproduction rates, i.e. leaves the overall population dynamics unchanged. We ignore a possible latency period and assume that the disease induces permanent immunity after recovery. To set up the model — following Dietz (1975) or Hoppensteadt (1974) — we subdivide the age-density $n(t,a)$ at time t into

$$n(t,a) = S(t,a) + I(t,a) + R(t,a) \tag{2.4}$$

with $S(t,\cdot), I(t,\cdot), R(t,\cdot)$ denoting the age-densities of the susceptible, infective and removed individuals. We assume that every neonate is susceptible. We ignore immunity induced in newborns by maternal antibodies:

$$\begin{aligned}
\left(\frac{\partial}{\partial t} + \frac{\partial}{\partial a}\right) S(t,a) &= -\lambda(t)S(t,a) - \mu(a)S(t,a) \\
\left(\frac{\partial}{\partial t} + \frac{\partial}{\partial a}\right) I(t,a) &= \lambda(t)S(t,a) - (\rho + \mu(a))I(t,a) \\
\left(\frac{\partial}{\partial t} + \frac{\partial}{\partial a}\right) R(t,a) &= \rho I(t,a) - \mu(a)R(t,a) \\
S(t,0) &= \int_0^\infty \beta(a)n(t,a)da \\
I(t,0) &= R(t,0) = 0 \\
\lambda(t) &= \int_0^\infty k(a)I(t,a)da.
\end{aligned} \tag{2.5}$$

$\lambda(t)$ is the rate of a susceptible individual to be infected at time t. Notice that this infection rate is the same for susceptibles of all ages. Actually we have made the simplifying assumption that the transmission of the disease — described by k — from an infective individual to a susceptible individual only depends on the age of the infective individual and not on the age of the susceptible individual.

ρ is the rate at which individuals are removed from the infective class. For simplicity the removal rate is assumed to be age-independent.

Notice that the equations (2.5) are consistent with (2.2), (2.3) if n is related to S, I, R by (2.4).

The model simplifies considerably if we introduce the fractions x of susceptible, y of infective, and z of removed individuals. To be precise we introduce the respective age densities

$$x(t,a) = \frac{S(t,a)}{n(t,a)}, \; y(t,a) = \frac{I(t,a)}{n(t,a)}, \; z(t,a) = \frac{R(t,a)}{n(t,a)}. \tag{2.6}$$

This has the effect that the dynamics of the population vanish from the equations almost completely with the only exemption that the population age-density appears in the equation for the per capita infection rate:

$$\begin{aligned}
\left(\frac{\partial}{\partial t} + \frac{\partial}{\partial a}\right) x(t,a) &= -\lambda(t)x(t,a) \\
\left(\frac{\partial}{\partial t} + \frac{\partial}{\partial a}\right) y(t,a) &= \lambda(t)x(t,a) - \rho y(t,a) \\
\left(\frac{\partial}{\partial t} + \frac{\partial}{\partial a}\right) z(t,a) &= \rho y(t,a) \\
x(t,0) &= 1 \\
y(t,0) = z(t,0) &= 0 \\
\lambda(t) &= \int_0^\infty k(a)n(t,a)y(t,a)da.
\end{aligned} \tag{2.7}$$

We note that the z equation is not actually needed in order to have a closed model.

In order to obtain a time-autonomous problem we assume that the population is in an equilibrium state, i.e. its size and age-distribution are independent of time. This is feasible if and only if

$$R_p := \int_0^\infty \beta(a)P(a)da = 1 \tag{2.8}$$

with

$$P(a) = \exp\left(-\int_0^a \mu(r)dr\right). \tag{2.9}$$

$P(a)$ is the probability to be still alive at age a. R_p is the *basic reproductive number* of the population, i.e. the mean number of offspring an average individual produces during its life. An easy computation shows that the equilibrium age density is given by

$$n(a) = \frac{NP(a)}{L} \tag{2.10}$$

with the stationary population size N and the life expectation

$$L = \int_0^\infty P(a)da.$$

Compare Aron (1989). The assumption of a stationary state for the population under (2.8) is justified by the celebrated renewal theorem — see Sharpe and Lotka (1911) for the formulation, Feller (1941) for the first rigorous proof, and Webb (1984) for a semigroup proof — which states under realistic conditions that the age distribution (2.10) is approached after sufficiently long time from whatever initial distribution the population has started. The equation for the infective force then takes the form

$$\lambda(t) = \int_0^\infty \frac{k(a)NP(a)}{L} y(t,a)da. \tag{2.11}$$

We make the assumption that

$$k(a)P(a) \leq \text{const } e^{-\epsilon a}$$

for some $\epsilon > 0$. This is not unrealistic, for the mortality rate $\mu(a)$ should be bounded away from zero for high age. This assumption in particular guarantees that (2.11) is well-defined because $0 \leq y(t,a) \leq 1$.

The model formulation in terms of the fractions x, y, z has the advantage that these entities are dimensionless. Before we start any serious manipulations of our model equations we make time and age dimensionless, too. To this end we introduce

$$\tau = \rho t, \alpha = \rho a.$$

This means that $\tau = 1$ corresponds to $t = \frac{1}{\rho}$ such that multiplying the dimensionless time by the length of the infectious period brings us back to real time. We set

$$\tilde{x}(\tau, \alpha) = x(t,a), \tilde{y}(\tau, \alpha) = y(t,a)$$

such that

$$\left(\frac{\partial}{\partial \tau} + \frac{\partial}{\partial \alpha}\right) \tilde{x}(\tau, \alpha) = \frac{1}{\rho}\left(\frac{\partial}{\partial t} + \frac{\partial}{\partial a}\right) x(t,a),$$

$$\left(\frac{\partial}{\partial \tau} + \frac{\partial}{\partial \alpha}\right) \tilde{y}(\tau, \alpha) = \frac{1}{\rho}\left(\frac{\partial}{\partial t} + \frac{\partial}{\partial a}\right) y(t,a). \tag{2.12}$$

We define

$$\tilde{\lambda}(\tau) = \frac{1}{\rho}\lambda(t), \quad \tilde{k}(\alpha) = \frac{N}{L\rho^2}P(a)k(a).$$

The equations (2.7) now take the following dimensionless form:

$$\begin{aligned}
\left(\frac{\partial}{\partial \tau} + \frac{\partial}{\partial \alpha}\right) \tilde{x}(\tau, \alpha) &= -\tilde{\lambda}(\tau)\tilde{x}(\tau, \alpha) \\
\left(\frac{\partial}{\partial \tau} + \frac{\partial}{\partial \alpha}\right) \tilde{y}(\tau, \alpha) &= \tilde{\lambda}(\tau)\tilde{x}(\tau, \alpha) - \tilde{y}(t, a) \\
\tilde{x}(\tau, 0) &= 1 \\
\tilde{y}(\tau, 0) &= 0 \\
\tilde{\lambda}(\tau) &= \int_0^\infty \tilde{k}(\alpha)\tilde{y}(\tau, \alpha)d\alpha.
\end{aligned} \tag{2.13}$$

The transformation

$$\breve{x}(\tau,\alpha)=\tilde{x}(\tau,\alpha)e^{\alpha},\quad \breve{y}(\tau,\alpha)=\tilde{y}(\tau,\alpha)e^{\alpha}$$

yields

$$\begin{aligned}\left(\frac{\partial}{\partial\tau}+\frac{\partial}{\partial\alpha}\right)\breve{x}(\tau,\alpha)&=(1-\tilde{\lambda}(\tau))\breve{x}(\tau,\alpha)\\ \left(\frac{\partial}{\partial\tau}+\frac{\partial}{\partial\alpha}\right)\breve{y}(\tau,\alpha)&=\tilde{\lambda}(\tau)\breve{x}(\tau,\alpha)\\ \breve{x}(\tau,0)&=1\\ \breve{y}(\tau,0)&=0\\ \tilde{\lambda}(\tau)&=\int_0^\infty \kappa Q(\alpha)\breve{y}(\tau,\alpha)d\alpha\end{aligned}\tag{2.14}$$

with

$$\kappa=\int_0^\infty \tilde{k}(\alpha)e^{-\alpha}d\alpha,\quad Q(\alpha)=\frac{1}{\kappa}\tilde{k}(\alpha)e^{-\alpha}.$$

Note that Q is a probability density, so it is make sense to look at its expected value

$$\gamma=\int_0^\infty \alpha Q(\alpha)d\alpha.$$

γ can be interpreted as some kind of mean age of infectivity — measured in dimensionless time — i.e. an average infective individual (provided it is infective since birth) spreads as much infectivity before as after that age. We scale time and age once more:

$$\tau=\gamma s,\quad \alpha=\gamma b.$$

We set

$$u(s,b)=\breve{x}(\tau,\alpha),\quad v(s,b)=\breve{y}(\tau,\alpha),\quad \nu(s)=\gamma\tilde{\lambda}(\tau),\quad T(b)=\gamma Q(\alpha).$$

T is normalized in a double way: it is not only a probability density, but also has mean 1. The equations (2.14) transform as follows:

$$\begin{aligned}\left(\frac{\partial}{\partial s}+\frac{\partial}{\partial b}\right)u(s,b)&=(\gamma-\nu(s))u(s,b)\\ \left(\frac{\partial}{\partial s}+\frac{\partial}{\partial b}\right)v(s,b)&=\nu(s)u(s,b)\\ u(s,0)&=1\\ v(s,0)&=0\\ \nu(s)&=\gamma\kappa\int_0^\infty T(b)v(s,b)db.\end{aligned}\tag{2.15}$$

In this form the problem is finally ready for being analyzed. Checking our transformations we find that the assumption $k(a)P(a)\leq \text{const}\ e^{-\epsilon a}$ implies that

$$0\leq T(b)\leq \text{ const } e^{-\tilde{\gamma}b},\quad b\geq 0,\quad \text{for some } \tilde{\gamma}>\gamma.\tag{2.16}$$

Keep in mind that $\gamma>0$ cannot be varied arbitrarily but is linked to T via (2.16).

3 Endemic equilibria

The equations (2.15) always have the disease-free equilibrium

$$u^\circ(b) = e^{\gamma b}, \quad v^\circ(b) = 0.$$

An endemic equilibrium satisfies

$$u^\star(b) = e^{(\gamma-\nu^\star)b}, \quad v^\star(b) = \nu^\star \int_0^b e^{(\gamma-\nu^\star)c} dc. \tag{3.1}$$

We substitute these relation into the ν equation and divide by $\nu^\star$. This yields

$$1 = \gamma\kappa \int_0^\infty T(b) \int_0^b e^{(\gamma-\nu^\star)c} dcdb. \tag{3.2}$$

We recognize that the right hand side of this equation is a strictly decreasing function of $\nu^\star$ unless T is identically zero. This implies that there is at most one endemic equilibrium. An endemic equilibrium is associated with a strictly positive solution $\nu^\star$ of (3.1) which exists if and only if the right hand side of (3.2) strictly exceeds 1 for $\nu^\star = 0$. So we obtain the following well-known threshold criterion — see Dietz and Schenzle (1985b), e.g.:

Theorem 3.1. *There exists an endemic equilibrium if and only if*

$$1 < \kappa \int_0^\infty T(b)(e^{\gamma b} - 1)db.$$

The endemic equilibrium — if it exists — is unique.

In order to study the stability properties of the endemic equilibrium we would like to consider (2.15) as a bifurcation problem. Rather than using κ it is more convenient to use $\nu^\star > 0$ as a parameter and to solve (3.2) for $\gamma\kappa$ and substitute the result into (2.15). To this end we rewrite (3.2) as

$$1 = \frac{\gamma\kappa}{\gamma - \nu^\star} \int_0^\infty T(b) \left(e^{(\gamma-\nu^\star)b} - 1\right) db = \frac{\gamma\kappa}{\gamma - \nu^\star} \left(\hat{T}(\nu^\star - \gamma) - 1\right).$$

Recall that T is a probability density, i.e. $\hat{T}(0) = 1$. Here we have used the *Laplace transform* notation

$$\hat{T}(w) = \int_0^\infty e^{-wr} T(r) dr.$$

Hence

$$\gamma\kappa = \frac{\nu^\star - \gamma}{1 - \hat{T}(\nu^\star - \gamma)}.$$

By (2.16), $\int_0^\infty e^{\gamma r} T(r) dr < \infty$ such that $\hat{T}(\nu^\star - \gamma)$ is defined for every $\nu^\star \geq 0$. We substitute $\gamma\kappa$ into (2.15):

$$\begin{aligned}
\left(\frac{\partial}{\partial s}+\frac{\partial}{\partial b}\right) u(s,b) &= (\gamma-\nu(s))u(s,b)\\
\left(\frac{\partial}{\partial s}+\frac{\partial}{\partial b}\right) v(s,b) &= \nu(s)u(s,b)\\
u(s,0) &= 1\\
v(s,0) &= 0\\
\nu(s) &= \frac{\nu^\star-\gamma}{1-\hat{T}(\nu^\star-\gamma)}\int_0^\infty T(r)v(s,r)dr.
\end{aligned} \tag{3.3}$$

It is convenient to shift the endemic equilibrium to zero. So we introduce

$$u(s,b)=u^\star(b)+U(s,b), \quad v(s,b)=v^\star(b)+V(s,b), \quad \nu(s)=\nu^\star+\phi(s).$$

(3.3) takes the form

$$\begin{aligned}
\left(\frac{\partial}{\partial s}+\frac{\partial}{\partial b}\right) U(s,b) &= (\gamma-\nu^\star)U(s,b)-\phi(s)u^\star(b)-\phi(s)U(s,b)\\
\left(\frac{\partial}{\partial s}+\frac{\partial}{\partial b}\right) V(s,b) &= \nu^\star U(s,b)+\phi(s)u^\star(b)+\phi(s)U(s,b)\\
U(s,0) &= 0\\
V(s,0) &= 0\\
\phi(s) &= \frac{\nu^\star-\gamma}{1-\hat{T}(\nu^\star-\gamma)}\int_0^\infty T(r)V(s,r)dr.
\end{aligned} \tag{3.4}$$

In order to save on notation we introduce

$$\eta=\nu^\star-\gamma. \tag{3.5}$$

Substituting (3.1) for $u^\star$ we obtain the final version of our equations:

$$\begin{aligned}
\left(\frac{\partial}{\partial s}+\frac{\partial}{\partial b}\right) U(s,b) &= -\eta U(s,b)-e^{-\eta b}\phi(s)-\phi(s)U(s,b)\\
\left(\frac{\partial}{\partial s}+\frac{\partial}{\partial b}\right) V(s,b) &= \nu^\star U(s,b)+e^{-\eta b}\phi(s)+\phi(s)U(s,b)\\
U(s,0) &= 0\\
V(s,0) &= 0\\
\phi(s) &= \frac{\eta}{1-\hat{T}(\eta)}\int_0^\infty T(b)V(s,b)db.
\end{aligned} \tag{3.6}$$

The stability of the endemic equilibrium is equivalent to the stability of the zero equilibrium associated with (3.6). We have to keep in mind that, by (3.5) the parameters $\nu^\star,\eta$ must satisfy

$$\nu^\star>0, \quad \nu^\star>\eta>-\gamma \tag{3.7}$$

in order to make epidemiologic sense. Further γ is linked to T by (2.16).

4 The stability of the endemic equilibrium and the characteristic equation

In order to decide about the stability or instability of the endemic equilibrium — or the zero equilibrium of (3.6) — we formally proceed as in the case of ordinary differential equations. We shall discuss later whether this procedure is justified.

We linearize (3.6) by dropping the higher order terms and look for special solutions of the form $U(s,b) = e^{ws}\tilde{U}(b), V(s,b) = e^{ws}\tilde{V}(b), \phi(s) = e^{ws}$:

$$\begin{aligned}
\left(w + \frac{d}{db}\right)\tilde{U}(b) &= -\eta\tilde{U}(b) - e^{-\eta b} \\
\left(w + \frac{d}{db}\right)\tilde{V}(b) &= \nu^\star\tilde{U}(b) + e^{-\eta b} \\
\tilde{U}(0) &= 0 \\
\tilde{V}(0) &= 0 \\
1 &= \frac{\eta}{1-\hat{T}(\eta)}\int_0^\infty T(b)\tilde{V}(b)db.
\end{aligned} \tag{4.1}$$

Lemma 4.1. *If* $\nu^\star > 0$, (4.1) *has no solution with* $w \in \mathbf{R}, w \geq 0$.

Proof. Let us assume that there is such a solution. Apparently $\tilde{U}(b) < 0, b > 0$. Then $\tilde{V}$ satisfies the differential inequality

$$\left(w + \frac{d}{db}\right)\tilde{V}(b) < e^{-\eta b}, \quad b > 0.$$

Hence

$$\tilde{V}(b) < \frac{1}{\eta}(1 - e^{-\eta b}).$$

Substituting this expression into the last equation in (4.1) yields the contradiction

$$1 < \frac{\hat{T}(0) - \hat{T}(\eta)}{1 - \hat{T}(\eta)} = 1.$$

Recall that T is a probability density. This completes the proof. □

We integrate the $\tilde{U}$ equation in (4.1):

$$\tilde{U}(b) = -e^{-\eta b}\int_0^b e^{-wr}dr.$$

We substitute $\tilde{U}$ into the $\tilde{V}$ equation in (4.1) and integrate:

$$\begin{aligned}\tilde{V}(b) &= \int_0^b e^{-\eta r} e^{w(r-b)} dr - \nu^\star \int_0^b e^{-\eta s} \left(\int_0^s e^{-wr} dr \right) e^{w(s-b)} ds \\ &= \int_0^b e^{-\eta r} e^{w(r-b)} dr - \nu^\star \int_0^b e^{-\eta s} \int_0^s e^{w(s-r-b)} dr ds \\ &= \int_0^b e^{-\eta r} e^{w(r-b)} dr - \nu^\star \int_0^b e^{-\eta s} \int_0^s e^{w(r-b)} dr ds.\end{aligned}$$

We change the order of integration:

$$\begin{aligned}\tilde{V}(b) &= \int_0^b e^{-\eta r} e^{w(r-b)} dr - \nu^\star \int_0^b e^{w(r-b)} \int_r^b e^{-\eta s} ds dr \\ &= \int_0^b e^{-\eta r} e^{w(r-b)} dr - \frac{\nu^\star}{\eta} \int_0^b e^{w(r-b)} \left(e^{-\eta r} - e^{-\eta b} \right) ds dr.\end{aligned}$$

Thus we obtain

$$\begin{aligned}\tilde{V}(b) &= \int_0^b e^{-wr} e^{\eta(r-b)} \left(1 - \frac{\nu^\star}{\eta} \left(1 - e^{-\eta r} \right) \right) dr \\ &= \frac{w - \nu^\star}{w(w-\eta)} e^{-\eta b} - \frac{\nu^\star}{w\eta} e^{-(\eta+w)b} + \frac{\nu^\star - \eta}{\eta(w-\eta)} e^{-wb}.\end{aligned} \tag{4.2}$$

We substitute $\tilde{V}$ into the last equation in (4.1):

$$\begin{aligned}1 =& \frac{\eta}{1 - \hat{T}(\eta)} \left(\frac{w - \nu^\star}{w(w-\eta)} \hat{T}(\eta) - \frac{\nu^\star}{w\eta} \hat{T}(w+\eta) + \frac{\nu^\star - \eta}{\eta(w-\eta)} \hat{T}(w) \right), \\ & 0 \neq w \neq \eta \neq 0.\end{aligned} \tag{4.3}$$

For $\eta = 0$, we take the limit of (4.3), $\eta \to 0$, or repeat the derivation of the characteristic equation from (4.1). Either way we obtain

$$1 = \frac{w - \nu^\star}{w^2} (1 - \hat{T}(w)) + \frac{\nu^\star}{w} \int_0^\infty r T(r) e^{-wr} dr, \quad \eta = 0, w \neq 0. \tag{4.4}$$

If $\eta \neq 0$, it is convenient to multiply (4.3) by $(1 - \hat{T}(\eta)) w (w - \eta)$ and rearrange terms:

$$0 = (w^2 - \eta\nu^\star)\hat{T}(\eta) + w(\nu^\star - \eta)\hat{T}(w) - \nu^\star(w-\eta)\hat{T}(w+\eta) - w(w-\eta), \quad \eta \neq 0. \tag{4.5}$$

We must be aware of the fact that, by the above multiplication, we have created roots $w = 0$ and $w = \eta$ of (4.5) which are not associated with solutions to (4.1) according to Lemma 4.1.

If T has *compact support*, i.e. if there is some $b^\natural > 0$ such that $T(b) = 0, b > b^\natural$, then the (in)stability of the endemic equilibrium can be linked with the position of the roots w of (4.2), (4.3) in the complex plane. See Inaba (1990), Lemmata 5.1 and 5.2.

Theorem 4.2. *Let T have support in $[0, b^\natural]$, $b^\natural < \infty$. Then the following holds: a) Let all roots w of (4.3), (4.4) with nonzero imaginary part have a strictly* *negative real part. Then the endemic equilibrium is locally asymptotically stable. In particular we have:*

(*i*) *For any* $\epsilon > 0$ *there is some* $\delta > 0$ such that

$$\int_0^{b^\natural} (|U(s,b)| + |V(s,b)|)db < \epsilon \quad \textit{for all } s > 0$$

whenever U, V *are solutions to* (3.6) *satisfying*

$$\int_0^{b^\natural} (|U(0,b)| + |V(0,b)|)db < \delta.$$

(*ii*) *There is some* $\delta_0 > 0$ *such that*

$$\int_0^{b^\natural} (|U(s,b)| + |V(s,b)|)db \to 0, \quad s \to \infty$$

whenever U, V *are solutions to* (3.6) *satisfying*

$$\int_0^{b^\natural} (|U(0,b)| + |V(0,b)|)db < \delta_0.$$

b) Let (4.3), (4.4) *have a root with strictly positive real and imaginary part. Then the endemic equilibrium is unstable. In particular we can find some* $\epsilon_0 > 0$ *with the following property: For any* $n \in \mathbf{N}$ *there are solutions* U_n, V_n *of* (3.6) *and times* $s_n > 0$ *such that*

$$\int_0^{b^\natural} (|U_n(0,b)| + |V_n(0,b)|)db < \frac{1}{n}$$

but

$$\int_0^{b^\natural} (|U_n(s_n;b)| + |V_n(s_n,b)|)db > \epsilon_0.$$

5 Stability of the endemic equilibrium for small disease prevalence

We show that the characteristic equation has no roots with non-negative real parts if $\nu^\star > 0$ is sufficiently small. This result follows from Inaba (1990), if T has compact support. We prove the result for the convenience of the reader. We include the case of non-compact support.

We first notice from (4.1) and (4.2) that

$$1 = \frac{\eta}{1 - \hat{T}(\eta)} \int_0^\infty T(b) \Re \tilde{V}(b) db. \tag{5.1}$$

and

$$\Re\tilde{V}(b) = \int_0^b \Re e^{-wr} e^{\eta(r-b)} W(r) dr$$
$$W(r) = \left(1 - \frac{\nu^\star}{\eta}\left(1 - e^{-\eta r}\right)\right). \tag{5.2}$$

We recall from (3.5), $\eta = \nu^\star - \gamma$, that $\eta < 0$ if $\nu^\star > 0$ is small. Define

$$b(\nu^\star) = -\frac{1}{\eta}\ln\left(1 - \frac{\eta}{\nu^\star}\right) = \frac{1}{\gamma - \nu^\star}\ln\left(\frac{\gamma}{\nu^\star}\right).$$

We realize that

$$b(\nu^\star) \to \infty, \quad \nu^\star \to 0$$

and

$$W(r) \geq 0 \iff r \leq b(\nu^\star).$$

Further we always have the estimate

$$|W(r)| \leq 1 + \frac{\nu^\star}{\eta}\left(1 - e^{-\eta r}\right).$$

In other words

$$|W(r)| \leq 1 - \text{ sign }\left(b(\nu^\star) - r\right)\frac{\nu^\star}{\eta}\left(1 - e^{-\eta r}\right). \tag{5.3}$$

It follows from (5.1), (5.2) that, for $\Re w \geq 0$,

$$1 \leq \frac{\eta}{1 - \hat{T}(\eta)} \int_0^\infty T(b) \int_0^b e^{\eta(r-b)} |W(r)| dr db.$$

As

$$1 = \frac{\eta}{1 - \hat{T}(\eta)} \int_0^\infty T(b) \int_0^b e^{\eta(r-b)} dr db,$$

this implies

$$0 \leq -\frac{\eta}{1 - \hat{T}(\eta)} \int_0^\infty T(b) \int_0^b e^{\eta(r-b)} \text{ sign }\left(b(\nu^\star) - r\right)\frac{\nu^\star}{\eta}\left(1 - e^{-\eta r}\right) dr\, db.$$

As

$$\text{sign }(b(\nu^\star) - r) \to 1, \quad \nu^\star \to 0, r \geq 0,$$

and T satisfies the estimate (2.16) and $\eta > -\gamma$, we find that the right hand side of this inequality converges to a strictly negative number as $\nu^\star \to 0, \Re w \geq 0$ by *Lebesgue's* theorem of dominated convergence. This contradiction implies

Theorem 5.1. *If $\nu^\star > 0$ is sufficiently small, the characteristic equation* (4.3), (4.4) *has no roots with non-negative real parts.*

6 Condition for stability change of the endemic equilibrium

In this section we prove that, for any η, the characteristic equation (4.3), (4.4) has roots with positive real part if $\nu^\star > \eta$ is chosen large enough and T is sufficiently concentrated at 1. The latter in particular involves (2.16) with $\gamma = \nu^\star - \eta$ — compare (3.5) and (3.7).

Equation (4.3), (4.4) is rather painful to analyze. So we try to take limits — hopefully without eliminating the parts of the equation which cause the existence of roots with strictly positive real part.

We take the limit of (4.5) for $\nu^\star \to \infty$ (we will incorporate (4.4) later.):

$$0 = -\eta\hat{T}(\eta) + w\hat{T}(w) - (w-\eta)\hat{T}(w+\eta). \tag{6.1}$$

We take another limit by considering the *Dirac* measure $T = \delta_1$ concentrated at 1 rather than a probability density T with mean 1:

$$0 = -\eta e^{-\eta} + we^{-w} - (w-\eta)e^{-w-\eta}. \tag{6.2}$$

Multiplying by $e^{\eta+w}$, dividing by η and rearranging terms we obtain

$$e^w = w\frac{e^\eta - 1}{\eta} + 1. \tag{6.3}$$

In order to include the case $\eta = 0$ we apply the same procedure to (4.4). Defining

$$\psi(\eta) = \begin{cases} \frac{e^\eta - 1}{\eta} & \text{if } \eta \neq 0 \\ 1 & \text{if } \eta = 0 \end{cases} \tag{6.4}$$

we obtain

$$e^w = w\psi(\eta) + 1.$$

We set $w = \delta + i\omega$ and separate real and imaginary part:

$$\begin{aligned} e^\delta \cos\omega &= \delta\psi(\eta) + 1 \\ e^\delta \sin\omega &= \omega\psi(\eta). \end{aligned} \tag{6.5}$$

We solve the second equation for δ:

$$\delta = \ln\frac{\omega}{\sin\omega} + \ln\psi(\eta). \tag{6.6}$$

We substitute this result into the first equation in (6.5) and rearrange terms:

$$\omega\cos\omega\psi(\eta) = \sin\omega\left(\left(\ln\frac{\omega}{\sin\omega} + \ln\psi(\eta)\right)\psi(\eta) + 1\right). \tag{6.7}$$

It follows from the intermediate value theorem that (6.6), (6.7) have solutions $\delta > 0, \omega$ with large ω for any η. Noting that $\ln\psi(\eta)$ with ψ defined by (6.4) is a strictly increasing function of η and $\ln\psi(0) = 0$ we even realize that (6.6), (6.7) has no roots with non-positive real part except 0 if $\eta = 0$ and no roots with non-positive real part at all if $\eta > 0$.

Rouché's theorem implies that (4.3), (4.4) has roots with strictly positive part, if $\nu^\star$ is sufficiently large and if T is a probability density with mean 1 that is sufficiently concentrated at 1.

We can now fix T and η and vary $\nu^\star$. If $\nu^\star$ is small, all roots of the characteristic equations strictly lie in the left complex half plane. If $\nu^\star$ becomes large enough, the characteristic equation has roots in the right half plane. It follows from the form of (4.3), (4.4) and *Rouché's* theorem that roots cross the imaginary axis from left to right — without passing through the origin — if $\nu^\star$ increases. When T has compact support this implies that the endemic equilibrium changes its stability — see theorem 4.2.

7 Discussion

This paper shows that an age-structured $S \to I \to R$ model has an unstable endemic equilibrium at certain parameter values, if the rate of a susceptible individual to be infected is independent of its age but, as for the age of the infective individual, is highly concentrated in a specific age class. Typically this happens if, in addition, the incidence of the disease and the mean age of infectivity are relatively high. Though our basic assumption — age-independent susceptibility and highly concentrated age-dependent infectivity — does not seem to be realistic, the above result is important in so far as it tells us that stability of the endemic equilibrium in an age-structured endemic model cannot be taken for granted but requires a sound analytical or numerical investigation. It would be interesting, e.g., to have the stability of the endemic equilibrium settled at least for the case that the infection rate is independent both of the age of the susceptible and the infective individual but that the survival probability $P(a)$ is arbitrary, i.e. $P(0) = 1$, $P(a)$ decreasing in a, $\int_0^\infty P(a)da < \infty$. This will still leave the problem open for age dependent susceptibility and infectivity — even in the case of proportionate mixing (1.1) and even assuming symmetry with respect to the age of the susceptible and the infective individual, i.e. $k_1 = k_2$ in (1.1). We have assumed k_1 to be constant in this paper because it simplifies the characteristic equation dramatically — though it remains complicated enough.

If the endemic equilibrium is unstable, it does not follow necessarily — though it is intuitive — that the disease dynamics oscillate in an undamped fashion. The author conjectures that, in the case of proportionate mixing, the function $\lambda(t)$ in (2.7) either converges to its endemic equilibrium value — provided that the endemic equilibrium exists — or oscillates around it in an undamped fashion. If the endemic equilibrium is unstable, the second should occur, though it cannot be excluded that, on one of their excursions away from the equilibrium, the disease dynamics hit the stable manifold and finally converge to the equilibrium.

Another unsettled problem is the existence of periodic solutions in case that the endemic equilibrium is unstable. The application of *local Hopf bifurcation* theorems requires checking nonresonance and transversal crossing of the roots of the characteristic equation over the imaginary axis. This seems to be a very difficult task in face of the complexity of the characteristic equation. The application of *global Hopf bifurcation* theorems does not require these properties, but still one has to find parameter values such that there are roots in the right complex half-plane and — unfortunately in addition — no roots on the imaginary axis — see, e.g., Fiedler (1986), assumption (1.10). Moreover one has to adapt the existing theory to the infinite-dimensional dynamical system generated by our endemic model. Actually we have been able to check the above condition in the limiting case $\mu^\star \to \infty$, T being the probability measure concentrated at 1, $\eta > 0$, but it is not clear whether this situation remains valid if T is a highly concentrated probability density rather than a point measure.

Finally, in view of the investigations of childhood disease data — see, e.g. Olsen, Truty, Schaffer (1988) and the references there — we mention the question whether more complicated than periodic oscillations can be generated by a deterministic age-structured endemic model.

Acknowledgment

This note presents parts of a systematic study of age-structured S → I → R endemic models by Stavros Busenberg, Mimmo Iannelli and the author which will be published elsewhere.

References

1. Anderson, R.M. Grenfell, B.T., May, R.M. (1984): Oscillatory fluctuations in the incidence of infectious diseases and the impact of vaccination: time series analysis. J. Hyg. Camb. **93**, 587-608
2. Anderson, R.M., May, R.M. (1982): Directly transmitted infectious diseases, Control by vaccination. Science **215**, 1053-1060
3. Anderson, R.M., May, R.M. (1983): Vaccination against rubella and measles: quantitative investigations of different policies. J. Hygiene **90**, 259-325
4. Anderson, R.M., May, R.M. (1985): Age-related changes in the rate of disease transmission: implication for the design of vaccination programmes. J. Hyg. Camb. **94**, 365-436
5. Andreasen, V. (1989a): Multiple time scales in the dynamics of infectious diseases. Mathematical Approaches to Problems in Resource Management and Epidemiology C. Castillo-Chavez, S.A. Levin, C.A. Shoemaker, eds., Lecture Notes in Biomathematics, **81**, 142-151, Springer-Verlag, Berlin-Heidelberg-New York
6. Andreasen, V. (1989b): Disease regulation of age-structured host populations. Theor. Pop. Biol. **36**, 214-239
7. Andreasen, V. (preprint a): Age-dependent host mortality in the dynamics of endemic infectious diseases.
8. Andreasen, V. (preprint b): SIR-models of the epidemiology and natural selection of co-circulating influenza virus with partial cross-immunity.
9. Aron, J.L. (1989): Simple versus complex epidemiological models. Applied Mathematical Ecology, S.A. Levin, T.G. Hallam, L. J. Gross, eds., Lecture Notes in Biomathematics, **18**, 176-192, Springer-Verlag, Berlin-Heidelberg-New York

10. Bernoulli, D. (1760): Essai d'une nouvelle analyse de la mortalité causeé par la petite vérole et des avantages de l'inoculation pour la prévenir. Mém. Math. Phys. Acad. Roy. Sci. Paris, 1-45
11. Busenberg, S., Cooke, K.L., Iannelli, M. (1988): Endemic thresholds and stability in a class of age-structured epidemics. SIAM J. Appl. Math. **48**, 1379-1395
12. Busenberg, S., Iannelli, M., Thieme, H.R. (to appear): Global behavior of an age-structured S–I–S epidemic model. SIAM J. Math. Anal.
13. Busenberg, S, Iannelli, M., Thieme, H.R. (preprint): Global behavior of an age-structured S–I–S epidemic model. The case of a vertically transmitted disease.
14. Castillo-Chavez, C. (1989): Some applications of structured models in population dynamics. Applied Mathematical Ecology, S.A. Levin, T.G. Hallam, L. J. Gross, eds., Lecture Notes in Biomathematics, **18**, 212-234, Springer-Verlag, Berlin-Heidelberg-New York
15. Castillo-Chavez, C., Hethcote, H.W., Andreasen, V., Levin, S.A., Liu, W.-M. (1989): Epidemiological models with age structure, proportionate mixing, and cross immunity. J. Math. Biol. **27**, 233-258
16. Dietz, K. (1975): Transmission and control of arbovirus diseases. Epidemiology, D. Ludwig, K.L. Cooke, eds., 104-121, SIAM
17. Dietz, K. (1981): The evaluation of rubella vaccination strategies. The Mathematical Theory of the Dynamics of Biological Populations II, Hiorns, R.W., Cooke, D., eds., 82-97, Academic Press
18. Dietz, K. (1988): The first epidemic model: A historical note on P.D. En'ko. Austral. J. Statist. **30 A**, 56-65
19. Dietz, K., Schenzle, D. (1985a): Mathematical models for infectious disease statistics. A Celebration of Statistics. The ISI Centenary Volume, Atkinson, A.C., Fienberg, S.E., eds., 167-204. Springer-Verlag, Belin-Heidelberg-New York
20. Dietz, K., Schenzle, D. (1985b): Proportionate mixing models for age-dependent infection transmission. J. Math. Biol. **22**, 117-120
21. Enderle, J.D. (1980): A stochastic communicable disease model with age specific states and application to measles. Ph.D dissertation, Rensselaer Polytechnic Institute
22. Feller, W. (1941): On the integral equation of renewal theory. Ann. Math. Stat. **12**, 243-267
23. Fiedler, B. (1986): Global Hopf bifurcation for Volterra integral equations. SIAM J. Math. Anal. **17**, 911-932
24. Greenhalgh, D. (1987): Analytical results on the stability of age-structured recurrent epidemic models. IMA J. Math. Appl. Med. Biol. **4**, 109-144
25. Greenhalgh, D. (1988a): Analytical threshold and stability results on age-structured epidemic models with vaccination. Theor. Pop. Biol. **33**, 266-290
26. Greenhalgh, D. (1988b): Threshold and stability results for an epidemic model with an age-structured meeting rate. IMA J. Math. Appl. Med. Biol. **5**, 81-100
27. Gripenberg, G. (1983): On a nonlinear integral equation modelling an epidemic in an age-structured population. J. Reine Angew. Math. **341**, 54-67
28. Hethcote, H.W. (1976): Qualitative analysis for communicable disease models. Math. Biosci. **28**, 335-356
29. Hethcote, H.W. (1988): Optimal ages of vaccination for measles. Math. Biosci. **89**, 29-52
30. Hethcote, H.W. (1989): Rubella. Applied Mathematical Ecology, S.A. Levin, T.G. Hallam, L. J. Gross, eds., Lecture Notes in Biomathematics, **18**, 212-234, Springer-Verlag, Berlin-Heidelberg-New York

31. Hethcote, H.W., Levin, S.A. (1989): Periodicity in epidemiological models. Applied Mathematical Ecology, S.A. Levin, T.G. Hallam, L. J. Gross, eds., Lecture Notes in Biomathematics, **18**, 193-211, Springer-Verlag, Berlin-Heidelberg-New York
32. Hethcote, H.W., Stech, H.W., van den Driessche, P. (1981): Periodicity and stability in epidemic models: a survey. Differential Equations and Applications in Ecology, Epidemics and Population Problems S. Busenberg, K.L. Cooke, eds., 65-82, Academic Press
33. Hoppensteadt, F. (1974): An age dependent epidemic model. J. Franklin Institute **297**, 325-333
34. Hoppensteadt, F. (1975): Mathematical Theories of Populations: Demographics, Genetics and Epidemics. Regional Conference Series in Applied Mathematics **20**. SIAM
35. Huang, W. (thesis): Studies in differential equations and applications. Ph.D dissertation, Claremont Graduate School, 1990
36. Inaba, H. (1990): Thresholds and stability results for an age-structured epidemic model. J. Math. Biol. **28**, 411-434
37. Katzmann, W., Dietz, K. (1984): Evaluation of age-specific vaccination strategies. Theor. Pop. Biol. **25**, 125-137
38. Kermack, W.O., McKendrick, A.G. (1927): A contribution to the mathematical theory of epidemics. Proc. Roy. Soc. **A 115**, 700-721
39. Kermack, W.O., McKendrick, A.G. (1932): Contributions to the mathematical theory of epidemics. II.– The problem of endemicity. Proc. Roy. Soc. **A 138**, 55-83
40. Kermack, W.O., McKendrick, A.G. (1933): Contributions to the mathematical theory of epidemics. III.– Further studies of the problem of endemicity. Proc. Roy. Soc. **A 141**, 94-122
41. Knox, E.G. (1980): Strategy for rubella vaccination. Int. J. Epidemiol. **9**, 13-23
42. Liu, W.-M., Hethcote, H.W., Levin, S.A. (1987): Dynamical behavior of epidemiological models with nonlinear incidence rates. J. Math. Biol. **25**, 359-380
43. McKendrick, A.G. (1926): Applications of mathematics to medical problems. Proc. Edin. Math. Soc. **44**, 98-130
44. McLean, A. (1986): Dynamics of childhood infections in high birthrate countries. Immunology and Epidemiology, G.W. Hoffman, T. Hraba, eds., Lecture Notes in Biomathematics, **65**, 171-197, Springer-Verlag, Berlin-Heidelberg-New York
45. May, R.M., Anderson, R.M., McLean, A.R. (1988): Possible demographic consequences of HIV/AIDS: I, assuming HIV infection always leads to AIDS. Math. Biosci. **90**, 475-505
46. May, R.M., Anderson, R.M., McLean, A.R. (1989): Possible demographic consequences of HIV/AIDS: II, assuming HIV infection does not necessarily lead to AIDS. Mathematical Approaches to Problems in Resource Management and Epidemiology, C. Castillo-Chavez, S.A. Levin, C.A. Shoemaker, eds., Lecture Notes in Biomathematics, **81**, 220-245, Springer
47. Olsen, L.F., Truty, G.L., Schaffer, W.M. (1988): Oscillations and chaos in epidemics: a nonlinear dynamic study of six childhood diseases in Copenhagen, Denmark. Theor. Popl. Biol. **33**, 344-370
48. Schenzle, D. (1984): An age-structured model of pre- and post- vaccination measles transmission. IMA J. Math. Appl. Med. & Biol. **1**, 169-191
49. Schenzle, D. (1985): Control of virus transmission in age-structured populations. Mathematics in Biology and Medicine, V. Capasso, E. Grosso, S.L. Paveri-Fontana, eds., Lecture Notes in Biomathematics, **57**, 171-178, Springer-Verlag, Berlin-Heidelberg-New York

50. Sharpe, F.R., Lotka, A.J. (1911): A problem in age-distribution. Phil. Mag. **21**, 435-438
51. Webb, G.F. (1984): A semigroup proof of the Sharpe-Lotka theorem. Infinite-Dimensional Systems, F. Kappel, W. Schappacher, eds., Lecture Notes in Mathematics, **1076**, 254-268, Springer-Verlag, Berlin-Heidelberg-New York

Part III

Ecology and Population Dynamics

Mathematical Model for the Dynamics of a Phytoplankton Population

E. Beretta and A. Fasano

Dipartimento di Matematica U. Dini, Università di Firenze, Florence, Italy

Introduction

The problem of modelling the vertical structure of a phytoplankton population in sea water has been treated in many papers (see e.g. A. Wörz-Busekros [9], N. Shigesada and A. Okubo [7], H. Ishii and I. Takagi [4]). In [9] the coupled dynamics of phytoplankton and nutrient is considered with a sinking term for the phytoplankton and both the algae and the nutrient diffusing with the same diffusion coefficient. A unique solution of the corresponding initial-boundary value problem was shown to exist. In [7] the self-shading effect of the biomass in a nutrient-saturated environment was taken into account (neglecting light absorption by water in order to have an autonomous system at equilibrium). The global stability of the unique positive stationary solution for an improved model was proved in [4].

More recently, S. Totaro [8] has considered a population of two different diffusing phytoplankton species (still under the assumption of saturating nutrients) with a sinking term and light influenced growth.

In our model we consider a phytoplankton population growing under controlled condition in a vertical test tube filled with sea water and with one nutrient having a limiting action on the algal growth.

The phytoplankton growth rate depends on the local values of the light intensity (exerting an activation or inhibition effect) and on the nutrient concentration. The light intensity decreases along the tube because of the water absorption and of the biomass self-shading effect. The death rate is given by a quadratic function of the biomass concentration (including the so-called overcrowding term) and the nutrient release by dead algae is described according to some delay profile and taking into account the sinking velocity. The question of the possible displacement of living algae is discussed, but for simplicity here we assume that they occupy a fixed position during their life.

The main scope of the paper is to investigate the steady state problem, analyzing the structure of nontrivial solutions.

The model is described in Sections 1 and 2 and the equilibrium problem is stated in Sect. 3 (and very briefly sketched for the case of absence of light in Sect.

4). Then the existence and the structure of positive solutions are illustrated in Sect. 5 (for the nutrient saturation case) and in Sect. 6 (including the nutrient dynamics), in both cases for different selections of the light activation-inhibition law.

In Sect. 7 we illustrate briefly the case in which not all the nutrient is given back to the solution by dead algae. In such a case the boundary conditions are of nonlocal type.

Some open problems are listed at the end of the paper.

1 A general description of the system

The phytoplankton population grows under controlled conditions in a vertical cylinder (typical size: 2 meters height and 20 cm diameter) filled with saline water and kept at a uniform temperature θ . Light comes from the top with a prescribed intensity I_0. Nutrients are dissolved in the water and the experimental conditions are such that only one of them (e. g. phosphorus) exerts a limiting action (in the sense that its concentration varies in a range influencing the dynamics of the system).

The basic quantities entering the model are:

- the concentration of living algae B as a function of time t and of the depth coordinate $x \in (0, h)$,
- the nutrient concentration $N(x,t)$,
- the local value of the light intensity I, expressed through the normalized variable $i = I/I_m$, I_m being some optimum reference value [1, 6].

Owing to absorption by water and to the photosynthesis action by the phytoplankton, the function $i(x,t)$ has the following form

$$i(x,t) = i_0 exp\left[-\left(\sigma_w x + \sigma_c \int_0^x B(\xi,t)\,d\xi\right)\right], \tag{1.1}$$

where $i_0 = I_0/I_m$ and σ_w, σ_c are positive coefficients (see Table 1 in the Appendix).

In order to formulate a mathematical model we have to describe the following features:

(I) the growth rate,

(II) the death rate,

(III) the nutrient dynamics,

(IV) the possible displacement of individuals.

(i) Growth rate.

The growth rate is governed by the classical nutrient uptake function

$$U(N) = N/(L+N), \tag{1.2}$$

L being a given positive constant, and by the light activation-inhibition function, which can be taken as in [1]

$$f(i) = ie^{-i+1} \tag{1.3}$$

(or any function with similar properties).

We will also consider two extreme cases, namely

$$f(i) = i, \tag{1.4}$$

i. e. no inhibition, and

$$f(i) = [2i - i^2]_+, \tag{1.5}$$

$[\,]_+$ denoting the positive part, i. e. total inhibition for $i > 2$.

It can be noticed that (1.5) can reasonably replace (1.3) in a range of values of i which is of practical interest.

Given $U(N)$ and $f(i)$, the population growth rate has the following expression

$$G(x,t) = Af(i)U(N)B(x,t), \tag{1.6}$$

where A is some known positive coefficient. For very low values of i the product $f(i)U(N)$ is sometimes replaced by $max[r(N), f(i)U(N)]$ where $r(N) > 0$ represents the residual nutrient absorption in the absence of light and can be expressed as $r(N) = min[r_0, U(N)]$, for some positive r_0.

(ii) Death rate.

Besides the usual term $\gamma B (\gamma > 0)$, we consider a quadratic term expressing an additional death rate due to overcrowding. Thus the death rate will be expressed by the product γM, with

$$M = B + \frac{a}{\gamma} B^2, \tag{1.7}$$

$a > 0$ being a measurable constant (see Table 1, Appendix).

(iii) Nutrient dynamics.

The limiting nutrient is absorbed at the rate $U(N)$ and it is released by dead cells with a delay described by the function

$$F(\tau) = \frac{1}{E}\alpha^2 \tau e^{-\alpha\tau}, \tag{1.8}$$

where τ is the time elapsed from death, α is a constant depending on the temperature θ and $E \geq 1$.

Since $\int_0^\infty F(\tau)\,d\tau = \frac{1}{E}$, the value of E determines whether the nutrient contained in a dead cell is completely returned to the solution for $\tau \to \infty$ $(E = 1)$, or not $(E > 1)$.

Thus, dead cells provide a distributed source of nutrient. In order to describe such a process in the correct way, we have to take into account the fact that dead cells fall through the cylinder at a constant velocity v. Indeed, the time scales of the cells motion and of the nutrient release are comparable. At each (x,t) we have to sum the contributions of the cells which died at higher locations ξ and which have travelled the distance $x - \xi$ over the time $(x - \xi)/v$. If Y represents the average nutrient content in a cell, the distributed nutrient source will be given by

$$S(x,t) = Y\gamma \int_{[x-vt]_+}^{x} M\left(\xi, t - \frac{x-\xi}{v}\right) F\left(\frac{x-\xi}{v}\right) \frac{d\xi}{v} \tag{1.9}$$

(iv) Cells migration.

Living cells can control their position in the cylinder, otherwise after a while all the population would be found at the botton of the cylinder, while experiments indicate the existence of nontrivial equilibrium distributions.

The more general way of describing cells displacements is to introduce a biomass current $q(x,t)$, due in part to diffusion and in part to the ability of the individuals of searching the optimal environment.

Since the quality of the environment is measured by the product $f(i)U(N)$, we can suppose that the cells tend to move in the direction of its gradient. Thus we conclude that

$$q(x,t) = -D_0 B_x + v_0 g(\zeta) B, \tag{1.10}$$

where D_0 is the diffusion coefficient of the population, v_0 is the maximum migration velocity, $\zeta = [f(i)U(N)]_x$, and $g(\zeta) = \frac{\zeta}{\lambda}$ for $|\zeta| < \lambda$, $g(\zeta) = 1$ for $\zeta > \lambda$, $g(\zeta) = -1$ for $\zeta < -\lambda$, λ being a positive parameter (however one could describe ascending and descending motion in different ways).

2 The complete evolution model

According to (I)-(IV) the nutrient and biomass balance are the following

$$N_t = DN_{xx} - YG + \gamma Y \int_{[x-vt]_+}^{x} M\left(\xi, t - \frac{x-\xi}{v}\right) F\left(\frac{x-\xi}{v}\right) \frac{d\xi}{v} \tag{2.1}$$

$$B_t = G - \gamma M - q_x, \tag{2.2}$$

where D is the nutrient diffusion coefficient and all other quantities have been described in (I)-(IV).

Remark 2.1 A more precise form of the absorption term YG in (2.1) is $YGH(N)$, with $H(N) = 1$ for $N > 0$ and $H(N) = 0$ otherwise. Of course such a correction becomes irrelevant if we can prove a-priori that $N(x)$ is positive.

The system above must be supplemented with initial and boundary conditions for N, B.

The boundary conditions for N need some explanation. At the botton of the cylinder, $x = h$, there is an accumulation of dead cells, continuously releasing nutrient. Therefore the entering flux at $x = h$ is

$$DN_x(h,t) = \gamma Y \int_0^t F(\tau) \int_{[h-v\tau]_+}^h M(x, t-\tau)\, dx\, d\tau. \tag{2.3}$$

In what follows we shall neglect the biomass current q, so that (2.2) simplifies to

$$B_t = G - \gamma M \tag{2.2'}$$

and we shall look for nontrivial equilibrium solutions.

3 The steady state equation

The biomass balance equation (2.2') reduces at equilibrium to $G = \gamma M$, i. e., recalling (1.6), (1.7) to

$$B[Af(i)U(N) - aB - \gamma] = 0, \quad B \geq 0, \tag{3.1}$$

while the equilibrium equation for the nutrient can be written as

$$\frac{D}{\gamma Y} N''(x) = M(x) - \int_0^x M(\xi) F\left(\frac{x-\xi}{v}\right) \frac{d\xi}{v}. \tag{3.2}$$

As we shall see in Proposition 3.1 below, the constraint $N \geq 0$ is automatically satisfied. Recalling (2.3) and (1.8), the entering flux at $x = h$ is

$$\begin{aligned} DN'(h) &= \gamma Y \int_0^\infty F(\tau) \int_{[h-v\tau]_+}^h M(x)\, dx\, d\tau \\ &= \gamma Y \int_0^h \left(\frac{1}{E} - \int_0^{\frac{h-x}{v}} F(\tau)\, d\tau\right) M(x)\, dx > 0. \end{aligned} \tag{3.3}$$

Therefore, integrating (3.2) along the cylinder, one obtains the cumulative nutrient balance at equilibrium

$$DN'(0) = -\left(1 - \frac{1}{E}\right) \gamma Y \int_0^h M(x)\, dx. \tag{3.4}$$

The remaining condition to be prescribed at $x = 0$ is

$$N(0) = N_0 > 0. \tag{3.5}$$

The problem consists in finding a pair of continuous functions (B, N) solving (3.1), (3.2) with the initial conditions (3.4), (3.5). The additional condition (3.3) will be satisfied automatically.

The first result we want to prove is the positivity of $N(x)$.

Proposition 3.1 *For any solution* (B, N) *of* (3.1)-(3.5) *the function* $N(x)$ *is positive everywhere in* $[0, h]$.

Proof. We assume that (B, N) is a solution and we suppose that $N(x)$ vanishes (for the first time) at some point $x_0 \in (0, h)$. We remark that in the vicinity of x_0 and everywhere N is nonpositive B has to be zero because of (3.1), and that on the other hand $B \equiv 0$ is consistent only with $N \equiv N_0$. Therefore there must be an interval on the left of x_0 where B, and hence M, is positive, implying from (3.2) that N'' is strictly negative in the zero set of B. Thus N cannot have a negative minimum (nor it can vanish identically over some interval) and we would have $N'(h) < 0$, contradicting (3.3).

The above result justifies the form in which (3.1), (3.2) are written, i. e. not inserting the constraint $N \geq 0$.

Remark 3.1 The system (3.1), (3.2), (3.4), (3.5) always admits the solution $B \equiv 0,\ N \equiv N_0$.

From now on we shall assume $E = 1$, thus simplifying (3.4) to

$$N'(0) = 0. \tag{3.6}$$

We will come back to the case $E > 1$ in the last section.

For $E = 1$ a first integration of (3.2) gives

$$\frac{D}{\gamma Y} N'(x) = \int_0^x M(\eta)[1 - \hat{F}(x, \eta)]\, d\eta$$

with $\hat{F}(x, \eta) = \int_\eta^x F\left(\frac{\xi - \eta}{v}\right) \frac{d\xi}{v}$, i. e.

$$\frac{D}{\gamma Y} N'(x) = \int_0^x M(\eta)\left(1 + \alpha \frac{x - \eta}{v}\right) exp\left(-\alpha \frac{x - \eta}{v}\right) d\eta. \tag{3.7}$$

Remark 3.2 Of course (3.7) is compatible with (3.3) for any h and for $E = 1$ we have a relevant simplification. Indeed for $E \neq 1$ the initial condition (3.4) is nonlocal, while for $E = 1$ $N(x)$ satisfies standard Cauchy conditions for $x = 0$ and there is no relevant influence of the parameter h.

A first consequence is

Proposition 3.4 *For* $E = 1$ *the function* $N(x)$ *is nondecreasing.*

Integrating (3.7) and recalling (3.1), we can write down the final form of the steady state equations for possible nontrivial solutions.

$$B(x) = \frac{1}{a}[Af(i)U(N) - \gamma]_+, \tag{3.8}$$

$$N(x) = N_0 + \frac{\gamma Y}{D} \int_0^x \int_0^\xi M(\eta)[1 - \hat{F}(\xi, \eta)]\, d\eta\, d\xi. \tag{3.9}$$

We recall from (1.1) that i contains the integral $\int_0^x B(\xi)\, d\xi$.

Remark 3.3 From (3.8) the value of $B(0)$ can be deduced, and it is positive if

$$Af(i_0)U(N_0) > \gamma. \tag{3.10}$$

4 Equilibrium without light

Before proceeding to the study of (3.8), (3.9) we consider briefly the case $i_0 = 0$, applying the modified growth rate expression

$$G = A\, min[r_0, U(N)]\, B \tag{4.1}$$

and assuming for simplicity that $U(N_0) \geq r_0$, so that (4.1) reduces to $G = Ar_0B$.

Provided $Ar_0 > \gamma$, we find a uniform positive equilibrium solution

$$B = \frac{(Ar_0 - \gamma)}{a}. \tag{4.2}$$

The expressions of M and N follow easily.

5 Eutrophication ($U \simeq 1$)

When $N_0/L \gg 1$ the uptake function $U(N)$ can be approximated by 1 in (3.8), thus obtaining a nonlinear Volterra integral equation for B:

$$B(x) = \frac{1}{a}[Af(i) - \gamma]_+, \tag{5.1}$$

$$i = i_0\, exp\left[-\sigma_w x - \sigma_c \int_0^x B(\xi)\, d\xi\right].$$

(A) Case $f(i) = i$.

The positivity condition (3.10) is in this case $Ai_0 > \gamma$.

Setting

$$\omega(x) = \sigma_w x + \sigma_c \int_0^x B(\xi)\, d\xi \tag{5.2}$$

and

$$y(x) = e^{\omega(x)}, \tag{5.3}$$

from (5.1) we obtain the differential equation

$$y' + \beta y = \alpha, \quad y(0) = 1 \tag{5.4}$$

with

$$\alpha = Ai_0\sigma_c/a, \quad \beta = \frac{\gamma}{a}\sigma_c - \sigma_w. \tag{5.5}$$

Hence we get $y(x)$ and eventually $B(x)$:

$$B(x) = \frac{1}{a}\left[Ai_0\frac{\beta e^{\beta x}}{\beta + \alpha(e^{\beta x} - 1)} - \gamma\right]_+. \tag{5.6}$$

Thus, if $Ai_0 > \gamma$ (implying $\alpha > \beta$), $B(x)$ is decreasing and positive up to the depth

$$\tilde{h} = \frac{1}{\beta}\,log\frac{\gamma}{Ai_0}\left(\frac{Ai_0 - \gamma}{a}\frac{\sigma_c}{\sigma_w} + 1\right) \tag{5.7}$$

if $h > \tilde{h}$, and positive everywhere if $h < \tilde{h}$.

Remark 5.1 $\tilde{h}$ is an increasing function of i_0, tending to $\tilde{h}_\infty = \frac{1}{\beta}\,log\frac{\gamma\sigma_c}{a\sigma_w}$ as $i_0 \to \infty$. For any $h < \tilde{h}_\infty$ there is one unique value of i_0 such that $\tilde{h} = h$, i. e. $B(x) > 0$ for $0 < x < h$.

(B) More general $f(i)$.

Both for (1.3) and for (1.5) we have $f(i) \leq 1$ with the only maximum for $i = 1$.
From (5.1) we deduce that where $B > 0$ B' has the opposite sign of f':

$$B'(x) = -\frac{A}{a}f'(i)(\sigma_w + \sigma_c B(x))i(x). \tag{5.8}$$

In particular B will reach a maximum at the point (if it exists) where $i = 1$. (and $f(i) = 1$). Obviously the maximum value of B will be $\frac{A-\gamma}{a}$.

It is easy to describe the dependence of the solution on i_0. Let $i_- < i_+$ be the roots of the equation $Af(i) = \gamma$ (for (1.5) $i_\pm = 1 \pm (1 - \gamma/A)^{1/2}$).

Then

(a) for $i_0 \leq i_-$ there exists only the trivial solution;

(b) for $i_- < i_0 < i_+$ we have one unique solution which remains positive until $i(x)$ reaches the value i_-, the same is true for $i_0 = i_+$, since $B(0) = 0$ but we can select a solution such that $B'(0) > 0$; from (5.1) it is easy to realize that varying i_0 in this range produces a rigid displacement of the biomass profile;

(c) for $i_0 > i_+$ we have $B \equiv 0$ up to the depth x^* such that $i(x^*) = i_+$, i. e. $x^* = \frac{1}{\sigma_w}\,log(i_0/i_+)$. Then, if $x^* < h$, we have $B'(x^*) > 0$, since $f'(i_+) < 0$, i. e. $B > 0$ in a right neighborhood of x^*. Since for $x > x^*$ we can write

$$i(x) = i_+\,exp\left[-\sigma_w(x - x^*) - \sigma_c\int_{x^*}^{x} B(\xi)\,d\xi\right],$$

it is immediately seen that if $B(x, x^*)$ denotes the solution above, then $B(x, x^* + y^*) = B(x - y^*, x^*)$. This means that also in this case the equilibrium profile of the biomass translates when i_0 varies.

All the profiles described above are illustrated in Fig. 1.

In both cases (b) and (c) B will vanish identically in the region where i drops below i_- (if h is large enough).

For the quadratic form (1.5) of $f(i)$, eq. (5.1) can be written in the form

$$i' = -i\left[\beta + \frac{\alpha}{i_0}(2i - i^2)\right]. \tag{5.9}$$

and can be integrated in the region $B > 0$.

For instance in the case (b), defining

$$\left[1 \pm \sqrt{\left(1 - \frac{\gamma}{A}\right) + \frac{a}{A}\frac{\sigma_w}{\sigma_c}}\right]\frac{1}{i_0} = p_\pm \tag{5.10}$$

we have for $p = i/i_0$ the implicit solution

$$(p_+ - p_-)log\, p(x) + p_- log\left|\frac{p(x) - p_+}{1 - p_+}\right| - p_+ log\left|\frac{p(x) - p_-}{1 - p_-}\right| = p_+ p_- (p_+ - p_-)\alpha i_0 x. \tag{5.11}$$

Remark 5.2 From (5.11) we can obtain the location of the maximum of B taking $p = 1/i_0$. Clearly $B(x)$ is expressed in terms of $i(x)$ by the formula

$$B = -\frac{1}{\sigma_c}\left(\sigma_w + \frac{i'}{i}\right). \tag{5.12}$$

In the case (c) eq. (5.9) must be integrated for $x > x^*$ with the condition $i(x^*) = i_+$. Therefore the result is similar to (5.11) but with i_+ replacing i_0 in the definition of p, p_+, p_- and with the product $i_0 x$ in the righ hand side of (5.11) substituted by $i_+(x - x^*)$.

6 Equilibrium in the presence of nutrient dynamics ($U < 1$)

We have to study the system (3.8), (3.9). Again we deal first with the case $f(i) = i$.

(A) Case $f(i) = i$

Following the procedure of Sec. 5(A) and supposing $U(N) > 0$ is known one can obtain the solution of (3.8) in the form

$$B(x) = \frac{1}{a}\left\{Ai_0 U(N(x))e^{\beta x}\left[1 + \alpha\int_0^x e^{\beta\xi}U(N(\xi))\,d\xi\right]^{-1} - \gamma\right\}_+ \tag{6.1}$$

(α, β given by (5.5)).

Therefore it is natural to look for the solutions as a fixed point of the mapping $M \to N \to B \to M$.

To this purpose we define the following subset of $C([0,h])$:

$$X(Z) = \{M \in C([0,h]) : 0 \le M \le Z\}$$

for any positive constant Z. The norm in $C([0,h])$ will be denoted by $\|\cdot\|$.

Taken $M \in X(Z)$ we define $N(x) > N_0$ using (3.9) and $B(x)$ using (6.1). Note that Proposition 3.4 applies to N even if M is not the solution, but is just a function in $X(Z)$. Then we have the mapping $\mathcal{F}: X(Z) \to C([0,h])$

$$\mathcal{F}M = \tilde{M} \equiv B + \frac{1}{\gamma}B^2. \tag{6.2}$$

Of course any fixed point of $\mathcal{F}$ is a solution to our problem.

Since $0 < U(N) < 1$, from (6.1) we obtain the a-priori bound

$$0 \le B(x) \le \frac{Ai_0}{a}e^{\beta h} \equiv \lambda, \tag{6.3}$$

implying

$$0 \le \tilde{M} \le \lambda + \frac{a}{\gamma}\lambda^2. \tag{6.4}$$

Therefore, if we take $Z_0 = \lambda + \frac{a}{\gamma}\lambda^2$, $\mathcal{F}$ maps $X(Z_0)$ into itself and in addition N, B and $\tilde{M}$ are twice continuously differentiable where $B > 0$. Moreover it can be seen immediately that N', B' and consequently $\tilde{M}'$ are bounded in $[0,h]$ uniformly with respect to the choice of M in $X(Z_0)$. Then the image $\mathcal{F}\ (X(Z))$ is relatively compact in $X(Z_0)$. Next we see that $\mathcal{F}$ is continuous. This follows from the following chain of inequalities, in which C_1, C_2, C_3 stand for computable constants, depending only on the characteristics parameters of the system, as long as M_1, M_2 are taken in $X(Z_0)$:

$$\begin{aligned}
\|N_1 - N_2\| &\le C_1\|M_1 - M_2\|,\\
\|B_1 - B_2\| &\le C_2\|N_1 - N_2\|,\\
\|\tilde{M}_1 - \tilde{M}_2\| &\le C_3\|B_1 - B_2\|.
\end{aligned}$$

Note that C_1 can be made as small as desired by reducing h.

We conclude that $\|\tilde{M}_1 - \tilde{M}_2\| \le c(h)\|M_1 - M_2\|$, with $c(h) \to 0$ as $h \to 0$. Thus $\mathcal{F}$ is completely continuous, and since $X(Z_0)$ is bounded, closed and convex, Schauder's theorem assures the existence of at least one fixed point.

Again we remark that if h is less than some h_0 then $\mathcal{F}$ is contractive and uniqueness follows. Such a property can be exploited in order to infer uniqueness for any value of h. Indeed we can solve the problem up to $x = h_0$ and reset the initial values of N and N': one can easily realize that the solution has a unique continuation up to $x = 2h_0$, and so on.

We have now to provide conditions ensuring the positivity of the solution.

Since $N(x)$ is increasing, we can say that $U(N_0) \le U(N) < 1$ and this enables us to estimate the first term in brackets in (6.1) from above and from below, obtaining

$$\frac{1}{a}[\varphi(x) - \gamma]_+ \le B(x) \le \frac{1}{a}[\psi(x) - \gamma]_+ \tag{6.5}$$

with

$$\varphi(x) = Ai_0 U(N_0) \frac{e^{\beta x}}{1 + \frac{\alpha}{\beta}(e^{\beta x} - 1)},$$
$$\psi(x) = Ai_0 \frac{e^{\beta x}}{1 + \frac{\alpha}{\beta} U(N_0)(e^{\beta x} - 1)}. \tag{6.6}$$

In this way not only we recover the obvious positivity condition

$$Ai_0 U(N_0) > \gamma, \tag{6.7}$$

but also we can find lower and upper estimates of the positivity interval of $B(x)$.

Finally, the smothness of B and N in the set $\{B > 0\}$ can be inferred by induction from the particular structure of the system (3.8), (3.9).

We can sumarize the results above in the following

Theorem 6.1 *Under the assumption* (6.7) *there exists one unique nontrivial solution* (B, N). *Moreover* $N \in C^\infty, B \in C^\infty$ *in the positivity interval of* B *and the estimates* (6.5) *hold true.*

(B) More general $f(i)$

For $f(i)$ given e. g. by (1.3) we no longer have the possibility of integrating (3.8). However we can observe that since $f(i) \leq 1$ and $U(N) < 1$, in any case from (3.8) we have the a-priori estimate

$$\|B\| \leq (A - \gamma)/a \equiv H \tag{6.8}$$

(we always have to suppose $A > \gamma$).

Therefore we introduce the set

$$\Sigma(H) = \{B(x) \in C([0, h]) : \|B\| \leq H\} \tag{6.9}$$

and for any $B \in \Sigma(H)$ we define $M(x)$ by means of (1.7), $N(x)$ by means of (3.9), and finally the mapping

$$JB = \tilde{B} = \frac{1}{a}[Af(i)U(N) - \gamma]_+, \tag{6.10}$$

where i is obtained by inserting the selected specification of B in (1.1).

Clearly, J maps $\Sigma(H)$ into itself. Since $i'(N), N'(x)$ are uniformly bounded as B varies in $\Sigma(H)$, $\tilde{B}$ is uniformly Lipschitz continuous. Moreover, for any pair B_1, B_2 in $\Sigma(H)$ we have the inequalities

$$\|M_1 - M_2\| \leq d_1 \|B_1 - B_2\|$$
$$\|N_1 - N_2\| \leq d_2 \|M_1 - M_2\|$$
$$\|\tilde{B}_1 - \tilde{B}_2\| \leq d_3 \|B_1 - B_2\| + d_4 \|N_1 - N_2\|,$$

with d_1, d_2, d_3, d_4 independent of (B_1, B_2) and d_3, d_4 proportional to h.

Therefore $||\tilde{B}_1 - \tilde{B}_2|| \leq d(h)||B_1 - B_2||$ and $d(h) = O(h)$.

As in the previous case we can deduce the existence of one unique smooth solution for any value of h.

Now we concentrate our attention on the conditions ensuring the positivity of the biomass B.

For any fixed N_0 let us define $j_- < j_+$ as the roots of

$$f(j_\pm) = \frac{\gamma}{AU(N_0)} \tag{6.11}$$

and we distinguish the usual three cases.

(a) For $i_0 \leq j_-$ the light intensity is not sufficient to sustain any equilibrium biomass throughout the column.

(b) For $j_- < i_0 < j_+$ we have a nontrivial solution taking the value $B(0) = \frac{1}{a}[Af(i_0)U(N_0) - \gamma]$; for $i_0 = j_+$ the initial value of B is zero, but $B'(0) > 0$ (discarding the trivial solution). We can say that for $j_- < i_0 \leq j_+$ a nontrivial solution exists until i becomes so small that the equation $Af(i)U(N) = \gamma$ is satisfied at some point $x = \bar{x}$. Since $U(N)$ is increasing, this necessarily happens for a value of $f(i)$ less than $f(j_\pm)$, i. e. for a value of i less than j_- being $i(x)$ decreasing. Calculating

$$B' = \frac{A}{a}[f'(i)i'U + f(i)U'(N)N'], \tag{6.12}$$

we realize that $B'(x) < 0$ and that B will remain identically zero for $x > \bar{x}$ (of course the argument applies if $h > \bar{x}$).

(c) For $i_0 > j_+$ B is zero and N is constant as long as i decreases to the value j_+. Then a positive solution starts with $B' = \frac{A}{a}|f'(j_+)|U(N_0)\sigma_w j_+ > 0$ which may vanish again as in the case (b). Increasing i_0 produces a shift of the biomass profile downwards.

The smoothness of the solution is as evident as in the case (A).

Therefore we can state the following theorem

Theorem 6.2. *If $j_- < i_0 \leq j_+$ there exists a unique nontrivial solution B, N. If $i_0 > j_+$ we may have a positivity interval for B, provided the column is long enough so that i can reach the value j_+ by water absorption. Varying i_0 in the range $i_0 > j_+$ produces a rigid displacement of the equilibrium profile of B. In any case $B, N \in C^\infty$ in the set $\{B > 0\}$, B is Lipschitz continuous and N has Lipschitz continuous second derivative in $[0, h]$.*

Of course more details can be obtained taking for $f(i)$ the quadratic form (1.5).

7 The case $E > 1$

When $E > 1$ we must replace the homogeneous condition (3.6) with the nonlocal condition (3.4).

For the sake of brevity we confine ourselves to the case treated in Sect. 6(B). The main change is due to the fact that we cannot guarantee the positivity of $N(x)$ for any choice of B in $\Sigma(H)$. However this difficulty can be overcome with a slight modification of the definition of J in (6.10):

$$\tilde{J}B = \frac{1}{a}[Af(i)U([N]_+) - \gamma]_+. \tag{7.1}$$

The rest of the fixed point argument needs minor changes. Thus we have found a solution to the modified problem (3.1), with N replaced by $[N]_+$, (3.2), (3.4), (3.5). The same arguments used in the proof of Proposition 3.1 show that such a solution has the property $N(x) > 0$, $\forall x \in [0, h]$ and therefore it is also a solution of the original problem. The existence of nontrivial solutions follows.

The only point which fails in Theorem 6.2 is the continuation argument (ensuring uniqueness for arbitrary h), because of the nonlocal condition (3.4).

Conclusions

The model proposed includes many of the relevant aspects of phytoplankton growth dynamics and leads to a description of positive equilibrium solutions reproducing the qualitative behaviour observed in the experiments. A further investigation will be concerned with a more detailed description of the cells migration.

Other problems to be studied are the stability of the equilibrium solutions, the time dependent case and the influence of temperature variations.

Appendix

The sea phytoplankton concentration is measured in 10^6 *cells/ℓ* or otherwise in the equivalent concentration of Clorophyll-a, i. e. $\mu gr.Clor.a/\ell$.

The conversion factor is:

$$4\ 10^{-6}\ \mu gr.Clor.a/cells$$

TABLE 1

Symbol	*Quantity*	*Value*	*Source*
θ	Environmental temperature	$22^0 C$	lab.
A	Maximum growth rate under optimal conditions of enlightening and saturating	$0.63\ day^{-1}$	lab.

	nutrient condition ($\theta = 22^0 C$)		
γ	Death rate constant:	$0.5\ day^{-1}$	lab.
H	Maximum biomass concentration in the text tube under optimal conditions	$150\ 10^6 cells/\ell$	lab.
a	$a = \frac{1}{H}(A - \gamma)$: overcrowding death rate	$0.22 \left(\frac{cm^3}{\mu gr.Clor.a}\right) day^{-1}$	lab.
α	$\alpha = \frac{2}{T}$, T mean decomposition time. At $\theta = 22^0 C\ T \simeq 1\ day$	$2 day^{-1}$	lab.
Y	average nutrient concentration in a cell (Phosphorus P) Y_P. $Y_N = 16Y_P,\ Y_C = 106Y_P$	$0.6125 \frac{10^3 \mu gr P/\ell}{10^6 cells/\ell}$	[7]
D	diffusion coefficient for the limiting nutrient (P)	$0.86 \frac{cm^2}{day}$	[8]
D_0	diffusion coefficient for the biomass		
v	Steady falling rate of dead microalgaes (diatoms)	$0.46 \frac{m}{day}$	p.
σ_w	Light extinction coefficient of water in the absence of biomass	$3\ 10^{-2} m^{-1}$	[9]
σ_c	Light extinction coefficient by Clorophyll-a in phytoplankton	$0.8 \left(\frac{cm^3}{\mu gr.clor.a}\right) m^{-1}$	lab.
I_m	the optimum of the light intensity	$3.6\ 10^4 \frac{erg}{cm^2 sec.}$	[5, 6]
L	Michaelis-Menten constant (or half saturation constant) for the phosphorus uptake function	$10 \mu gr\ P/\ell$	[5]

lab: Centro Universitario per le Risorse Biologiche e Marine - Cesenatico - Regione Emilia-Romagna - Italy.

p: Estimation of the authors assuming a radius for diatoms cells of $5\ 10^{-4} cm$ (see [5]).

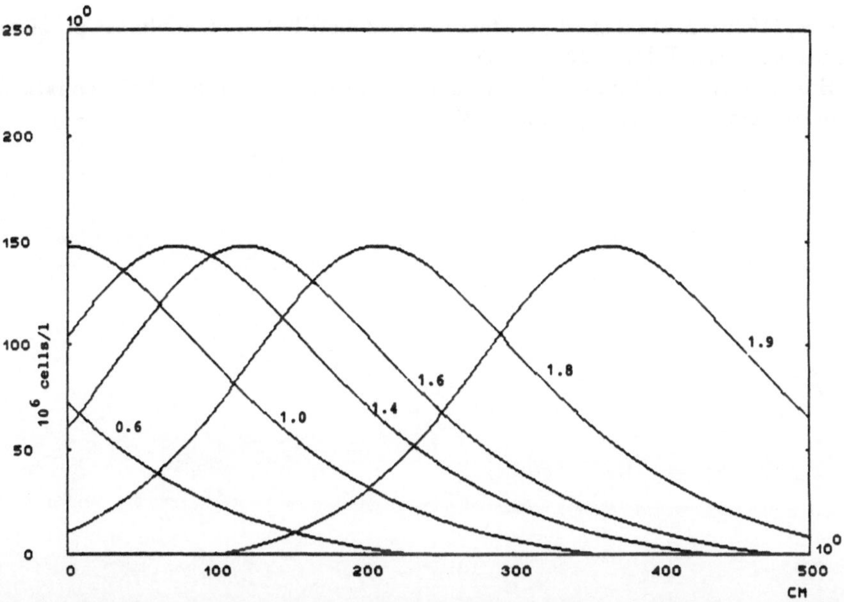

Fig. 1. Case $U = 1$, $f(i)$ as in (1.3) (see. Sect 3). Biomass profiles for i varying from 0.6 to 1.9. Required parameters from Table 1. $j_- = 0.465$. $j_+ = 1.842$.

Acknowledgment

Work partially supported by the Italian National Project "Equazioni di evoluzione e applicazioni fisico-matematiche" (M.U.R.S.T.).

References

1. Di Toro, D.M. , O'Connor, D.J., Thoman, R.V. (1985): Phytoplankton-Zooplankton-Nutrient Interaction Model for Western Lake Erie. System Analysis And Simulation in Ecology. Patten ed., Academic Press, 423-474
2. Handbook of Enviromental Data and Ecological Parameters. Editor S.E. Jorgensen. Distributed by: International Society for Ecological Modeling, Pergamon Press (1980)
3. Hull, H., Lagonegro, M., Puccia, C.J. (1988): Modellizzazione di Sistemi Ecologici Complessi. Ed. C.L.U.P., Milano
4. Ishii, H. , Takagi, I., (1982): Global stability of stationary solutions to a nonlinear diffusion equation in phytoplankton dynamics. J. Math. Biol. **16**, 1-24
5. Mitchell, J.G., Okubo, A., Furham, J.A., (1985): Microzones surrounding phytoplankton form the basis for a stratified marine microbial ecosystem. Nature **316**, 58-59
6. Powell, T., Richerson, P.J., (1985): Temporal variation, spatial heterogeneity and competition for resources in plankton systems: a theoretical model. The American Naturalist **125**, 431-464
7. Shigesada, N., Okubo, A., (1981): Analysis of the self-shading effect on algal vertical distribution in natural waters. J. Math. Biol. **12**, 311-326

8. Totaro, S., (1989): Mutual shading effect on algal distribution: a nonlinear problem. Nonlinear Analysis, T.M.A. **13**, 969-986
9. Wörz-Busekros A., (1976): Solutions to a degenerate system of parabolic equations from marine biology. J. Math. Biol. **3**, 393-406

Some Delay Models for Juvenile vs. Adult Competition

J. M. Cushing

Department of Mathematics, University of Arizona, Tuscon, Arizona 85721

1 Introduction

Models of competition have played a central role in theoretical population dynamics and ecology. The vast majority of mathematical models of competitve interactions that have been formulated have been done so with regard to highly aggregate state variables at the total population level and have ignored differences between individual organisms, in effect treating all individuals of a species as identical. Biological populations generally consist, however, of individuals with diverse physiological characteristics, such as age, body size or weight, life cycle stages, etc., (with intra-species variances that in fact can exceed inter-specific variances amongst competing species) and it has become widely recognized that this diversity can have a significant influence upon population level dynamics (Werner and Gilliam (1984), Ebenman and Persson (1988), Metz and Diekmann (1986)). Models that ignore individual level physiological variances cannot, except in the simplest of cases, adequately account for the mechanisms that result in competition between individual organisms for limited resources. In particular, intra-specific competition can be accounted for in such models only in highly qualitative ways at best.

One common type of intra-specific competition found in species with overlapping generations occurs between different age or size classes within the population, in particular between (younger or smaller) juveniles and (older or larger) adults. In species with complex life cycles during which individuals undergo radical changes in morphology, physiology, or behavior (e.g. holometabolous insects, amphibians, and many marine invertebrates), there are often significant changes in resource utilization during the life cycle. For such species, these ontogenetic niche shifts lessen or even exclude possibility of competition between adults and juveniles. However, the potential for juvenile vs. adult competition is increased in species that have simpler life cycles, during which individuals undergo fewer changes during their life history. This is the case for most fishes, birds, mammals, many plants, and for hemimetabolous insects (Ebenman (1987, 1988)). Some questions that arise concerning such species included: what effects do strong competitive interactions between juveniles and adults have on the dynamics of the population? Do such interactions stabilize or destabilize the population and in what sense?

In an early paper, May et al. (1974) use a simple two-age class difference equation model to conclude that juvenile vs. adult competition has a destabilizing effect on population equilibria. This conclusion was also reached by Tschumy (1982) using a simple differential equation model. Ebenman (1987), however, uses a simple two-age difference model to argue that destabilization was not the only possibile effect of strong juvenile vs. adult competition (also see Ebenman (1988), Ebenman and Persson (1988), Cushing and Li (1989)). Ebenman distinguishes between suppressed adult fertility due to strong competition from juveniles and increased juvenile mortality due to competition with adults and he tabulates "all the studies known to (him), where even crude estimates of relevant parameters can be made" and demonstrates two opposing trends (Table 2, Ebenman (1988)). First, in species that possess adult density dependent juvenile survival, population level dynamics progress from "unstable" oscillatory dynamics to stable equilibrium dynamics as one passes from species with weak to species with strong juvenile vs. adult competition. Conversely, in species that possess juvenile density dependent adult fertility, the opposite is true, i.e. population dynamics is destabilized from equilibration to oscillations as one passes from weak to strong juvenile vs. adult competition. These observations were commensurate with the conclusions drawn from his simple model.

Cushing and Li (1989) study Eberman's model in more mathematical detail and argue, in agreement with Eberman, that strong juvenile vs. adult competition can be either stabilizing or destabilizing. They point out, however, that it is difficult to make sweeping conclusions over wide parameter ranges and that any conclusions drawn depend very much upon what is meant by "stabilization" or "destabilization". Different criteria can lead to contradictory conclusions. May et al. and Eberman study the stabilization/destabilization question with regard to the relative sizes of stability regions in certain parameter spaces. Tschumy (1982), on the other hand, uses a criterion based upon linearized eigenvalue sensitivity or changes in equilibrium levels. Loreau (1990) criticizes Eberman's analysis (and implicitly that of May, et al.) and, amongst other things, points out that if the linearized eigenvalues are analyzed for his model one finds that increaed competition is always destabilizing (also see Cushing and Li (1989)). We will see here, however, that for more sophisticated continuous time-age variable models, stabilization can indeed occur under the eigenvalue sensitivity criterion.

In this paper we will study the stabilization/destabilization question by models derived using the now standard modeling methodology for age-structured popultion dynamics (Metz and Diekmann (1986)). In Sect. 2 we will derive, under certain simplifying assumptions, differential delay equations for the dynamics of populations exhibiting the two types of juvenile vs. adult competition distinguised by Eberman. These two cases, namely juvenile density dependent adult fertility and adult density dependent juvenile survivability, will be considered independently in Sect. 3. The existence and stability of positive equilibrium states will be studied and eigenvalue sensitivity to changes in competition coefficients will be used to investigate the stabilization/destabilization question.

2 Model derivation

Using what is now standard methodology for modeling age-structured poulations (Hoppensteadt (1975), Gurtin and MacCamy (1974), Metz and Diekmann (1986)), we consider a population described by an age specific density $\rho = \rho(t,a)$ whose dynamics are governed by the equations

$$\partial_t \rho + \partial_a \rho + \mu\rho = 0, \tag{2.1a}$$

$$\rho(t,0) = \int_0^\infty f\rho da, \tag{2.1b}$$

$$\rho(0,a) = \phi(a) \tag{2.1c}$$

where $t > 0$ is time, $a > 0$ is age, $\mu \geq 0$ and $f \geq 0$ are the unit density mortality and fertility rates respectively, and $\phi \geq 0$ is the initial age distribution. To model the density dependent dynamics of a population one must prescribe how the vital rates μ and f depend on ρ.

In this paper we are interested in populations which consist of immature juveniles and mature adults. It will be assumed that maturity is determined strictly by age, i.e. there exists an age $m > 0$ such that individuals of age $a < m$ are immature and individuals of age $a > m$ are mature. We can then structure the population into two subgroups, namely juveniles and adults whose total numbers are given by the integrals

$$J(t) = \int_0^m \rho(t,a)da, A(t) = \int_m^\infty \rho(t,a)da,$$

respectively. As discussed in Sect. 1, we are interested in the case when competition between these two subgroups is expressed through either increased juvenile mortality and/or decreased adult fertility. Therefore we assume

$$\begin{gathered} 0 \leq \mu = \mu(a, W(t)), W(t) = J(t) + \gamma A(t) \\ 0 \leq f = r\beta(a, V(t)), V(t) = \alpha J(t) + A(t) \\ \partial_W \mu(a,W) \geq 0, \quad \partial_V \beta(a,V) \leq 0, \quad \beta(a,V) = 0 \quad \text{for} \quad 0 \leq a \leq m. \end{gathered} \tag{2.2}$$

It is also assumed that the normalization

$$\int_0^\infty \beta(a,0)\exp\left(-\int_0^a \mu(s,0)ds\right)da = 1 \tag{2.3}$$

is satisfied so that the constant $r \geq 0$ is the "inherent net reproductive rate", i.e. the expected number of offspring per individual per lifetime in the absence of density effects (Cushing (1985)). Here the competition coefficient γ measures the relative effect that an adult individual has on juvenile survival as compared to that of a juvenile. Similarly the coefficient α measures the relative effect that a juvenile individual has on adult fertility as compared to that of an adult individual. The last line in (2.2) expresses the fact that the death rate increases and the fertility rate decreases with increases in population numbers and that no individual of age less than m is fertile.

In this paper we will be interested in the existence and stability of (non-negative) equilibrium solutions of equations (2.1) and how equilibrium stability depends upon the competition coefficients γ and α. Obviously these equations possess the trivial equilibrium $\rho = 0$ ($J = A = 0$). It follows from general results of Cushing (1985) that there exists an unbounded continuum of equilibrium pairs (ρ, r) that bifurcates from the trivial solution pair $(0,1)$ and consists of positive equilbria (except for $(0,1)$). Moreover, because of conditions (2.2), these positive equilibria are locally asymptotically stable at least for r greater than, but close to the critical value 1. In a forthcoming paper, Cushing and Li (1990) study the sensitivity of the stability of these positive equilibria on the competition coefficients γ and α. Here we will study this question for some specializations of equations (2.1) from which can be derived differential delay equations for A and J, or more precisely for A and the total population size $P = J + A = \int_0^\infty \rho(t,a)da$.

We will study the effects of density on juvenile mortality and on adult fertility separately by considering the following two models. For the case of density dependent adult fertility we set

$$\mu = \begin{Bmatrix} \mu_J, 0 < a < m \\ \mu_A, m < a \end{Bmatrix}, \beta = \begin{Bmatrix} 0, 0 < a < m \\ \beta_A(V), m < a \end{Bmatrix}, V = \alpha J + A \tag{2.4}$$

where $\mu_A > 0$, $\mu_J > 0$ are constants and $\beta_A(V) \geq 0$, $\partial_V \beta_A(V) \leq (0)$ for all $V \geq 0$. (For technical reasons we define $\beta_A(V) = \beta_A(0)$ for $V < 0$.) Thus, in this case, the death rate is assumed independent of weighted population size V. Moreover the death rates of juveniles and of adults are assumed constant (independent of an individual juvenile's or adult's age), although these two death rates need not necessarily be the same. The normalization (2.3) requires that

$$\beta_A(0) = \mu_A \exp(\mu_J m). \tag{2.5}$$

For the second case of density dependent juvenile mortality we set

$$\mu = \begin{Bmatrix} \mu_J(W), 0 < a < m \\ \mu_A, m < a \end{Bmatrix}, \ \beta = \begin{Bmatrix} 0, 0 < a < m \\ \beta_A, m < a \end{Bmatrix}, \ W = J + \gamma A \tag{2.6}$$

where $\mu_A > 0$, $\beta_A > 0$ are constants and $\partial_W \mu_J(W) \geq 0$ for all $W \geq 0$, $\mu_J(0) > 0$. (For technical reasons we define $\mu_J(W) = \mu_J(0)$ for $W < 0$.) In this case the juvenile death rate is independent of age, but dependent on population size W. The adult fertility rate is a constant, independent of both age and population size. The normalization (2.3) requires

$$\beta_A = \mu_A \exp(\mu_J(0)m). \tag{2.7}$$

Under these assumptions we can derive differential equations for the rates of change of $J(t)$ and $P(t)$ by integrating (2.1a) with repect to a from 0 to m and 0 to ∞ respectively. Before doing this we point out that an integration of (2.1a) along characteristics, when $\mu = \mu(t,a)$ and $\beta = \beta(t,a)$ yields

$$\rho(t,a) = \begin{cases} \phi(a-t)\exp(-\int_0^t \mu(s,a-t+s)ds), 0 \le t \le a \\ \rho(t-a,0)\exp(-\int_0^a \mu(a-t+s,s)ds), a < t \end{cases}$$

which implies for the adult fertility problem (2.4) that

$$\rho(t,m) = \begin{cases} \phi(m-t)\exp(-\mu_J t), 0 \le t \le m \\ r\beta_A(V(t-m))A(t-m)\exp(-\mu_J m), m < t \end{cases} \tag{2.8}$$

and for the juvenile mortality problem (2.6) that

$$\rho(t,m) = \begin{cases} \phi(m-t)\exp(-\int_0^t \mu_J(W(s)ds), 0 \le t \le m \\ r\beta_A A(t-m)\exp(-\int_0^m \mu_J(W(t-m+s))ds), m < t \end{cases} \tag{2.9}$$

These formulas describe maturation rates for the two problems, i.e. the rate at which juveniles reach age m. For example, in (2.8) the maturation rate during the interval $0 \le t \le m$ is simply the initial density $\phi(m-t)$ of individuals of age $m-t$ times the probability $\exp(-m_J t)$ of surviving t units of time. Afterwards, for $t > m$, the maturation rate is the birth rate $r\beta_A(V(t-m))A(t-m)$ at time $t-m$ multiplied by the probability $\exp(-\mu_J m)$ of surviving to maturation age m. (2.9) has a similar interpretation.

Under the assumption that ρ vanishes at $a = \infty$, integrations of (2.1a) with respect to a from 0 to ∞ and from m to ∞ yield

$$P' - \rho(t,0) + \int_0^\infty \mu\rho da = 0, \quad A' - \rho(t,m) + \mu_a A = 0.$$

These equations together with (2.1b) and (2.8)-(2.9) in turn yield the following system of delay equations

$$P' = -\mu_J P + \mu_d A + r\beta_A(V)A \text{ for } t > 0 \text{ where } \mu_d = \mu_J - \mu_A \tag{2.10a}$$

$$A' = -\mu_A A + \phi(m-t)\exp(-\mu_J t) \quad \text{for} \quad 0 \le t \le m \tag{2.10b}$$

$$A' = -\mu_A A + r\beta_A(V(t-m))A(t-m)\exp(-\mu_J m) \text{ for } t > m \tag{2.10c}$$

for the density dependent adult fertility problem and

$$P' = -\mu_J(W)P + (r\beta_A - \mu_A + \mu_J(W))A \quad \text{for} \quad t > 0 \tag{2.11a}$$

$$A' = -\mu_A A + \phi(m-t)\exp(-\int_0^t \mu_J(W(s)ds) \quad \text{for} \quad 0 \le t \le m \tag{2.11b}$$

$$A' = -\mu_A A + r\beta_A A(t-m)\exp(-\int_0^m \mu_J(W(t-m+s))ds) \text{ for } t > m \tag{2.11c}$$

for the density dependent juvenile mortality problem. Note that

$$V(t) = \alpha P(t) + (1-\alpha)A(t) \quad \text{and} \quad W(t) = P(t) + (\gamma - 1)A(t).$$

Both problems have the initial conditions (see (2.1c))

$$P(0) = \int_0^\infty \phi(a)da \geq 0 \quad \text{and} \quad A(0) = \int_m^\infty \phi(a)da \geq 0. \tag{2.12}$$

Equations (b) and (c) in (2.10) and (2.11) have straightforward interpretations. (b) They state that the rate of change of the adult population is determined by the loss $-\mu_A A$ due to deaths plus the gain due to maturation of juveniles. On the interval $0 \leq t \leq m$ this maturation rate is given simply by the survivors, over t units of time, from the initial density $\phi(m-t)$ of juvenile individuals of age $m-t$. For times $t > m$ this maturation rate is given by the birth rate at time $t-m$ multiplied by the probability of surviving the m time units to maturation. Equations (a) express the net change in population size due to deaths and births. For example, the right hand side of (2.10a) can be rewritten as $-\mu_J J - \mu_A A + r\beta_A(V)A$ whereby the change in population size P' is seen to be given by the loss of both juveniles and adults due to death plus the gain due to births from the adult class. A similar interpretation holds for (2.11a).

The problems (2.10)-(2.12) are somewhat nonstandard in that the initial conditions for the delay equations (a) and (c) for $t > m$, rather than being given over the "initial" interval $0 \leq t \leq m$ as is usually the case for delay equations, are determined by the initial conditions (2.12) at $t = 0$ and the equations in parts (a) and (b) over the interval $0 \leq t \leq m$. We will not dwell on this point here, but instead turn to the asymptotic dynamics by restricting our attention to $t > m$.

3 Equilibria and stability

In this section we will analyze the equilibrium solutions of equations (2.10a,c) and (2.11a,c) and their stability properties, including the trivial zero state solution $P = A = 0$. Four theorems will be proved. Theorem 1 shows that if the inherent net reproductive rate r is less than one then the trivial equilibrium is globally attracting and consequently the population asymptotically dies out. The condition $r < 1$ means that at low population densities, when the adverse effects on survival and fertility are minimal, individuals do not on average replace themselves. This "reasonable" result is a consequence of the monotonicity assumptions in (2.2) on the death and fertility rates as functions of population size, assumptions which rule out depensation or Allee effects that in nonlinear problems can lead to viable populations with $r < 1$.

Theorems 2-4 deal with the case when $r > 1$. In this case, the trivial solution is unstable and there exists, for both problems, a unique positive equilibrium which at least for r sufficiently close to 1 is locally asymptotically stable (Theorem 2). It is frequently the case in population dynamical models that equilibria destabilize (usually through a Hopf-type bifurcation to a limit cycle) as r or some other measure of reproductive output is increased sufficiently. We cannot expect, in the generality assumed, that the positive equilibria of equations (2.10a,c) and 92.11a,c) remain stable for all $r > 1$. We do not study this question here, however, and turn instead in Theorems 3 and 4 to the question of the dependence of the equilibria and their stability properties on the competition coefficients α

and γ when r is close to 1 (when we are assured by Theorem 2 that they are indeed stable).

Theorem 3 indicates that increased juvenile vs. adult competition is destabilizing in both models in the sense that equilibrium levels for both the juvenile and adult classes decreases with increased competition. As Theorem 4 indicates, however, this is not always the case if the strength of the equilibrium stability is measured by the effect of the competition on the magnitude of the linearized, stability-determining eigenvalue λ (often referred to as a measure of the "resilience" of the equilibrium). Theorem 4 deals with the change in this eigenvalue caused by changes in α and γ by computing the sign of the derivative $\partial_\alpha \lambda$ for (2.10) and $\partial \lambda_\gamma$ for (2.11). Theorem 2 guarantees that $\lambda < 0$ for $r > 1$ sufficiently close to 1. If the sign of one of these derivatives is positive then an increase in the strength of the intra-specific competition is represented by an increase in the corresponding competitive coefficient is destabilizing since the negative eigenvalue is thereby increased (causing decreased resilience). Similarly a negative derivative implies a stabilizing effect due to increased competition (due to an increase in resilience). Theorem 4 shows that the two problems (2.10) and (2.11) have opposite effects on equilibrium stability. Specifically, in the first case when adult fertility is dependent on population size, increased competition from juveniles is destabilizing. On the other hand, in the second case when juvenile survival is dependent on population size, increased competition is stabilizing.

It is not difficult to see from the equations (2.1)-(2.2) that non-negativity of the initial age distribution $\phi(a) \geq 0$ implies that the density $\rho(t,a) \geq 0$ is non-negative for all time t. Thus the solutions $P(t)$ and $A(t)$ of (2.10) and (2.11) are non-negative for all time t. (A direct proof from equations (2.10) and (2.11) that non-negative initial conditions imply non-negative solutions for all time doesn't appear to be trivial.) Both systems have the trivial zero equilbrium $(P, A) = (0,0)$ (corresponding to $\phi \equiv 0$).

Theorem 1 *If* $r < 1$ *and* $P(0) \geq 0$, $A(0) \geq 0$, *then* $(P(t), A(t)) \rightarrow (0,0)$ *as* $t \rightarrow +\infty$ *for solutions of* (2.10) *and* (2.11).

Proof. For both problems (2.10) and (2.11) it is easy to see that

$$A'(t) \leq -\mu_A(t) + r\mu_A A(t-m), \quad t > m$$

from which follows that $0 \leq A(t) \leq x(t)$ where $x(t)$ is the solution of

$$x'(t) = -\mu_A x(t) + r\mu_A x(t-m), \quad t > m \tag{3.1}$$

$$x(t) = A(t), \quad 0 \leq t \leq m.$$

Since $r \leq 1$ it follows from a result of Hayes (Theorem 13.8, Chapter 13, of Bellman and Cooke (1963)) that $x(t) \rightarrow 0$ and hence $A(t) \rightarrow 0$ as $t \rightarrow +\infty$.

For (2.10) we note that $P' = -\mu_J P + \psi(t)$ where

$$\psi(t) = (\mu_d + r\beta_A(V(t)))A(t) \rightarrow 0 \quad \text{as} \quad t \rightarrow +\infty.$$

Thus $P(t) \rightarrow 0$ as $t \rightarrow +\infty$.

For (2.11) we define $Q(t)$ as the solution of

$$Q' = -\mu_J(W)Q + (r\beta_A - \mu_A)A, \quad Q(0) = P(0) \geq 0$$

and let $R(t) = P(t) - Q(t)$ so that

$$R' = -\mu_J(W)R + \mu_J(W)A, \quad R(0) = 0.$$

Since $\mu_J(W) \geq \mu_J(0) > 0$ and $A(t) \to 0$ we see that $Q(t) \to 0$ as $t \to +\infty$. Using an integrating factor we find from an integration by parts that

$$R(t) = A(t) - A(0)\exp\big(-\int_0^t \mu_J(W)d\sigma\big) - \exp\big(-\int_0^t \mu_J(W)d\sigma\big)\int_0^t \exp\big(\int_0^s \mu_J(W)d\sigma\big)A'(s)ds.$$

The first two terms tend to 0 as $t \to +\infty$ as does the last term since from (2.11c) we see that $A'(t) \to 0$. Thus $R(t)$ and hence $P(t)$ tend to 0 as $t \to +\infty$. □

Theorem 2 *Suppose* $r > 1$. *Then (0,0) is unstable for both* (2.10) *and* (2.11) *and both systems have a unique positive equilibrium* (P_e, A_e) *which is aysmptotically stable for at least* r *close to 1.*

Proof. The linearization of part (c) of both (2.10) and (2.11) at the trivial equilibrium (0,0) yields the scalar delay equation (3.1), all of whose nontrivial solutions are unbounded (again see Theorem 13.8, Chapter 13, of Bellman and Cooke (1963)) since $r > 1$. Thus (0,0) is unstable for $r > 1$.

Next we look at solutions $(P_e, A_e) \neq (0,0)$ of (2.10) and (2.11) that are constant for $t > m$ by looking for nonzero equilibrium solutions of (a) and (c). (Parts (b) then serve to define the initial equilibrium age distribution ϕ.) For (2.10) this leads to the equations

$$0 = -\mu_J P_e + \big(r\beta_A(V_e) + \mu_d\big)A_e$$

$$0 = -\mu_A + r\beta_A(V_e)$$

where $V_e = \alpha P_e + (1-\alpha)A_e$. Using $r > 1$, the normalization (2.5) and the monotonicity of β_A we can rewrite these two equations as

$$\mu_J P_e - \big(\beta_A(0) + \mu_d\big)A_e = 0$$

$$\alpha P_e + (1-\alpha)P_e = \beta_A^{-1}(\beta_A(0)/r)$$

from which follows the existence of the positive equilibria

$$P_e = V_e(r)\frac{\beta_A(0) - \mu_A + \mu_J}{\alpha(\beta_A(0) - \mu_A) + \mu_J}, \quad A_e = V_e(r)\frac{\mu_J}{\alpha(\beta_A(0) - \mu_A) + \mu_J} \tag{3.2}$$

for $r > 1$ where $V_e(r) = \beta_A^{-1}(\beta_A(0)/r)$.

Similar manipulations with (2.11) lead to the positive equilibria

$$P_e = W_e(r)\frac{r\beta_A - \mu_A + \mu_J(W_e)}{r\beta_A - \mu_A + \gamma\mu_J(W_e)}, \quad A_e = W_e(r)\frac{\mu_J(W_e)}{r\beta_A - \mu_A + \gamma\mu_J(W_e)} \tag{3.3}$$

for $r > 1$ where $W_e(r) = \mu_J^{-1}(\frac{1}{m}\ln r + \mu_J(0))$.

The calculation of the characteristic functions for (2.10) and (2.11) at these two respective equilibria is straightforward. This exercise yields equations

$$p(\lambda;\alpha,r) = 0 \quad \text{and} \quad q(\lambda;\gamma,r) = 0$$

for complex λ where

$$p(\lambda;\alpha,r) = (\lambda + \mu_J)\left(\lambda + \mu_A - \mu_A e^{-\lambda m}\right) - rV_e(r)\mu_J\beta_A'(V_e(r))\frac{a(\lambda)\alpha + b(\lambda)}{c\alpha + d} \tag{3.4}$$

$$a(\lambda) = (\lambda + \mu_A)(1 - \exp(-\mu_J m - \lambda m)), \quad b(\lambda) = (\lambda + \mu_J)\exp(-\mu_J m - \lambda m)$$

$$c = \beta_A(0) - \mu_A, \quad d = \mu_J$$

and where

$$q(\lambda;\gamma,r) = (\lambda + \mu_J(W_e(r)))\left(\lambda + \mu_A - \mu_A e^{-\lambda m}\right) + \mu_J'(W_e(r))W_e(r)\frac{r\beta_A - \mu_A}{r\beta_A - \mu_A + \gamma\mu_J(W_e)}\left(\lambda + \mu_A - \mu_A e^{-\lambda m}\right) + \mu_J'(W_e(r))W_e(r)\mu_A\frac{1 - e^{-\lambda m}}{\lambda}c(r)\frac{a(\lambda,r)\gamma + b(\lambda,r)}{c(r)\gamma + d(r)} \tag{3.5}$$

$$a(\lambda,r) = \lambda + \mu_J(W_e(r)), \quad b(\lambda,r) = -\lambda + r\beta_A - \mu_A$$

$$c(r) = \frac{1}{m}\ln r + \mu_J(W_e(r)), \quad d(r) = r\beta_A - \mu_A.$$

Here we have indicated the dependence of p and q on r and the competition coefficients α and γ.

The following partial derivatives evaluated at the critical values $r = 1$ and $\lambda = V_e = W_e = 0$ are easily computed:

$$\partial_\lambda p(0;\alpha,1) = \mu_J(1 + m\mu_A) > 0$$

$$\partial_\lambda q(0;\gamma,1) = \mu_J(0)(1 + \mu_A) > 0 \tag{3.6}$$

More tedious calculations show that

$$\partial_r p(0;\alpha,1) > 0, \quad \partial_r q(0;\gamma,1) > 0$$

and hence in both cases $\partial_r\lambda < 0$ at $r = 1$. It follows that $\lambda < 0$ for r greater than, but close to 1. □

The effect of increased juvenile vs. adult competition on the positive total and adult reproductive levels P_e and A_e is clear from (3.2) and (3.3). Moreover from the formulas

$$J_e = P_e - A_e = V_e(r)\frac{\beta_A(0) - \mu_A}{\alpha(\beta_A(0) - \mu_A) + \mu_J}$$

$$J_e = P_e - A_e = W_e(r)\frac{r\beta_A - \mu_A}{r\beta_A - \mu_A + \gamma\mu_J(W_e)}$$

for the juvenile equilibrium levels in the two cases respectively, we obtain, with a reference to (2.5) and (2.7), the following result.

Theorem 3 *The unique, positive equilibrium levels for total population size P_e, adult population size A_e, and juvenile population size J_e, all decrease with increased competition coefficients α or γ. The relative proportions A_e/P_e and J_e/P_e, however, remain unaffected.*

Finally we consider the stability sensitivity of the positive equilibria guaranteed by Theorem 2 for r greater then, but close to 1. This we do by computing the change in the stability-determining eigenvalue λ near 0, or more specifically by determining the sign of the derivatives

$$\partial_\alpha \lambda = -\partial_\alpha p(\lambda;\alpha,r)/\partial_\lambda p(\lambda;\alpha,r)$$

$$\partial_\gamma \lambda = -\partial_\gamma q(\lambda;\gamma,r)/\partial_\lambda q(\lambda;\gamma,r)$$

for r greater than, but close to 1. By (3.6) these signs are the opposites of those of the derivatives $\partial_\alpha p(\lambda;\alpha,r)$, $\partial_\gamma q(\lambda;\gamma,r)$ for r greater than, but close to 1.

Theorem 4 *For r greater than, but close to 1, $\partial_\alpha\lambda > 0$ for* (2.10) *and $\partial_\gamma\lambda < 0$ for* (2.11).

Proof. Consider first (2.10) with the characterisitic equation (3.4). It is not difficult to see that $\partial_\alpha p(0;\alpha,1) = 0$ and therefore we must consider p for r close to but not equal to 1, i.e. for $\lambda < 0$, but close to 0. To determine the monotonicity of p in the variable α we need to determine that of the linear fractional expression in (3.4). Because $\beta_A' < 0$, this linear fractional expression and p have the same monotonicity, which is determined by the sign of the determinant-like expression $\Delta(\lambda) = a(\lambda)d - b(\lambda)c$ for $\lambda < 0$, but near 0. Since $\Delta(0) = 0$ and

$$\partial_\lambda \Delta(0) = \mu_J(1 - \exp(-\mu_J m)) + \mu_A(-1 + m\mu_J + \exp(-\mu_J m)) > 0$$

we find that $\Delta(\lambda) < 0$ for $\lambda < 0$, but near 0, and hence $r > 1$, but near 1. This implies that the linear fractional expression in p and hence p itself is decreasing in α, i.e. $\partial_\alpha p(\lambda;\alpha,r) < 0$, for $r > 1$ near 1.

A similar analysis must be carried out on q defined by (3.5) with respect to γ in order to determine the sign of $\partial_\gamma q(\lambda;\gamma,r)$ for $r > 1$, but near 1. The first term in q is independent of γ and the second term clearly increases as a function of γ. The monotonicity of the third term is the opposite of that of the linear fractional expression appearing there, which is determined by

$$\begin{aligned}\Delta(\lambda) &= a(\lambda,r)d(r) - b(\lambda,r)c(r)\\ &= \Big(\frac{1}{m}\ln r + \mu_J - \mu_A(1 - r\exp(\mu_J m))\Big)\lambda + \mu_A(1 - r\exp(\mu_J m))\frac{1}{m}\ln r.\end{aligned}$$

For $r > 1$, but near 1, and $\lambda < 0$, but near 0, we see that $\Delta(\lambda) < 0$ and hence the third term in q is also increasing in γ. □

4 Concluding remarks

From the general model equations (2.1) for age-structured population dynamics, we derived the two model systems of differential delay equations (2.10) and (2.11) to describe the dynamics of a population whose intra-specific competition between juveniles and adults leads to juvenile density dependent adult fertility and adult density dependent juvenile survival respectively. Under the assumption that the density dependencies are monotonic functions of weighted total population sizes (and that within the juvenile and adult age classes survival and fertility rates are age independent), it is shown that these model equations possess a positive equilibria if and only if the inherent net reproductive rate r is greater than 1, that these positive equilibria are locally asymptotically stable at least for r near 1, and that all solutions tend to 0 if $r < 1$.

With regard to the stabilization/destabilization question we reached conclusions that in one case disagree and in another agree with those of Ebenman(1988). First of all, the effect on equilibrium levels of increased juvenile vs. adult competition is always adverse (Theorem 3) and in this sense such competition is always destabilizing, in agreement with May et al.(1974) and Tschumy(1982) and in disagreement with Ebenman. However, the effect of increased competition on the linearized eigenvalue leads us to the same conclusion as Ebenman, although for different reasons. Namely, Theorem 4 implies that increases in the strength of competition (as measured by the coefficient α) in the juvenile density dependent adult fertility problem is a destabilizing influence, while increases in the strength of competition (as measured by the coefficient γ) in the adult dependent juvenile survival problem is a stabilizing influence. Ebenman reached these same conclusions using stability region arguments, a procedure criticized by Loreau(1990). To further confuse matters, Ebenman's model predicts destabilization in both cases if linearized eigenvalue sensitivity is used as the criteria. This contradiction between the eigenvalue analysis of Ebenman's model and ours here can be explained by the fact that Ebenman's model is a 2×2 matrix model which is degenerate in the sense that it is not "primitive", i.e. does not have a strictly dominant linearized eigenvalue. (In fact, an unusual double bifurcation of both equilibria and 2-cycles occurs at $r = 1$ because both +1 and -1 are linearized eigenvalues; see Cushing and Li (1989).) This peculiarity is purely a mathematical artifact that occurs because Ebenman's discrete time and age matrix equation has exactly two age categories, only one of which is adult, and disallows adult survival after reproduction. The number of age categories and the time scale is arbitrary as far as the modeling is concerned and even a similar model that utilizes just two adult age classes eliminates this degeneracy with the result that an eigenvalue analysis is then in agreement with those in Theorem 4. Loreau's conclusion that stabilization is an unlikely result of juvenile vs. adult competition should perhaps be re-evaluated, both in view of our result here and because his arguments are also based upon Ebenman's highly simplified model.

Aknowledgment

The author gratefully acknowledges the support of the Applied Mathematics Division and the Population/Ecology Division of the National Science Foundation under grant No. DMS-8902508.

References

1. Bellman, R., Cooke, K. (1963): Differential-Difference Equations. Academic Press, New York
2. Cushing, J.M. (1985): Equilibria in structured populations. J. Math. Biology **23**, 15-39
3. Cushing, J.M., Li, J.(1989): On Ebenman's model for the dynamics of a population with competing juveniles and adults. Bull. Math. Biol. **6**, 687-713
4. Cushing, J.M., Li, J.(1990): Juvenile vs. adult competition. To appear in J. Math. Biol.
5. Ebenman, B.(1987): Niche differences between age classes and intraspecific competition in age-structured populations. J. Theor. Biol. **124**, 25-33
6. Ebenman, B.(1988): Competition between age classes and population dynamics. J. Theor. Biol. **131**, 389-400
7. Ebenman, B., Persson, L.(1988): Size-Structured Populations: Ecology and Evolution. Springer-Verlag, Berlin-Heidelberg-New York
8. Gurtin, M.E., MacCamy, R.C.(1974): Nonlinear age-dependent population dynamics. Arch. Rat. Mech. Anal. **54**, 281-300
9. Hoppensteadt, F.C.(1975): Mathematical Theories of Populations: Demographics, Genetics, and Epidemics. Regional Conference Series in Applied Mathematics, **20**, SIAM, Philadelphia
10. Loreau, M.(1990): Competition between age classes, and the stability of stage-structured populations: a re-examination of Ebenman's model. J. Theor. biol. **144**, 567-571
11. May, R.M., Conway, G.R., Hassell, M.P., Southwood, T.R.E. (1974): Time delays, density-dependence and single-species oscillations. J. Animal Ecol. **43**, 747-770
12. Metz, J.A.J., Diekmann, O.(1986): The Dynamics of Physiologically Structured Populations. Lec. Notes in Biomath. **68**, Springer-Verlag, Berlin-Heidelberg-New York
13. Tschumy, W. (1982): Competition between juveniles and adults in age-structured populations. Theor. Pop. Biol. **21**, 255-268
14. Werner, E., Gilliam, J.(1984): The ontogenetic niche and species interactions in size-structured populations. Ann. Rev. Ecol. Syst. **15**, 393-425

McKendrick Von Foerster Models for Patch Dynamics

Alan Hastings

Division of Environmental Studies and Institute of Theoretical Dynamics, University of California, Davis, CA 95616

Introduction

For many years, Lotka-Volterra equations served as the basic paradigm in ecological modelling. These equations, and models based on them, were responsible for the embodiment of ideas such as the competitive exclusion principle, and even quite recently, many ideas concerning the dynamics of food webs. However, as has long been recognized, these models represent a vast simplification of ecological reality. In particular, they ignore the consequences of structure within the populations modelled. This can be spatial structure, age structure, physiological structure, genetic or phenotypic structure, or possibly other kinds of structure. My goal in this paper will be to describe how models based on systems of first order partial differential equations, known as the McKendrick or Von Foerster equation, can be used to describe interacting spatially structured populations.

In particular, I will describe how such models can be used to begin understanding several questions of biological interest which have been difficult to understand using more conventional models. I will focus on models for competition in a patchy environment. Here a primary goal is to infer the action of competition based on patterns of species abundances, assuming that the system is at equilibrium. I will present a general model for competition and present the determination of information about the equilibrium in a special case. Similar models can also be used to consider the explanation of local 'outbreaks' of pests by predator prey systems in a patchy environment. The models can be used to study the role local (in space) dynamics play in the interaction between a predator and a patchily distributed prey.

Before describing models appropriate for answering these questions, I will first describe analogous models for simpler situations involving single species.

Modelling approach

The Von Foerster or McKendrick model has long been used as a descriptor of age structured populations, beginning with the pioneering work of McKendrick (1926). An excellent review of both classical and important recent work on this model and its extensions is contained in the volumes by Webb (1985) and by Metz and Diekmann (1986). The initial work, and much that followed until recently, was based on linear models, i.e., models which ignored the role of density dependence. The first models considered only the role of age as an internal variable. Let $n(t,a)$ denote the age distribution in the population at time t. This age distribution satisfies the McKendrick - Von Foerster equation:

$$\frac{\partial n(t,a)}{\partial t} + \frac{\partial n(t,a)}{\partial a} = -\mu(t,a)n(t,a), \tag{1}$$

where $\mu(t,a)$ is the time and age specific death rate. To complete the formulation a boundary condition for the birth rate must be appended:

$$n(t,0) = b(t). \tag{2}$$

Typically the birth rate $b(t)$ would depend on the number of individuals of different ages, as in:

$$b(t) = \int_0^\infty m(a)n(a,t)da, \tag{3}$$

where $m(a)$ is the fecundity of an individual of age a. (Note that some care may be needed to take into account that only females have offspring.) At this point one can determine asymptotic information, but a full description would also require an initial condition, $n(0,a)$. This model can also be written in an equivalent, integral equation form which is easier to use in some applications. I will describe this formulation below in the context of some specific applications.

In an important paper Sinko and Streifer (1967) emphasized the role of variables other than age as structuring variables, leading to the following model. Denote by $g(t,a,\sigma)$ the function describing the rate of change of σ with respect to age and time. Then the population satisfies the following equation:

$$\frac{\partial n(t,a,\sigma)}{\partial t} + \frac{\partial n(t,a,\sigma)}{\partial a} + \frac{\partial g(t,a,\sigma)n(t,a,\sigma)}{\partial \sigma} = -\mu(t,a,\sigma)n(t,a), \tag{4}$$

where now $n(t,a,\sigma)$ is a density function on age a and size σ for the number of individuals at time t. Similarly, $\mu(t,a,\sigma)$ is the time, age and size specific death rate. Note that the variable σ could represent other structuring variables besides size, and there could be more than one such variable.

The models I have described at this point are quite simple to manipulate, but they ignore density dependent interactions among individuals. A great advance was made by Gurtin and MacCamy (1974) (see also Hoppensteadt 1974, 1975) who formalized the mathematical study of density dependent (nonlinear) versions of the McKendrick model. Here, both the death rate $\mu(a,t)$ and the birth rate $b(t)$ can be taken as nonlinear functionals (depending on all ages) of the number of individuals at time t. Many advances since this time in understanding the behavior of these nonlinear models are described in Metz and Diekmann (1986).

The use of Von Foerster type models to describe spatially structured populations is reviewed in Metz and Diekmann (1986) and Hastings and Wolin (1989). The general form of the model for a single species is as follows. Let $p(n,t)$ denote the number (or fraction) of patches with n individuals in them at time t. Thus, the patch plays the same role as the individual in the structured models just described above. Formulation of a model of this kind requires making a number of important assumptions (cf. Hastings and Wolin, 1989):

1 Ignore any stochastic effects due to small numbers of patches.

2 Ignore any underlying differences in the patches, although this can be taken into account in more complex models.

3 Ignore any influence of the spatial arrangement of the patches on the dynamics of the model.

4 Assume that the rate of change of individuals within a patch of size n at time t is given by $g(n,p(\cdot,t),t)$. The dependence on $p(\cdot,t)$ allows for the effects of immigration from or emigration to, other patches.

5 Within any patch, assume that population growth is deterministic, except for events (disasters) which remove all individuals within a patch, returning it to empty. This rate is given by $\mu(n,p(\cdot,t),t)$. Here, the dependence on $p(\cdot,t)$ may allow for the effect of a forager.

6 An assumption is needed to formulate a term analogous to the birth term given in (3). The production of new propagules, which colonize empty patches, is assumed to depend on the population sizes in the occupied patches. The analysis is greatly simplified if all propagules are of the same size, so I will assume that.

An equation similar to (4) results, namely:

$$\frac{\partial p(n,t)}{\partial t} + \frac{\partial g(n,p(\cdot,t),t)p(n,t)}{\partial n} = -\mu(n,p(\cdot,t),t)p(n,t), \tag{5}$$

The boundary condition is of the form

$$p(n_0,t) = \int_0^\infty M(n,p(\cdot),t)p(n,t)dn, \tag{6}$$

where n_0 is the initial propagule size, and the function M gives the rate of settling of empty patches per patch with n individuals, which may depend on the population characteristics as a whole. One of the first uses of an approach like this was in Levin and Paine (1974), where the patches represented gaps in the mussel bed in the intertidal, and actually equation (4) was used to describe the dynamics of these patches.

The assumptions imply that knowing the time since a patch was colonized, one can determine the population size within the patch, since one can use the fixed propagule size and deterministic dynamics. Thus, we will call the time since a patch was colonized the age, a, of the patch, and formulate an age structured model to describe the dynamics of the entire ensemble of patches (cf. Metz and Diekmann, 1986, Hastings and Wolin, 1989). This device avoids the use of the derivatives with respect to n and makes numerical solutions much easier. Let $p(a,t)$ be (a density function for) the fraction of patches of age a at time t . Describe the dynamics within a patch by the local growth law: $dn/da = f(n,a)$ with initial condition $n(0) = n_0$, which thus provides a way to compute a given n. Thus, equation (5) is replaced by:

$$\frac{\partial p(a,t)}{\partial t} + \frac{\partial p(a,t)}{\partial a} = -\mu(a, p(\cdot,t), t)p(a,t), \tag{7}$$

This is just the basic nonlinear model used in age dependent population dynamics. Following the standard analysis of such equations, one can compute the distribution of patches of different sizes, assuming that the extinction rate μ depends only on the 'age', a:

$$p(a,t) = p(0, t-a)e^{-\int_0^a \mu(s)ds}. \tag{8}$$

If the function M is such that substituting (8) into (6) yields a unique time independent solution, then the model has a unique solution. This would be the case, for example, if M was a monotonically decreasing function of p(0) when p(a) was given by (8). A sufficient condition for stability can be determined from convexity of the extinction probability as a function of age (Rorres, 1979). Using this result, Hastings and Wolin (1989) showed that, under the assumption that the extinction probability is a decreasing function of the number of individuals within a patch, if the equilibrium of (7) is feasible then the equilibrium is locally stable. It is also shown that the equilibrium is unique. Thus, any feasible equilibrium is locally stable.

I have recently extended this model to include the possibility of disasters which do not remove all individuals within a patch (Hastings, 1990).

Biological questions

There has been an extensive literature on the role of spatial structure in the dynamics of biological populations, beginning with the seminal work of Skellam (1951). Summaries are given in Levin (1976) and Okubo (1980) and DeAngelis and Waterhouse (1987). There have been two general approaches, an approach based on reaction-diffusion equations, and one based on the dynamics of 'patches' which are in different states. As noted by Chesson (1981), the latter approach is inherently stochastic, although the number of patches is typically assumed to be so large that the models focus only on the mean number of patches in various states.

Kareiva in a series of papers (summarized in Kareiva 1986) has examined using an experimental situation the dynamics of aphids and a major predator, the ladybug beetle, on goldenrod stands (see also Kareiva, 1987). He has obtained data on local outbreaks, and considered how models based on reaction diffusion equations might explain the observed dynamics. He has also considered the role played by dispersal as a stabilizing influence as indicated in patch models such as Hastings (1977). However, there have been very few patch models developed which include the dynamics of numbers within patches, which would be needed to better understand the interaction among dispersal, local dynamics, and predation in producing outbreaks.

Another important situation where the paradigm of patch models is useful is in understanding competition in patchy environments. One major controversy during the past decade has been the relative importance of predation and competition in structuring natural communities (e.g., Diamond and Case, 1986). One aspect of this question has concerned the inference of competition from observations of co-occurrence data for the putative competitors. I have shown (Hastings, 1987) that inferences of competition are very difficult to make based on presence or absence data, in a system with a large number of patches.

Consider a system of n species which all behave similarly, and affect each other only by increasing extinction rates in patches where species co-occur. A full dynamic model for this system is difficult to write down. However, one can easily write down the equations describing the equilibrium which is symmetric in the sense that the frequency of patches with each unique combination of species depends only on the number of species, and not their identity (Hastings, 1987). Thus, use as a variable the frequency of each unique i species combination, p_i. For example, the fraction of patches with a single species is given by np_1 since there are n different one species combinations. Denote by e_i the extinction rate per species in a patch with i species. Let M be the rate at which each species colonizes patches. Then, by counting the rate at which different transitions take place, the equations for the symmetric equilibrium are found to be:

$$0 = iMp_{i-1} - ie_i p_i - (n-i)Mp_i + (n-i)e_{i+1}p_{i+1} \text{ for } i = 0 \text{ to } n, \quad (9)$$

if one defines

$$p_{-1} = p_{n+1} = e_0 = e_{n+1} = 0. \quad (10)$$

Under either the assumption that the colonization rate M is a constant m, or that M is proportional to the frequency of patches occupied by a given species one can show that there is a unique equilibrium. In the case that M is constant, an explicit formula for the equilibrium results:

$$p_0 = \sum_{i=0}^{n} \left\{ \frac{\binom{n}{i} m^i}{\prod_{k=1}^{i} e_k} \right\} \quad (11)$$

$$p_i = \frac{m^i p_0}{\prod_{k=1}^{i} e_k}. \quad (12)$$

Results on the statistical analysis of systems of finite number of patches near this equilibrium (Hastings, 1987) show that deviations from random assortment are very difficult to detect with sample sizes (number of patches) typical of those obtained in natural systems. Simulations show that the deterministic results described here are relevant for determining the statistical behavior of systems with as few as 30 patches. No determination of the stability of this equilibrium was undertaken, and the stability analysis remains an open question.

Using only presence or absence data ignores the information available about numbers of individuals of each species within a patch. A number of investigators (e.g., Toft and Schoener, 1983; Gutzweiller and Anderson, 1988) have obtained this more complete data including numbers of individuals within patches. Thus, a theory relating competition in a patchy environment to numbers within patches is vitally needed which can incorporate this extensive additional information. Note that at least for the system studied by Toft and Schoener (1983), extinctions are an important event, so the underlying stochastic nature of the patch models is important in their use as a descriptor of the dynamics.

Modelling approaches for multiple species

I will now describe in more detail modelling approaches which can be used to answer the questions which I have posed above. These will lead to a system of integro-partial differential equation. Techniques described in Metz and Diekmann (1986) may prove helpful in the analysis. I will now present the formulation of a model for competition in a patchy environment. The model will be restricted to two species, although extensions are possible. All the assumptions outlined above for the single species case will be used. However, now instead of just having two general classes of patches, empty and occupied, there are four: empty, species one alone, species two alone, and both species. Again, I will focus only on the last three of these categories.

Following the approach given above, I will present a model given in terms of three density functions. The first density function, p_1, will be for the fraction of patches which have n_1 individuals of species 1 and none of species 2. The second density function, p_2, will be for the fraction of patches that have n_2 individuals of species 2. The final density function p_{12}, will be a function of both n_1 and n_2, the number of individuals of both species 1 and 2. Let g_{11}, which is a function of $n_1, p_1(\cdot), p_{12}(\cdot), t$, give the growth rate of populations with n_1 individuals of species 1 in patches which only have species 1. The extinction rate of species 1 in patches which only have species 1 is given by μ_1, which depends on $n_1, p_1(\cdot), p_{12}(\cdot), t$. Let γ_1 describe the rate of invasion of patches with species 1 alone by species 2, which will in general depend on $n_1, p_1(\cdot), p_2(\cdot), p_{12}(\cdot)$ and t. The functions g_2, μ_2, and γ_2 are defined similarly. The functions $g_{12;i}$ will give the growth rate of species i in patches with both species, as a function of $n_1, n_2, p_1(\cdot), p_2(\cdot), p_{12}(\cdot)$ and t. Finally, let $\mu_{12;12}$ describe the rate of simultaneous extinction of both species in a patch with both species, and let $\mu_{12;i}$ be the rate of extinction of species i in a patch with both species, where again these functions depend on all the appropriate variables. A general model thus takes the form:

$$\frac{\partial p_1}{\partial t} + \frac{\partial g_{11} p_1}{\partial n_1} = -\mu_1 p_1 - \gamma_1 p_1 + \int_0^\infty \mu_{12;2} p_{12} dn_2 \qquad (13)$$

$$\frac{\partial p_2}{\partial t} + \frac{\partial g_{22} p_2}{\partial n_2} = -\mu_2 p_2 - \gamma_2 p_2 + \int_0^\infty \mu_{12;1} p_{12} dn_1 \tag{14}$$

$$\frac{\partial p_{12}}{\partial t} + \frac{\partial g_{12;1} p_{12}}{\partial n_1} + \frac{\partial g_{12;2} p_{12}}{\partial n_2} = -\mu_{12;12} p_{12} - \mu_{12;1} p_{12} - \mu_{12;2} p_{12} \tag{15}$$

To this system one must append boundary conditions describing the results of invasions. For simplicity, I will assume that independent of the kind of patch, the initial propagule size for species i will be $n_{0;i}$. Let $I_{k;i}$ be the rate of invasion of patches of type k (with $k = 0$ for empty patches) by species i. I use a capital letter here to distinguish this case from the single species model where the invasion conditions are given in terms of per patch functions. The boundary conditions are now:

$$p_1(n_{0;1}, t) = I_{0;1} \tag{16}$$

$$p_2(n_{0;2}, t) = I_{0;2} \tag{17}$$

$$p_{12}(n_{0;1}, n_2, t) = I_{2;1} \tag{18}$$

$$p_{12}(n_1, n_{0;2}, t) = I_{1;2} \tag{19}$$

This is now a complete model which can be solved, at least numerically, once all the functions are specified, using standard techniques from age structured models (see e.g., Diekmann and Metz, 1986).

A symmetric case of the competition model

I will describe how in a special case, substantial analytical progress can be made towards the determination of the equilibrium of this system. This special case also illustrates the substantial difficulties involved in an analytical study of this system. I will assume that each species affects the extinction rate and colonization rate, but not the growth rate of the other species. I will also assume that the two species are identical, in that the extinction rates, initial propagule sizes, and colonization rates are the same. Finally, I will only look for an equilibrium where the distribution of population sizes of the two species are the same, and only examine the distribution of patch sizes of occupied patches. Note that the distribution of patch sizes of occupied patches is the important biological question, since this is what is easily observed. In this case, the system can be reduced to the study of a single Fredholm integral equation. If one further restricts attention to the case discussed above where colonization rates are independent of population sizes or occupancies of other patches, the integral equation is linear. An algorithm for determining the equilibrium can be presented.

The device used above for the single species models of changing from a density function defined in terms of numbers of individuals within a patch to a density defined in terms of an 'age' of the patch will prove useful. Thus classify all patches by an 'age' variable: the time to reach a population of that size starting from the initial propagule size. Also, it is easier to study the model in integral equation form, which can be derived by integrating the system (13)-(15) along characteristics, or by reasoning directly from the biological assumptions. There are two ways a one-species patch can be of age a at time t: an empty patch was colonized at time t and has not been colonized by the other species nor gone extinct, or from the extinction of one species in a two species patch at the appropriate time in the past. (One need only consider colonizations or the last extinction event.) This leads to the following equation for species 1:

$$\begin{aligned} p_1(t,a_1) =& p_1(t-a_1,0)e^{-\int_0^{a_1} \mu_1(t-s,a_1-s)+\gamma_1(t-s,a_1-s)ds} \\ &+ \int_0^{a_1}\int_0^{\infty} e^{-\int_{a_1-\alpha}^{a_1} \mu_1(t-a_1+s,s)+\gamma_1(t-a_1+s,s)ds} \\ &\mu_{12;2}(t-\alpha,a_1-\alpha,a_2)p_{12}(t-\alpha,a_1-\alpha,a_2)da_2d\alpha, \end{aligned} \tag{20}$$

with a similar equation for species p_2. By similar reasoning, one arrives at the following equation for p_{12}:

$$p_{12}(t,a_1,a_2) = \begin{cases} p_{12}(t-a_1,0,a_2-a_1)e^{-\int_0^{a_1} \mu_{12}(t-s,a_-s,a_2-s)ds} & \text{for } a_2 > a_1 \\ p_{12}(t-a_2,a_1-a_2,0)e^{-\int_0^{a_2} \mu_{12}(t-s,a_1-s,a_-s)ds} & \text{for } a_2 < a_1 \end{cases} \tag{21}$$

where $\mu_{12} = \mu_{12;12} + \mu_{12;1} + \mu_{12;2}$. Observe that

$$p_{12}(t-a_1,0,a_2-a_1) = p_2(t-a_1,a_2-a_1)\gamma(t-a_1,a_2-a_1) \text{ for } a_2 > a_1 \tag{22}$$

and

$$p_{12}(t-a_1,a_1-a_2,0) = p_1(t-a_2,a_1-a_2)\gamma(t-a_2,a_1-a_2) \text{ for } a_1 > a_2. \tag{23}$$

Now assume that both competitors are identical, and that colonization rates and extinction rates are independent of time, and the overall distribution of numbers within patches. Also, look for the equilibrium, or time independent, solutions of (20)-(23). Thus set $\gamma_1 = \gamma_2(a) = \mu_2(a)$, and $\mu_{12;1}(a_1,a_2) = \mu_{12;2}(a_2,a_1)$. I will focus on the 'symmetric' equilibrium, where the distribution of the two species are the same, as I did in the analysis of the presence absence model above. The time independent 'stable age' density function for the symmetric equilibrium is given by (here stable does not refer to return to equilibrium, but is used in the sense of demography):

$$g(a) = p_1(a)/p_1(0) = p_2(a)/p_2(0). \tag{24}$$

Using equations (20)-(23) one finds a single linear nonhomogeneous Fredholm integral equation for $f(a) = \mu_1(a)g(a)$ of the form:

$$f(a) = l(a) + \int_0^{\infty} K(\alpha,a)f(\alpha)d\alpha, \tag{25}$$

where the kernel $K(\alpha,a)$ is nonnegative and continuous. This integral equation can be analyzed using the methods applied to a single species model in Hastings (1990). The reason for the change to f is to allow integration by parts in estimating the spectrum of K. Consider the homogeneous eigenvalue equation:

$$f(a) = \lambda \int_0^\infty K(\alpha, a) f(\alpha) d\alpha. \tag{26}$$

One can show that

$$\int_0^\infty K(\alpha, a) da \leq \nu < 1, \tag{27}$$

which implies that the smallest positive eigenvalue of (26) is larger than one. Thus the integral equation has a unique solution which can be obtained using the iterative method of successive approximations (pp. 73-74, Zabreyko et al., 1975). This solution provides the biologically important distribution of population sizes of occupied patches at equilibrium. The stability of this equilibrium is still an open question.

Discussion

The formulation of tractable models which include the effects of population structure is a substantial challenge to the mathematical population biologist. I have indicated here how such models can be formulated for one important case of biological interest, competition with only two species. One can use this model as well to describe predation. Simply denote one of the species as a prey and the other as a predator. In this case it may make more sense to ignore the possibility of patches which contain only the predator, and thus concentrate only on three kinds of patches: empty, prey alone, or both predator and prey. Again, standard techniques can be used to find solutions of this model.

For a special case of the competition model with two identical species, I have sketched the steps which can be used to show that the 'symmetric' equilibrium, with both species having identical distributions of population sizes in patches, is unique because the solution reduces to the study of a linear integral equation. There is an explicit algorithm for determining this equilibrium, and thus the forms of the distributions generated can be contrasted with available data. It is interesting to note that in this model if the colonization rates do depend on population sizes other than in the patch being colonized this approach no longer applies. Consequently, solutions may not be unique. This indicates the substantial complexity found in even this simple case. Even for the case where I did determine the 'symmetric' equilibrium, I have not ruled out the possibility of alternate equilibria with two identical species. Another important open question concerns the stability of the 'symmetric' equilibrium.

Unfortunately, mathematical tractability is not the same as biological tractability. The models as formulated, require specification of numerous functions describing invasion rates, and extinction rates, and give answers in terms of joint densities of the fractions or numbers of patches with the two species. This is both more input data and more output data than the field or experimental biologist is likely to have or want. The next challenge is to describe approximations to these models in terms of simpler descriptions, such as means, variances and covariances of specified distributions. However, this approach is likely to be most successful if it proceeds from mathematical approximations of the models described here, rather than verbal approximations leading directly to simpler models.

Acknowledgment

This research was supported by grant DE - FG03 - 89ER60886/A000 from the Ecological Research Division, Office of Health and Environmental Research, U.S. Department of Energy. This support does not constitute an endorsement of the views expressed here. I thank Stavros Busenberg for comments on an earlier version.

References

1. DeAngelis, D. L., Waterhouse, J. C. (1987): Equilibrium and nonequilibrium concepts in ecological models. Ecological Monographs **57**, 1-21
2. Diamond, J., Case,T. J. (1986): Community Ecology. New York: Harper and Row.
3. Gurtin, M. E., MacCamy, R. C. (1974): Nonlinear age dependent population dynamics. Arch. Rat. Mech. Anal. **54**, 281-230
4. Gutzwiller, K. J., Anderson, S. H. (1988): Co-occurrence patterns of cavity-nesting birds in Wyoming cottonwood-willow stands. Manuscript
5. Hastings, A. (1977): Spatial heterogeneity and the stability of predator-prey systems. Theoretical Population Biology **12**, 37-48
6. Hastings, A. (1987): Can competition be detected using species co-occurrence data. Ecology **68**, 117-124
7. Hastings, A. (1990): A metapopulation model with local disasters of varying sizes. Submitted to Journal of Mathematical Biology
8. Hastings, A., Wolin, C.L. (1989): Within patch dynamics in a metapopulation. Ecology **70**, 1261-1266
9. Hoppensteadt, F. (1974): An age dependent epidemic model. J. Franklin Institute **297**, 325-333
10. Hoppensteadt, F. (1975): Mathematical Theories of Populations; Demographics, Genetics, and Epidemics. SIAM Regional Conference Series in Applied Mathematics
11. Kareiva, P. (1986): Patchiness, dispersal, and species interactions: consequences for communities of herbivorous insects. In: Diamond, J. and Case, T.J. (editors) Community Ecology. New York:Harper and Row
12. Kareiva, P. (1987): Habitat fragmentation and the stability of predator-prey interactions. Nature **326**, 388-390
13. Levin, S. A. (1976): Population dynamics in heterogeneous environments. Ann. Rev. Ecol. Syst. **7**, 287-310.
14. Levin, S. A., Paine, R. T. (1974): Disturbance, patch formation and community structure. Proc. Nat. Acad. Sci. (USA) **71**, 2744-2747
15. McKendrick, A. G. (1926): Application of mathematics to medical problems. Proc. Edinb. Math. Soc. **44**, 98-130
16. Metz, J. A. J., Diekmann, O. (1986): The Dynamics of Physiologically Structured Populations. Springer-Verlag, Berlin-Heidelberg-New York

17. Okubo, A. (1986): Diffusion and Ecological Problems: Mathematical Models. Springer-Verlag, Berlin-Heidelberg-New York
18. Rorres, C. (1979): Local stability of a population with density-dependent fertility. Theoretical Population Biology **16**, 283-300
19. Sinko, J. W., Streifer, W. (1967): A new model for age-size structure of a population. Ecology **48**, 910-918
20. Toft, C. A., Schoener, T. W. (1983): Abundance and diversity of orb spiders on 106 Bahamanian islands: biogeography at an intermediate trophic level. Oikos **41**, 411-426
21. Webb, G. (1985): Theory of Nonlinear Age-Dependent Population Dynamics. New York: Marcel Dekker, Inc.
22. Zabreyko, P.P., Koshlev, A.I., Krasnosel'skii, Mikhlin, S.G., Rakovshcik, L. S., Stet'senko, V. Ya. (1975): Integral equations – a reference text. Levden, The Netherlands: Noordhoff International Publishing

Generic Failure of Persistence and Equilibrium Coexistence in a Model of m-species Competition in an n-vessel Gradostat when $m > n$

Willi Jäger[1], *Hal Smith*[2] *and Betty Tang*[2]

[1] Universität Heidelberg, Institut für Angewandte Mathematik, D-6900 Heidelberg 1, Germany
[2] Department of Mathematics, Arizona State University, Tempe, AZ 85287, USA

Abstract

A mathematical model of competition between m species for a single limiting resource in an n vessel gradostat is studied. It is shown that the system has no steady state coexistence in general if $m > n$, and there is always a saturated equilibrium which must be on the boundary of the state space. Thus the system does not exhibit any form of persistence.

1 Introduction

The population dynamics of m species in an ecosystem can often be described by a system of ordinary differential equations of the form

$$x_k' = x_k g_k(x_1, ..., x_m), \quad k = 1, ..., m, \tag{1.1}$$

on $\mathbb{R}_+^m$, where x_k is the density of species k and $g_k(x_1, ..., x_m)$ its specific growth rate. Competitive coexistence or exclusion of the m species when there are less than m resources or limiting factors in the ecosystem is an intriguing question and has no simple answer. A good review of this is given by Armstrong and McGehee [1].

Coexistence can occur in different forms including steady state and oscillation. The precise definition of coexistence is quite delicate, as illustrated by the example of May and Leonard [10]. Several concepts related to coexistence of interacting populations as modeled by (1.1) have been developed. These include *weak persistence*, *persistence* (or *strong persistence*) and *uniform persistence*, (or *permanence*). (See Butler, Freedman and Waltman [3] or Hofbauer and Sigmund [7].) Weak persistence means

$$\limsup_{t\to\infty} x_k(t) > 0,$$

persistence means

$$\liminf_{t\to\infty} x_k(t) > 0,$$

and uniform persistence means there exists $\delta > 0$, independent of the initial data, such that

$$\liminf_{t\to\infty} x_k(t) \geq \delta,$$

for all k, whenever $x_k(0) > 0$. If (1.1) is dissipative, uniform persistence can be defined equivalently as the existence of a compact set K in $\overset{\circ}{\mathbb{R}}{}^m_+$ such that all trajectories end up in K. (K is called an *attractor*.) Obviously uniform persistence is the most desirable situation as all m species coexist at positive densities uniformly bounded away from zero indefinitely. On the other hand, the absence of any form of persistence does not automatically imply exclusion since the species could still coexist for certain initial conditions.

In this paper we study whether m species competing for a single substrate in an n vessel gradostat, where $m > n$, can exhibit some form of coexistence or persistence. The gradostat is a concatenation of chemostats simulating an ecosystem where there is a substrate gradient [14]. Mathematical models of microbial competition in the gradostat, assuming Monod growth model, have been studied previously [9], [11], [12]. It was shown that uniform persistence, in fact steady state coexistence, of two species competing for a single substrate in the gradostat is possible. Moreover, the phenomenon is robust in the sense that there is a set of nonzero measure in the parameter space corresponding to which the two species coexist. This result is different from the situation in a homogenous environment, as simulated by a chemostat, where at most one species lives indefinitely [8].

We show that the set of parameters in the parameter space corresponding to which all the m species coexist in steady state is of measure zero, thus in general no steady state coexistence is possible. In addition, there exist some trajectories which originate from the interior of the state space and eventually approach a boundary equilibrium point, so the system does not exhibit any form of persistence.

This paper is organized as follows. The model equations are given in Section 2. The case of one species growth is summarized in Section 3. In Section 4 we show that in general m species do not coexist at steady state in the gradostat. The main mathematical tool used is Sard's Theorem. A degree theory argument is used in Section 5 to show that the system has a saturated equilibrium on the boundary and therefore does not exhibit any form of persistence.

2 The model

Denote by $\mathbb{R}^k_+$ the k-dimensional closed nonnegative cone and $\overset{\circ}{\mathbb{R}}{}^k_+$ its interior. We sometimes write $x \geq y$ when $x - y \in \mathbb{R}^k_+$. Let $S = (S_1, ..., S_n)$ and $u^j = (u^j_1, ..., u^j_n)$, where $S_i(t)$ and $u^j_i(t)$ are respectively the concentration of the limiting substrate and of the j-th species in the i-th vessel of the gradostat at time t, $1 \leq i \leq n$, $1 \leq j \leq m$. Competition between the species for the limiting substrate, using Monod growth model, can be described by the following system of equations:

$$S' = e_1 + AS - \sum_{j=1}^{m} F_j(S)u^j$$

$$u^{j\prime} = Au^j + F_j(S)u^j, \tag{2.1}$$

$$S(0) \in \mathbb{R}^n_+, \quad u^j(0) \in \mathbb{R}^n_+, \quad 1 \leq j \leq m,$$

where e_1 is the n dimensional vector $(1, 0, ..., 0)$, and A and $F_j(S)$ are $n \times n$ matrices:

$$A = \begin{pmatrix} -2 & 1 & 0 & 0 & \dots & 0 \\ 1 & -2 & 1 & 0 & \dots & 0 \\ 0 & 1 & -2 & 1 & \dots & 0 \\ \vdots & \vdots & \vdots & \vdots & \ddots & \vdots \\ 0 & \dots & \dots & \dots & 1 & -2 \end{pmatrix}$$

and $F_j(S)$ is a diagonal matrix with the i-th diagonal element being

$$f_j(S_i) = \frac{m_j S_i}{a_j + S_i}.$$

(Some parameters have been scaled from the equations. See [9],[12] for more details on the derivation of the equations and the appropriate scaling.) The quantities m_j and a_j are properties of the j-th species and are related to respectively the maximal growth rate and the Michaelis-Menten constant of the j-th species with respect to the limiting substrate.

Lemma 2.1 *Solutions of* (2.1) *with nonnegative initial data exist and are nonnegative and bounded for* $t \geq 0$. *Moreover,*

$$\lim_{t\to\infty} S(t) + \sum_{j=1}^{m} u_j(t) = z$$

where $z = (z_1, ..., z_n)$, $z_i = 1 - i/(n+1)$, *is the unique solution of*

$$Az + e_1 = 0.$$

The proof of this lemma is essentially the same as in [12], [13].

Thus on the omega limit set, solutions of (2.1) satisfy

$$u^{j\prime} = \left[A + F_j \left(z - \sum_{j=1}^{m} u^j \right) \right] u^j, \quad u^j(0) \geq 0, \quad 1 \leq j \leq m \tag{2.2}$$

on

$$\Omega = \left\{ (u^1, ..., u^m) | (u^1, ..., u^m) \in \mathbb{R}_+^{nm}, \quad \sum_{j=1}^{m} u^j \leq z \right\}.$$

It is system (2.2) that we analyze.

3 Growth without competition

Single population growth in the gradostat is governed by the equation

$$u^{j\prime} = \left[A + F_j(z - u^j) \right] u^j, \quad u^j(0) = u^j_o. \tag{3.1}$$

Let $\sigma(M)$ be the spectrum of a matrix M. The *stability modulus* of M, $s(M)$, is defined as

$$s(M) = \max\{Re(\lambda) | \lambda \in \sigma(M)\}.$$

Theorem 3.1 ([12], [13]) *If $s(A + F_j(z)) \leq 0$ then $u^j(t) \to 0$ as $t \to \infty$ for every $u^j_o \in \mathbb{R}^n_+$. If $s(A + F_j(z)) > 0$ there exists a unique equilbrium $\hat{u}^j \in \mathbb{R}^n_+$, such that $u^j(t) \to \hat{u}^j$ as $t \to \infty$ for every nonzero $u^j_o \in \mathbb{R}^n_+$.*

Let E_0 be the equilbrium point of (2.2) corresponding to $u^j = 0$ for all j, and E_p the equilibrium point corresponding to $u^j = 0$ if $j \neq p$ and $u^p = \hat{u}^p$. If $s(A + F_p(z)) \leq 0$, E_p does not exist and $u^p(t) \to 0$ as $t \to \infty$ for every solution of (2.2) by simple comparison [5, Ch. 1, Thm. 10]. In that case the omega limit set of any solution of (2.2) is in the $n \times (m - 1)$ dimensional set $\Omega \cap \{u^p = 0\}$. If $s(A + F_j(z)) \leq 0$ for all j, all solutions of (2.2) tend to E_0.

4 Coexistence equilibrium

It has been shown in [9],[12] that two species can coexist in equilibrium in the gradostat. In particular, if m_i and a_i, $i = 1, 2$, are such that

$$\begin{aligned} s(A + F_2(z - \hat{u}^1)) &> 0, \\ s(A + F_1(z - \hat{u}^2)) &> 0, \end{aligned} \tag{4.1}$$

then the system is uniformly persistent and there exists a coexistence equilibrium $(u^{1*}, u^{2*}) \in \overset{\circ}{\mathbb{R}}{}^{n2}_+$. If n=2, (u^{1*}, u^{2*}) is unique and $(u^1(t), u^2(t)) \to (u^{1*}, u^{2*})$ as $t \to \infty$ whenever $(u^1(0), u^2(0)) \in \overset{\circ}{\mathbb{R}}{}^4_+$ [9]. Numerical simulations suggest that the set of (m_1, a_1, m_2, a_2) in the parameter space for which (4.1) is satisfied is of nonzero measure. Thus if $m = 2$, the existence of an equilibrium point of (2.2) in $\overset{\circ}{\mathbb{R}}{}^{n2}_+$ is a robust feature.

We study here equilibrium points E^* of (2.2) corresponding to the coexistence of all m species in the n-vessel gradostat, where $m > n$. Since there is obviously no coexistence equilbrium if any E_j does not exist, we assume henceforth the following:

$$s(A + F_j(z)) > 0, \quad 1 \leq j \leq m, \tag{H1}$$

so that system (2.2) has at least the following nontrivial single species equilibrium points

$$E_1, ..., E_m$$

on the boundary of $\overset{o}{\mathbb{R}}{}^{nm}_{+}$.

Rather than considering solutions $E^* = (u^{1*}, ..., u^{m*})$ of the system of algebraic equations

$$\left[A + F\left(z - \sum_{j=1}^{m} u^j\right)\right] u^j = 0, \quad 1 \leq j \leq m \tag{4.2}$$

in $\overset{o}{\mathbb{R}}{}^{nm}_{+}$ directly, we will first consider the problem of finding $S = (S_1, ..., S_n) \in \overset{o}{\mathbb{R}}{}^{n}_{+}$, such that

$$s(A_j(S, m_j, a_j)) = 0, \quad 1 \leq j \leq m, \tag{4.3}$$

where $A_j(S, m_j, a_j)$ is the matrix

$$\begin{pmatrix} D_1 & a_j + S_1 & 0 & \dots & 0 \\ a_j + S_2 & D_2 & a_j + S_2 & \dots & 0 \\ \vdots & \vdots & \vdots & \ddots & \vdots \\ 0 & 0 & \dots & a_j + S_n & D_n \end{pmatrix},$$

where

$$D_i = m_j S_i - 2(a_j + S_i).$$

Since positive solutions of (4.2) exist only if (4.3) has a solution S, by the Perron-Frobenius Theorem [2], conditions for insolvability of (4.3) also apply to (4.2).

Denote $s(A_j(S, m_j, a_j))$ by $r_j(S, m_j, a_j)$ and let $m = (m_1, ..., m_m)$ and $a = (a_1, ..., a_m)$. Note the distinction between m as a subscript or superscipt and m as a vector.) Define

$$r : \overset{o}{\mathbb{R}}{}^{n}_{+} \times \overset{o}{\mathbb{R}}{}^{m}_{+} \times \overset{o}{\mathbb{R}}{}^{m}_{+} \to \mathbb{R}^m$$

to be a map such that

$$r(S, m, a) = (r_1(S, m_1, a_1), ..., r_m(S, m_m, a_m)).$$

Since $r_j = r_j(S, m_j, a_j)$ is smooth for all $S \in \overset{o}{\mathbb{R}}{}^{n}_{+}$, $m_j > 0$ and $a_j > 0$, r is also smooth.

Lemma 4.1 *The $m \times m$ matrix $\frac{\partial r}{\partial m}$ is nonsingular at every $(S, m, a) \in r^{-1}(0)$, hence $0 \in \mathbb{R}^m$ is a regular value or r.*

Proof. Let $(S,m,a) \in r^{-1}(0) \subset \overset{\circ}{\mathbb{R}}{}^n_+ \times \overset{\circ}{\mathbb{R}}{}^m_+ \times \overset{\circ}{\mathbb{R}}{}^m_+$. Observe that

$$\frac{\partial r}{\partial m}(S,m,a) = diag(\alpha_j)$$

where

$$\alpha_j = \frac{\partial r_j}{\partial m_j}(S,m_j,a_j), \quad 1 \le j \le m,$$

thus $\frac{\partial r}{\partial m}(S,m,a)$ is nonsingular if $\alpha_j \neq 0$ for all j. Let $w^j = (w^j_1,...,w^j_n) \in \overset{\circ}{\mathbb{R}}{}^n_+$ be the unit eigenvector of $A_j(S,m_j,a_j)$ corresponding to the eigenvalue $r_j(S,m_j,a_j) = 0$, and $\delta m_j > 0$. Then

$$\begin{aligned} A_j(S,m_j+\delta m_j,a_j)w^j &= \delta m_j(S_1w^j_1,...,S_nw^j_n) \\ &\ge \delta m_j\left(\min_k S_k\right)w^j. \end{aligned}$$

Hence $r_j(S,m_j+\delta m_j,a_j) \ge \delta m_j\,(\min_k S_k)$ [2, p.28, Thm. 1.11] and therefore

$$\begin{aligned} \alpha_j &= \lim_{\delta m_j \to 0} \frac{r_j(S,m_j+\delta m_j,a_j) - r_j(S,m_j,a_j)}{\delta m_j} \\ &> 0. \end{aligned}$$

□

Corollary 4.2 [4, p. 57, Thm. 10.3]. *The set*

$$Q = \{(m,a) \in \overset{\circ}{\mathbb{R}}{}^{2m}_+ \mid 0 \quad \textit{is a critical value of} \quad r(\cdot,m,a)\}$$

is of measure zero in $\mathbb{R}^{2m}$.

Let P be the set $\{(m,a) \in \overset{\circ}{\mathbb{R}}{}^{2m}_+ | (4.3)$ has a solution $S \in \overset{\circ}{\mathbb{R}}{}^n_+$ corresponding to $(m,a)\}$. The main result of this section is given in the next theorem.

Theorem 4.3 *If* $m > n$, $P \subseteq Q$ *and therefore* P *also has measure zero in* $\overset{\circ}{\mathbb{R}}{}^{2m}$.

Proof. Let $(m,a) \in P$, then there exists $S \in \overset{\circ}{\mathbb{R}}{}^n_+$ such that $r(S,m,a) = 0$. Since $m > n$, the $m \times n$ matrix $\frac{\partial r}{\partial S}$ is not a surjective map from $\mathbb{R}^n$ to $\mathbb{R}^m$. Therefore 0 is a critical value of the map $r(\cdot,m,a)$ and $(m,a) \in Q$. □

The last theorem implies that if $m > n$, an equilibrium E^*, which corresponds to steady state coexistence of all m species in a gradostat with n vessels, can exist only if (m,a) belongs to a set of measure zero and therefore steady state coexistence is not a robust phenomenon.

5 Saturated equilibrium

While in general all the m species cannot coexist in the n-vessel gradostat in equilibrium if $m > n$, there is still the possibility that the system (2.2) is persistent. We will show that such possibility does not exist by establishing the existence of a boundary saturated equilibrium. The notion of a saturated equilibrium was first introduced by Hofbauer [6], but we use a slightly different definition here.

Definition. A steady state $\tilde{u} = (\tilde{u}^1, ..., \tilde{u}^m) \in \mathbb{R}_+^{nm}$ of (2.2) is a *saturated equilibrium* if $s(A + F_j(\tilde{S})) \leq 0$ whenever $\tilde{u}^j = 0$, where $\tilde{S} = z - \sum_{j=1}^m \tilde{u}^j$.

Remark. Note that any equilibrium point $\tilde{u} = (\tilde{u}^1, ..., \tilde{u}^m)$ of (2.2) is such that either $\tilde{u}^j = 0$ or $\tilde{u}^j \in \overset{\circ}{\mathbb{R}}{}_+^n$ for all j. If $\tilde{u}^j \in \overset{\circ}{\mathbb{R}}{}_+^n$ for all j, $\tilde{u}$ is trivially saturated. E_0 is not saturated because of hypothesis (H1).

Theorem 5.1 *There exists a saturated equilibrium of system* (2.2).

The proof is similar to the arguments in [6]. An immigration term is introduced into the equations and then a degree theory argument is used.

We first consider the following system which is a perturbation of system (2.1):

$$\begin{aligned} S' &= e_1 + AS - \sum_{j=1}^{m} F_j(S)u^j \\ u^{j\prime} &= A u^j + F_j(S)u^j + \epsilon\, e \\ &1 \leq j \leq m, \end{aligned} \tag{5.1}$$

where $e = (1, ..., 1)$ and $\epsilon > 0$. The term ϵe can be interpreted as immigration. Let $v(t) = S(t) + \sum_{j=1}^m u^j(t)$, then

$$v' = Av + e_1 + m\epsilon\, e$$

and $v(t) \to Z(\epsilon) = -A^{-1}(e_1 + m\epsilon\, e)$ as $t \to \infty$. The analogous system to (2.2) is therefore

$$u^{j\prime} = [A + F_j(S(\epsilon))]\, u^j + \epsilon\, e, \quad 1 \leq j \leq m,$$

where $S(\epsilon) = Z(\epsilon) - \sum_{j=1}^m u^j$.

Let $F^j(u;\epsilon) = (F_1^j(u;\epsilon), ..., F_n^j(u;\epsilon))$ be the vector field

$$F^j(u;\epsilon) = [A + F_j(S(\epsilon))]\, u^j + \epsilon\, e,$$

and $F(\cdot;\epsilon) : \Omega_\epsilon \to \mathbb{R}^{nm}$ the vector field defined by

$$F(u;\epsilon) = (F^1(u;\epsilon), ..., F^m(u;\epsilon))$$

and $\Omega_\epsilon = \{u = (u^1, ..., u^m) \in \mathbb{R}_+^{nm} | \sum_{j=1}^m u^j \leq Z(\epsilon)\}$.

Lemma 5.2 *For $\epsilon > 0$, the vector field $F(\cdot;\epsilon)$ points inward on $\partial\Omega_\epsilon$, the boundary of Ω_ϵ, and $F(\cdot;\epsilon)^{-1}(0) \cap \partial\Omega_\epsilon = \emptyset$.*

Proof. Let $u \in \partial\Omega_\epsilon$, then $u_i^j = 0$ for some (i,j) or $\sum_{j=1}^m u_i^j = Z_i(\epsilon)$ for $i \in I \subseteq \{1,...,n\}$. In the former case $F_i^j(u;\epsilon) > 0$. In the latter case $S_i(\epsilon) = 0$, therefore

$$\begin{aligned}\sum_{j=1}^m F_i^j(u;\epsilon) &= \sum_{j=1}^m u_{i-1}^j - 2Z_i(\epsilon) + \sum_{j=1}^m u_{i+1}^j + m\epsilon \\ &\leq Z_{i-1}(\epsilon) - 2Z_i(\epsilon) + Z_{i+1}(\epsilon) + m\epsilon \\ &= \begin{cases} 0, & i \neq 1 \\ -1, & i = 1, \end{cases}\end{aligned}$$

if $i \in I$. If $1 \in I$, $\sum_{j=1}^m F_i^j(u;\epsilon) < 0$. If $1 \notin I$, there exists i such that $i \in I$ but $i-1 \notin I$, hence $\sum_{j=1}^m F_i^j(u;\epsilon) < 0$. □

We now show that $deg(F(\cdot;\epsilon)0) = (-1)^{nm}$ so there exists $u_\epsilon \in \mathring{\Omega}_\epsilon$ such that $F(u_\epsilon;\epsilon) = 0$. Let $G(\cdot,\epsilon) : \Omega_\epsilon \to \mathbb{R}^{nm}$ be the map

$$G(u;\epsilon) = (A\,u^1 + \epsilon\, e, ..., A\,u^m + \epsilon\, e)$$

and define a homotopy between F and G: $H(\cdot,\cdot;\epsilon) : \Omega_\epsilon \times [0,1] \to \mathbb{R}^{nm}$,

$$H(u,\lambda;\epsilon) = \lambda F(u;\epsilon) + (1-\lambda)G(u;\epsilon).$$

Lemma 5.3 *For $0 \leq \lambda < 1$, $H(\cdot,\lambda;\epsilon)^{-1}(0) \cap \partial\Omega_\epsilon = \emptyset$.*

Proof. If $H(u,0;\epsilon) = 0$, $u^j = -\epsilon\, A^{-1} e \in \mathring{\mathbb{R}}_+^n$ for all j. For $0 < \lambda < 1$, similar arguments as in the proof of the last lemma show that $H(u,\lambda;\epsilon) \neq 0$ if $u \in \partial\Omega_\epsilon$. □

Direct calculation shows that $deg(G(\cdot;\epsilon)) = sgn(\det A)^m = (-1)^{nm}$, hence $deg(F(\cdot;\epsilon)) = (-1)^{nm}$ by homotopy invariance.

Proof of Thm 5.1. Let $\{\epsilon_k\}$ be a sequence and $\epsilon_k \to 0$ as $k \to \infty$. By degree theory it follows that there exists $u_{\epsilon_k} \in \mathring{\Omega}_{\epsilon_k}$ such that $F(u_{\epsilon_k};\epsilon_k) = 0$ for all k. Since that sequence $\{u_{\epsilon_k}\}$ lies in a compact set in $\mathbb{R}_+^{nm}$, we may as well assume $u_{\epsilon_k} \to \tilde{u}$ as $k \to \infty$, where $\tilde{u} \in \Omega_0 = \Omega$, and $F(\tilde{u};0) = 0$ for continuity. This shows that $\tilde{u}$ is an equilibrium point of (2.2). If $\tilde{u} \in \mathring{\Omega}$ it must be saturated. Suppose $\tilde{u}_i^j = 0$ for some point (i,j), then $\tilde{u}^j = 0$. Let $\sigma_j(\epsilon_k) = s(A + F_j(S(\epsilon_k)))$, where

$$S(\epsilon_k) = Z(\epsilon_k) - \sum_{j=1}^m \tilde{u}_{\epsilon_k}^j.$$

The Perron-Frobenius theorem asserts the existence of $w^j(\epsilon_k) = (w_1^j(\epsilon_k), ..., w_n^j(\epsilon_k)) \in \mathring{\mathbb{R}}_+^n$ such that

$$[A + F_j(S(\epsilon_k))]\, w^j(\epsilon_k) = \sigma_j(\epsilon_k) w^j(\epsilon_k).$$

Since $F^j(u^j_{\epsilon_k};\epsilon_k) \bullet w^j(\epsilon_k) = 0$,

$$\begin{aligned}[A + F_j(S(\epsilon_k))]u_{\epsilon_k} \bullet w^j(\epsilon_k) + \epsilon_k e \bullet w^j(\epsilon_k) \\ = \sigma(\epsilon_k)\,(u_{\epsilon_k} \bullet w^j(\epsilon_k)) + \epsilon_k e \bullet w^j(\epsilon_k) \\ = 0,\end{aligned}$$

so $\sigma_j(\epsilon_k) < 0$. Upon taking limits we obtain $s(A + F_j(\tilde{S})) \leq 0$. □

Remark. The existence of a saturated equilibrium is independent of m and n.

Corollary 5.4 *Suppose all equilibrium points of* (2.2) *are nondegenerate and* $m > n$, *then system* (2.2) *is not weakly persistent.*

Proof. Let $\tilde{u} = \tilde{u}(m,a)$ be a saturated equilibrium point of (2.2). Suppose $\tilde{u} \in \overset{\circ}{\mathbb{R}}{}^{nm}_{+}$, i.e, $\tilde{u}$ is a coexistence equilibrium, then by the Implicit Function Theorem there exists an open set U in $\overset{\circ}{\mathbb{R}}{}^{2m}_{+}$ containing (m,a) such that for each $(m,a) \in U$ there is a coexistence equilibrium of u of (2.2). This contradicts Theorem 4.3 and so $\tilde{u}$ must be on a portion of the boundary $B = \prod_{j \in J} \{u^j = 0\}$ of Ω, where $J \subset \{1, ..., m\}$, and $s(A + F_j(\tilde{S})) < 0$ for $j \in J$ by our nondegeneracy assumption. The tangent space of the stable manifold of $\tilde{u}$ (denoted by $W^+(\tilde{u})$) at $\tilde{u}$ contains vectors transverse to B. Thus $W^+(\tilde{u}) \cap \overset{\circ}{\Omega} \neq \emptyset$.

Remark. The saturated equilibrium $\tilde{u}$ may not be stable on B, but there is a nonempty subset of $\overset{\circ}{\Omega}$ which is attracted to $\tilde{u}$. Ecologically it means that the introduction of a small concentration of the missing species $j \in J$ to the vessels of the gradostat will result in their decay to zero. Thus the species $j \in J$ are unable to invade the coexistence steady state $\tilde{u}$ of the species $k \in J^c$.

Acknowledgment

The work of Hal Smith was partially supported by NSF Grant DMS8722279. Betty Tang's work was partially supported by NSF Grant DMS8905482.

References

1. Armstrong, R.A., McGehee, R. (1980): Competitive exclusion. Am. Nat. **115**, 151-170
2. Berman, A., Plemmons, R.J. (1979): Nonnegative Matrices in the Mathematical Sciences. Academic Press, New York
3. Butler, G., Freedman, H.I., Waltman, P. (1986): Uniformly persistent systems. Proc. Amer. Math. Soc. **96**, 425-430

4. Chow, S.N., Hale, J.K. (1982): Methods of Bifurcation Theory. Springer-Verlag, Berlin-Heidelberg-New York
5. Coppel, W.A. (1965): Stability and Asymptotic Behavior of Differential Equations. Heath, Boston
6. Hofbauer, J. (1988): Saturated equilibria, permanence, and stability for ecological systems. In Mathematical Ecology, L.J. Gross, T.G. Hallam, S.A. Levin, eds., World Scientific
7. Hofbauer, J., Sigmund, K. (1988): The Theory of Evolution and Dynamical Systems. University Press, Cambridge
8. Hsu, S.B., Hubbell, S.P., Waltman, P. (1977): A mathematical theory for single nutrient competition in continuous cultures of microorganisms. SIAM J. Appl. Math. **32**, 366-383
9. Jäger, W., So, J., Tang, B., Waltman, P. (1987): Competition in the gradostat. J. Math. Biol. **25**, 23-42
10. May, R.M., Leonard, W. (1975): Nonlinear aspects of competition between three species. SIAM J. Appl. Math. **29**, 243-252
11. Smith, H.L., Tang, B. (1989): Competition in the gradostat: the role of the communication rate. J. Math. Biol. **27**, 139-165
12. Smith, H.L., Tang, B., Waltman, P.: Competition in an n-vessel gradostat. SIAM J. Appl. Math., to appear
13. Tang, B. (1986): Mathematical investigations of growth of microorganisms in the gradostat. J. Math. Biol. **23**, 319-339
14. Wimpenny, J.W.T., Lovitt, R.W. (1984): The investigation and analysis of heterogeneous enviroments using the gradostat. In Microbiological Methods for Environmental Biotechnology, J.M. Grainger, J.M. Lynch, eds., Academy Press, Orlando

Boundedness of Solutions in Neutral Delay Predator-Prey and Competition Systems

Y. Kuang

Department of Mathematics, Arizona State University, Tempe, AZ 85287

Abstract

In this paper, we establish conditions under which solutions of

$$\begin{cases} \dot{x}(t) = rx(t)[1 - \int_0^{\tau_1} x(t-s)d\mu(s) - \rho\dot{x}(t-\tau_2) - y(t-\tau_3)g(x(t))] , \\ \dot{y}(t) = y(t)[a + bx(t-\tau_4)g(x(t-\tau_4)) - cy(t-\tau_5)] , \end{cases} \tag{2.1}$$

will be bounded. This partially answers the open questions proposed by this author in his recent works on neutral predator-prey and competition systems.

1 Introduction

The autonomous logistic delay differential equation

$$\dot{x}(t) = rx(t)[1 - x(t-\tau)/K] , \tag{1.1}$$

where "$\cdot$"$= d/dt, r, K, \tau$ are positive constants, has been widely used as a model equation capable of showing oscillations of single-species population sizes in constant environments closed to both immigration and emigration (see Cushing (1977), Gopalsamy and Zhang (1988), Hale (1977), Kuang and Feldstein (1991) and Pielou (1977)). It has been the object of intensive analysis by numerous authors (see the references cited in Gopalsamy and Zhang (1988)). Indeed, it is a natural generalization of the following well-known logistic single species population equation

$$\dot{x}(t) = rx(t)[1 - x(t)/K] . \tag{1.2}$$

Here r is called the intrinsic growth rate of the species x, K is interpreted as the environment capacity for x, and $r[1 - x(t)/K]$ is the per capita growth rate of x at time t. Based on his investigation on laboratory populations of Daphnia magna, F.E. Smith (1963) argued that the per capita growth rate in (1.2) should be replaced by $r[1 - (x(t) + \rho\dot{x}(t))/K]$ (for details see Pielou (1977)). This leads to the following equation

$$\dot{x}(t) = rx(t)[1 - (x(t) + \rho\dot{x}(t))/K]. \tag{1.3}$$

We may think of x as a species grazing upon vegetation, which takes time τ to recover. In this case, it will be even more realistic to incorporate a single discrete delay τ in the per capita growth rate, which results in the following neutral delay logistic equation

$$\dot{x}(t) = rx(t)[1 - (x(t-\tau) + \rho\dot{x}(t-\tau))/K]. \tag{1.4}$$

This equation was introduced and investigated by Gopalsamy and Zhang (1988). Subsequently, it was studied by Freedman and Kuang (1991), Kuang and Feldstein (1991). The focus of these works is the qualitative behavior of the solutions, such as boundedness, asymptotic stability and oscillation.

Assume the population $x(t)$ described by (1.4) is a prey species, and suppose there exist a predatory species $y(t)$ prey on species $x(t)$, then it is natural to proposed the following mathematical model to describe their interaction,

$$\begin{cases} \dot{x}(t) = rx(t)[1 - (x(t-\tau) + \rho\dot{x}(t-\tau))/K] - y(t)p(x(t)), \\ \dot{y}(t) = y(t)[-\alpha + \beta p(x(t-\sigma))]. \end{cases} \tag{1.5}$$

Here α, β, σ are all positive constants, and $p(x)$ is the predator response function for the predator species y with respect to the prey species x. A slightly more general version of (1.5) was introduced and studied in Kuang (1991a). Where the focus of the study was the local stability and oscillation analysis of system (1.5).

Assume $x(t)$ described by (1.4) is the population of a species competing with another species with population $y(t)$ for a shared limited resource—space or a nutrient, for example, then the following system may model their interaction,

$$\begin{cases} \dot{x}(t) = r_1x(t)[1 - k_1x(t) - \alpha x(t-\tau_1)) - \beta\dot{x}(t-\tau_0) - c_1y(t-\tau_2)] \\ \dot{y}(t) = r_2y(t)[1 - c_2x(t-\tau_3) - k_2y(t-\tau_4)]. \end{cases} \tag{1.6}$$

Here all parameters except β are assumed to be positive constants. We have included $k_1x(t)$ into the per capita growth rate of $x(t)$, which may reflect the possible instantaneous interference within species x. System (1.6) was introduced and studied in Kuang (1991b). Again the focus of that work was the local stability and oscillatory analysis of system (1.6).

For both systems (1.5) and (1.6), the question on the boundedness of solutions was left open in the previous works. This will be the focus of our present study.

2 Preliminaries

In this paper, we will consider the following more general system:

$$\begin{cases} \dot{x}(t) = rx(t)[1 - \int_0^{\tau_1} x(t-s)d\mu(s) - \rho\dot{x}(t-\tau_2) - y(t-\tau_3)g(x(t))], \\ \dot{y}(t) = y(t)[a + bx(t-\tau_4)g(x(t-\tau_4)) - cy(t-\tau_5)], \end{cases} \tag{2.1}$$

where $\tau_i \geq 0, i = 1, \ldots 5$; $\rho > 0, r > 0, c \geq 0$; a and b are real numbers, μ is nondecreasing and $\int_0^{\tau_1} d\mu(s) < +\infty$. When $a < 0$, $b > 0$, $\tau_1 = \tau_2$, $\tau_3 = 0$ and $c = 0$, (2.1) reduces to a slightly more general form of (1.5). When $a > 0, b < 0$, (2.1) has (1.6) as a special case. Let $p(x) = xg(x)$. We always assume:

(H1) $p(x)$ is continuously differentiable, $p(0) = 0$, and $p'(x) > 0$ for $x > 0$,

(H2) $\lim_{x\to+\infty} p(x) > |a/b|$.

Let $\tau_0 = \max\{\tau_i, i = 1, 2, \ldots, 5\}$. We always assume that the initial conditions for (2.1) are of the type:

$$x(s) = \phi_1(s) \geq 0, s \in [-\tau_0, 0], \phi_1(0) > 0 \text{ and } \phi_1 \in C^1([-\tau_0, 0], R^+),$$
$$y(s) = \phi_2(s) \geq 0, s \in [-\tau_0, 0], \phi_2(0) > 0 \text{ and } \phi_2 \in C^1([-\tau_0, 0], R^+).$$

We say $(x(t), y(t))$ is a solution of (2.1) on $[-\tau_0, \infty)$, if both $x(t)$ and $y(t)$ are positive continuously differentiable functions and satisfy both the above initial conditions and system (2.1). It is not difficult to see that solutions of (2.1) corresponding to the initial conditions of the above type exist and are unique, and they are always positive and are defined on $[0, \infty)$.

We say a function $x(t)$ defined on $[-\tau, \infty)$ is *oscillatory about* x^*, if there exists a sequence $\{t_n\} \to +\infty$, as $n \to +\infty$, for which $x(t_n) = x^*$, $n = 1, 2, \ldots$.

Without loss of generality, we will *always assume* $\tau_1 > 0$, $\int_0^{\tau_1} d\mu(s) = 1$, and $\tau_2 \leq \tau_1$.

In the rest of this paper, we denote $||\phi_1(s)|| = \max\{\phi, (s), s \in [-\tau_0, 0]\}$.

3 Boundedness of $x(t)$

In this section, our main object is to establish conditions under which $x(t)$ will be bounded. To this end, we need the following lemma.

Lemma 3.1 *If* $0 < \alpha \leq e^{-1}$, *then there exists a* $\lambda(\alpha)$, $\lambda(\alpha) > 1$, *such that* $e^{\alpha\lambda(\alpha)} = \lambda(\alpha)$, *and* $e^{\alpha x} > x$, *for* $x < \lambda(\alpha)$.

Proof. If $\alpha = e^{-1}$, then $e^{\alpha x} \geq x$, for all $x \in (-\infty, +\infty)$, and $e^{\alpha x} = x$ if and only if $x = e$. Since $e^{\alpha x}$ is strictly increasing with respect to α (when $x > 0$), and $e^{\alpha x} > 0$, for all $x \in (-\infty, +\infty)$, we see that if $0 < \alpha < e^{-1}$, then $e^{\alpha x}$ will intersect with x at two points, say $x_1(\alpha)$ and $x_2(\alpha)$, $1 < x_1(\alpha) < x_2(\alpha)$. Let $\lambda(\alpha) = x_1(\alpha)$, then it is easy to see that $e^{\alpha x} > x$, for $x < \lambda(\alpha)$. This proves the lemma. □

Now we are ready to state and prove our main result of this section.

Theorem 3.1 *Let* $\alpha = r(\tau_1 + \rho) \le e^{-1}$, *and* $\lambda(\alpha)$ *be defined as in* Lemma 3.1. *If* $\|\phi_1\| < 1$, *then* $x(t) < \lambda(\alpha)$, *for* $t \ge -\tau_0$.

Proof. It is clear that we have

$$\dot{x}(t) \le rx(t)\left[1 - \int_0^{\tau_1} x(t-s)d\mu(s) - \rho\dot{x}(t-\tau_2)\right]. \tag{3.1}$$

If $x(t)$ is not bounded by $\lambda(\alpha)$, then there exist $t^* > t_0 > 0$, such that $x(t^*) = \lambda(\alpha)$, $x(t_0) = 1$, $1 < x(t) < \lambda(\alpha)$, for $t_0 < t < t^*$, and $x(t) < \lambda(\alpha)$ for $t \in [-\tau_0, t^*)$. It is easy to see that (3.1) implies

$$x(t) \le x(t_0)e^{r\int_{t_0}^{t}[1-\int_0^{\tau_1} x(\tau-s)d\mu(s)]\,d\tau} e^{r\rho[x(t-\tau_2)-x(t_0-\tau_2)]}, \text{ for } t \ge t_0 \tag{3.2}$$

Since $x(t^* - \tau_2) \ge 0$, $x(t_0 - \tau_2) < \lambda(\alpha)$, we have

$$e^{-r\rho[x(t^*-\tau_2)-x(t_0-\tau_2)]} < e^{r\rho\lambda(\alpha)}. \tag{3.3}$$

If $t^* \le t_0 + \tau_1$, then

$$e^{r\int_{t_0}^{t^*}[1-\int_0^{\tau_1} x(\tau-s)d\mu(s)]\,d\tau} \le e^{r\int_{t_0}^{t^*}d\tau} \le e^{r\tau_1}.$$

If $t^* > t_0 + \tau_1$, then

$$\begin{aligned} & e^{r\int_{t_0}^{t^*}[1-\int_0^{\tau_1} x(\tau-s)d\mu(s)]\,d\tau} \\ = & e^{r\int_{t_0+\tau_1}^{t^*}[1-\int_0^{\tau_1} x(\tau-s)d\mu(x)]\,d\tau} \cdot e^{r\int_{t_0}^{t_0+\tau_1}[1-\int_0^{\tau_1} x(\tau-s)d\mu(s)]\,d\tau} \\ \le & e^{r\int_{t_0}^{t_0+\tau_1}[1-\int_0^{\tau_1} x(\tau-s)d\mu(s)]\,d\tau} \le e^{r\tau_1}, \end{aligned}$$

since

$$\begin{aligned} & \int_{t_0+\tau_1}^{t^*}\left[1 - \int_0^{\tau_1} x(\tau-s)d\mu(s)\right]d\tau \\ = & \int_{t_0+\tau_1}^{t^*}\left(\int_0^{\tau_1}[1 - x(\tau-s)]\,d\mu(s)\right)d\tau \le 0. \end{aligned}$$

Thus in both cases, we have

$$e^{r\int_{t+0}^{t^*}[1-\int_0^{\tau_1} x(\tau-s)d\mu(s)]\,d\tau} \le e^{r\tau_1} < e^{r\tau_1\lambda(\alpha)}. \tag{3.4}$$

Hence

$$\lambda(\alpha) = x(t^*) < e^{r\tau_1\lambda(\alpha)} \cdot e^{r\rho\lambda(\alpha)} = e^{r(\tau_1+\rho)\lambda(\alpha)} = e^{\alpha\lambda(\alpha)}.$$

Clearly, this contradicts to the definition of $\lambda(\alpha)$. Therefore $x(t^*)$ must be less than $\lambda(\alpha)$,and the theorem is proved. □

Remark 3.1 Theorem 3.1 seems to be more general and applicable than the ones obtained in Kuang and Feldstein (1991) in the following senses:

(i) The method used in the proof of Theorem 3.1 can be applied to the more general system (2.1), while the method developed in Kuang and Feldstein (1991) can only be applied to

$$\dot{x}(t) = rx(t)[1 - x(t-\tau) - \rho\dot{x}(t-\tau)]. \tag{3.5}$$

However, their method can be applied to the nonautonomous version of (3.5).

(ii) The conditions obtained in [12] for the boundedness of solutions of (3.5) is rather complicated and thus difficult to verify. Clearly, the conditions stated in Theorem 3.1 is easy to check.

If $x(t)$ is unbounded, then the following simple theorem may characterize its behavior.

Theorem 3.2 *In system* (2.1), *if $x(t)$ is unbounded, then $x(t)$ is oscillatory about* 1.

Proof. Assume $x(t)$ is unbounded and not oscillatory about 1, then there exists a $t_0 > 0$, such that for $t \geq t_0$, $x(t) > 1$. Since $x(t)$ is unbounded, there exists a $t^* > t_1 = t_0 + \tau_1$, such that

$$x(t^* - \tau_2) > x(t_1 - \tau_2) + (r\rho)^{-1}[1 + \ell n(x(t_1))].$$

Since $y(t - \tau_3)g(x(t)) \geq 0$, we have

$$\begin{aligned} x(t^*) &\leq x(t_1)e^{r\int_{t_1}^{t^*}[1-\int_0^{\tau_1} x(\tau-s)d\mu(s)]\,d\tau} \cdot e^{-r\rho[x(t^*-\tau_2)-x(t_1-\tau_2)]} \\ &\leq x(t_1)e^{-r\rho[x(t^*-\tau_2)-x(t_1-\tau_2)]} \\ &\leq x(t_1)e^{-1-\ell n(x(t_1))} = e^{-1} < 1. \end{aligned}$$

This contradicts our assumption. Thus the theorem is proved. □

4 Main results

We consider first the case when $c > 0$ in system (2.1). This will include the neutral competition system (1.6) as a special case. When $a < 0, b > 0$, system (2.1) can be used to model the predator-prey interaction with self crowding effect on predator.

Theorem 4.1 *Assume $c > 0$, $\alpha = r(\tau_1 + \rho) \leq e^{-1}$, $\lambda(\alpha)$ is defined as in* Lemma 3.1, *and $\|\phi_1\| < 1$, then solutions of (2.1) are bounded. Moreover, $x(t) < \lambda(\alpha)$, for $t \geq -\tau_0$, and $\lim_{t\to+\infty} y(t) \leq c^{-1}\beta e^{\beta\tau_5}$, where $\beta = |a| + |b|\lambda(\alpha)g(\lambda(\alpha))$.*

Proof. The assertion on $x(t)$ follows from Theorem 3.1. Thus we have

$$\dot{y}(t) \leq y(t)[|a| + |b|\lambda(\alpha)g(\lambda(\alpha)) - cy(t - \tau_5)], \tag{4.1}$$

since $p(x) = xg(x)$ is increasing and $y(t) > 0$, for $t \geq 0$.

Let $\beta = |a| + |b|\lambda(\alpha)g(\lambda(\alpha))$. From (4.1), we have

$$\dot{y}(t) \leq \beta y(t). \tag{4.2}$$

Thus

$$y(t) \leq y(t_0)e^{\beta(t-t_0)}. \tag{4.3}$$

In particular,

$$y(t) \leq y(t-\tau_5)e^{\beta\tau_5}, \tag{4.4}$$

which implies

$$-y(t-\tau_5) \leq -y(t)e^{-\beta\tau_5}. \tag{4.5}$$

Substituting (4.5) into (4.1), we obtain

$$\dot{y}(t) \leq y(t)(\beta - ce^{-\beta\tau_5 y(t)}). \tag{4.6}$$

Clearly, solutions of

$$\dot{y}(t) = \beta y(t)(1 - c\beta^{-1}e^{-\beta\tau_5}y(t)) \tag{4.7}$$

satisfies

$$\lim_{t\to\infty} y(t) \leq c^{-1}\beta e^{\beta\tau_5}. \tag{4.8}$$

Therefore solutions of (4.6) must satisfy (4.8), which proves the theorem. □

In the rest of this section, we assume $c = 0$, $a = -\delta < 0$, $b > 0$. System (2.1) thus reduces to

$$\begin{cases} \dot{x}(t) = rx(t)\left[1 - \int_0^{\tau_1} x(t-s)d\mu(s) - \rho\dot{x}(t-\tau_2) - y(t-\tau_3)g(x(t))\right], \\ \dot{y}(t) = y(t)[-\delta + bx(t-\tau_4)g(x(t-\tau_4))]. \end{cases} \tag{4.9}$$

When delays are absent from (4.9), it may model the so- called intermediate type predator-prey interactions. For (4.9), we have the following theorem.

Theorem 4.2 *Assume $\alpha(\tau_1 + \rho) \leq e^{-1}$, $\lambda(\alpha)$ is defined as in* Lemma 3.1, *and $\|\phi_1\| < 1$. Let $x^* > 0$ be the unique solution of $bx^*g(x^*) = \delta$, $\gamma = \min\{g(x), x \in [0, \lambda(\alpha)]\}$, $y^* = \max\{y(0), \gamma^{-1}[(\rho+1)r\lambda(\alpha) + \ell n(\lambda(\alpha)/x^*)]\}$, and $\Delta = y^* \exp\{[b\lambda(\alpha)g(\lambda(\alpha))](\tau_3 + \tau_4 + 1)\}$. Then the solution $(x(t), y(t)$ of (4.9) is bounded. In particular, $x(t) < \lambda(\alpha)$ for $t \geq -\tau_0$, and $y(t) < \Delta$ for $t \geq 0$.*

Proof. Again, the assertion on $x(t)$ follows from Theorem 3.1. In the following, we assume $y(t)$ is not bounded by Δ. Clearly in this case, x^* must be less than $\lambda(\alpha)$, otherwise $\dot{y}(t) \leq 0$, thus $y(t) \leq y(0) < \Delta$.

The first equation of system (4.9) gives us

$$x(t) = x(t_0)\exp\left\{\int_{t_0}^{t}\left[1 - \int_0^{\tau_1} x(\tau - s)d\mu(s)\right]d\tau - r\rho[x(t-\tau_2) - x(t_0 - \tau_2)]\right\}$$

$$\cdot \exp\left\{-\int_{t_0}^{t} y(s-\tau_3)g(x(s))ds\right\},$$

which implies

$$x(t) \le \lambda(\alpha)\exp\{r\rho\lambda(\alpha)+r\lambda(\alpha)(t-t_0)\}\exp\left\{-\gamma\int_{t_0-\tau_3}^{t-\tau_3} y(s)ds\right\}. \tag{4.10}$$

The second equation of (4.9) implies

$$\dot{y}(t) < b\lambda(\alpha)g(\lambda(\alpha))y\,, \tag{4.11}$$

which leads to

$$y(t) < y(t_0)\exp\{b\lambda(\alpha)g(\lambda(\alpha))(t-t_0)\}\,. \tag{4.12}$$

Since $y(t)$ is not bounded by Δ, there are $t_2 > t_1 \ge 0$, such that $y(t_1) = y^*$,

$$y(t_2) = \Delta = y^*\exp\{[b\lambda(\alpha)g(\lambda(\alpha))](\tau_3+\tau_4+1)\} \tag{4.13}$$

and

$$y(t) \ge y^* \quad for \quad t \in [t_1, t_2]\,.$$

From (4.12), we see that $t_2 - t_1 > \tau_3 + \tau_4 + 1$. In (4.10), let $t_0 = t_1 + \tau_3$ and $t_0 + 1 \le t \le t_2$, then $y(s - \tau_3) \ge y^*$ for $s \in [t_0, t_2]$, thus

$$\begin{aligned} x(t) &\le \lambda(\alpha)\exp\{r\rho\lambda(\alpha)+r\lambda(\alpha)(t-t_0)\}\exp\{-\gamma y^*(t-t_0)\} \\ &\le \lambda(\alpha)\exp\{r\rho\lambda(\alpha)+r\lambda(\alpha)(t-t_0)\} \\ &\qquad \exp\{-[(\rho+1)r\lambda(\alpha)+\ell n(\lambda(\alpha)/x^*)](t-t_0)\} \\ &\le \lambda(\alpha)e^{r\rho\lambda(\alpha)}\cdot e^{-r\rho\lambda(\alpha)+\ell n(\lambda(\alpha)/x^*)}e^{[-\rho r\lambda(\alpha)+\ell n(\lambda(\alpha)/x^*)](t-t_0-1)}\,. \end{aligned}$$

Clearly, the last inequality implies that for $t \in [t_0 + 1,\, t_2]$

$$x(t) \le x^*\,. \tag{4.14}$$

From the second equation of (4.9), we have

$$y(t) = y(t_0)\exp\left\{\int_{t_0}^{t}[-\delta + bx(t-\tau_4)g(x(t-\tau_4))]\,ds\right\}. \tag{4.15}$$

Let $t_0 = t_1$ in (4.15), then for $t_1 + \tau_3 + \tau_4 + 1 \le t \le t_2$,

$$\begin{aligned} y(t) = y^*\exp&\left\{\int_{t_1}^{t_1+\tau_3+\tau_4+1}[-\delta + bx(t-\tau_4)g(x(t-\tau_4))]\,ds\right\} \\ &\cdot\exp\left\{\int_{t_1+\tau_3+\tau_4+1}^{t}[-\delta + bx(t-\tau_4)g(x(t-\tau_4))]\,ds\right\}, \end{aligned}$$

hence

$$y(t) < y^*\exp\{[b\lambda(\alpha)g(\lambda(\alpha)](\tau_3+\tau_4+1)\}\,. \tag{4.16}$$

In particular,

$$y(t_2) < y^*\exp\{[b\lambda(\alpha)g(\lambda(\alpha)](\tau_3+\tau_4+1)\} = \Delta\,.$$

This clearly contradicts our assumption (4.13). Thus $y(t)$ must be bounded, and

$$y(t) < y^*\exp\{[b\lambda(\alpha)g(\lambda(\alpha))](\tau_3+\tau_4)\},\ \text{for } t \ge 0\,.$$

This proves the theorem. □

5 Discussion

The present work is complemental to the author's previous work in [11] and [13]. In [11], this author studied the local asymptotic stability of the (assumed to exist) positive steady state of (4.9), and obtained some results on the oscillatory and non-oscillatory characteristics of its positive solutions. In [13], similar analysis was carried out for neutral delay competitive system. One of the fundamental tools used in [11] and [13] is the stability switching theory introduced in [1], and further developed in [2] and [5]. Generally speaking, the role played by the neutral term $\rho\dot{x}(t-\tau_2)$ in the per capita growth rate for x is very complex. It may serve as both a stabilizing and a destabilizing factor, dependent on the specific situation. At the very least we know its presence causes various difficulties in the mathematical analysis of these models.

One of the important questions still left open is, *what is the criterion for the existence of periodic solutions in system (2.1)*? In fact, so far, we do not even know what is the condition for a Hopf bifurcation to occur in system (2.1). Although our numeric results strongly indicates that a similar Hopf bifurcation theorem (as the retarded case) may be true for general nonlinear neutral system, we cannot provide any rigorous proof for that yet. Hopefully, the integral manifold technique developed by Hale [8,9,10] may contribute to the solution of this general problem.

Another interesting question remaining to be investigated is, *what are the conditions for the global stability of the unique positive steady state (if any) in system (2.1)*?

Finally, we would like to admit that the way we have introduced the delays in system (2.1) is not the most realistic one. The best way to do so is probably to replace those discrete delays by continuously distributed ones. Indeed, by virtue of the proofs, we can see that all the results in this paper are valid for distributed delays.

References

1. Cooke, K.L., Grossman, Z. (1982): Discrete delay, distributed delay and stability switches. J. Math. Anal. Appl. **86**, 529–627
2. Cooke, K.L., van den Driessche, P. (1986): On zeros of some transcendental equations. Funkcialaj Ekvacioj **29**, 77-90
3. Cushing, J.M. (1977): Integrodifferential Equations and Delay Models in Population Dynamics. Lecture Notes in Biomathematics **20**, Springer-Verlag, Berlin-Heidleberg-New York
4. Freedman, H.I. (1980): Deterministic Mathematical Models in Population Ecology, Marcel Dekker, New York
5. Freedman, H.I., Kuang, Y. (1991): Stability switches in linear scalar neutral delay equations. Funkcialaj Ekvacioj, **34**, to appear
6. Gopalsamy, K., Zhang, B.G. (1988): On a neutral delay logistic equation. Dynamics and Stability of Systems **2**, 183–195
7. Gopalsamy, K.: Equations Mathematical Ecology, Part 1. Autonomous Systems, preprint

8. Hale, J.K. (1971): Critical cases for neutral functional differential equations. J. Differential equations **10**, 59–82
9. Hale, J.K. (1974): Behavior near constant solutions of functional differential equations. J. of Differential Equations **15**, 278–294
10. Hale, J.K. (1977): Theory of Functional Differential Equations, Springer-Verlag, Berlin-Heidelberg-New York
11. Kuang, Y. (1991a): On neutral delay logistic Gause-type predator-prey systems. Dynamics and Stability of Systems, **6**, to appear
12. Kuang, Y., Feldstein, A. (1991): Boundedness of solutions of a nonlinear nonautonomous neutral delay equation. J. Math. Anal. Appl., **156**, to appear
13. Kuang, Y. (1991b): On neutral delay two species Lotka-Volterra competitive systems, J. Austral. Math. Soc. Ser. B **32**, 311-326
14. Pielou, E.C. (1977): Mathematical Ecology. Wiley Interscience, New York
15. Smith, F.E. (1963): Population dynamics in Daphnia magna. Ecology **44**, 651–653
16. Waltman, P. (1986): A Second Course in Elementary Differential Equations. Academic Press, New York

Some Examples of Nonstationary Populations of Constant Size

Fabio A. Milner[1] *and Tanya Kostova*[2]

[1] Department of Mathematics, Purdue University, West Lafayette, Indiana 47907, U.S.A., and Dipartimento di Matematica, IIa. Università di Roma, 00133 Roma, Italy.

[2] Institute of Mathematics, Bulgarian Academy of Sciences, 1113 Sofia, Bulgaria.

Dedicated to Kenneth Cooke on the occasion of his 65th birthday

Abstract

Examples of nonstationary populations of constant size are explicitly constructed. Necessary and sufficient conditions for a constant birth function are established under mild restrictions on the death rate and the initial distribution and examples are provided.

Introduction

An important issue in demography is that of establishing conditions, in terms of the vital dynamics rates (natality and mortality), which lead to constant populations. In the simplest setting of constant birth and death rates a necessary and sufficient condition is, obviously, that those rates be equal. In the case of age-structured models, the issue is considerably more complicated and has only been partially explored. The simplest example of an age-structured population of constant size is that of a steady state distribution (that is, a time-independent one) [2]. Prior to the work of the authors [3] no other examples were known. One of the objectives of the present work is to show numerically produced age-distributions which are not steady state but are of constant size. The computations were done based on the theory developed in [3]. Also, sufficient conditions for a birth function to be constant are given, and the asymptotic stability of the equilibrium distribution is demonstrated.

1 Background

We consider the following equations describing linear age-dependent population dynamics with an age-distribution $u = u(a,t)$ satisfying

$$\begin{cases} u_a + u_t = -\mu(a)u, & a,t > 0, \\ u(0,t) = B(t), & t \geq 0, \\ u(a,0) = \phi(a), & a \geq 0, \end{cases} \tag{1.1}$$

where a is the age, t is the time, μ is the age specific mortality and B is the birth function. The total population size is given by

$$P(t) = \int_0^\infty u(a,t)da. \tag{1.2}$$

Definition 1.1 We say $u(a,t)$ is a *distribution of constant size* P_0 if $P(t) \equiv P_0$.

If the birth function is given by

$$B(t) = \int_0^\infty b(a)u(a,t)da \tag{1.3}$$

where the non-negative function b is the age specific fertility, then the resulting model is the classical McKendrick–von Foerster's [4,5]. Assume that $b(\cdot)\exp(-\int_0^\cdot \mu(s)ds) \in L^1(R^+)$. Then, a necessary and sufficient condition for the steady state equations

$$\begin{cases} u_a = -\mu(a)u, & a > 0, \\ u(0) = \int_0^\infty b(a)u(a)da & , \end{cases}$$

to have a solution is that

$$R = \int_0^\infty b(a)\Pi(a)\, da = 1,$$

where we introduced the *net reproduction rate of the population* R and the *survival function*

$$\Pi(a) = e^{-\int_0^a \mu(\tau)d\tau}. \tag{1.4}$$

In this case $u(a)$ is given explicitly as the *equilibrium* distribution

$$p_\infty(a) = \frac{P_0 \Pi(a)}{\int_0^\infty \Pi(\tau)\, d\tau} = p_\infty(0)\,\Pi(a). \tag{1.5}$$

We shall also assume that no individual is immortal, that is,

$$\lim_{a\to\infty} \Pi(a) = 0.$$

This means that $\int_0^\infty \mu(a)\, da = +\infty$. An even more realistic requirement is that every individual die before reaching a fixed age, A, say. This is guaranteed by assuming that $\int_0^A \mu(a)\, da = +\infty$, that is, $\Pi(a) = 0$ for $a \geq A$.

Equilibrium distributions have been known for quite some time and play an important role in the determination of the asymptotic behavior of solutions of (1.1) [2]. They obviously provide an example of populations of constant size but, for demographic purposes are relatively impractical since, for any given population (for which data is available), either $R \neq 1$ or the present distribution is not given by (1.5). On the other hand, it is of demographic interest to know under what conditions the size of the population will remain constant and how the age-distribution will change in time for such populations. We shall give some insight into these issues through specific examples.

Let us assume that μ, and ϕ in (1.1) are piecewise continuous non-negative functions, $\mu \not\equiv 0$, and ϕ is compactly supported in $[0, +\infty)$. We shall also assume the following compatibility condition on the initial and boundary data of (1.1):

$$\phi(0) = B(0). \tag{1.6}$$

The first (and completely trivial) necessary and sufficient condition for a population to be of constant size P_0 is the following:

$$B(t) = \int_0^\infty \mu(a)u(a,t)da, \quad t \geq 0, \tag{1.7}$$

which simply restates that the number of births must equal the number of deaths at all times.

The following (implicit) representation of the solution of (1.1) will be useful in the sequel:

$$u(a,t) = \begin{cases} B(t-a)\exp\left(-\int_0^a \mu(\tau)d\tau\right), & t \geq a, \\ \phi(a-t)\exp\left(-\int_0^t \mu(a-t+\tau)d\tau\right), & t < a. \end{cases} \tag{1.8}$$

Necessary and sufficient conditions for the existence of a solution $u(a,t)$ of (1.1) with constant size were derived in [3]. A necessary condition is that the mean value (or, equivalently, the improper integral $\int_0^\infty g(a)\,da$) of the function

$$g(a) = \phi'(a) + \phi(a)\mu(a), \qquad 0 \leq a < +\infty, \tag{1.9}$$

be zero (this results from the combination of the compatibility condition (1.6) with the equality between the number of births and the number of deaths (1.7) at time $t = 0$). We also find in that paper an explicit way of constructing examples of populations of constant size as follows. Let $\phi \in C^0([0,+\infty))$ be such that $\phi\mu \in L^1((0,+\infty))$ and assume that g given by (1.9) has mean value zero. Let

$$F(t) = \int_0^\infty [\phi'(s) + \phi(s)\mu(s)]e^{-\int_0^t \mu(s+\tau)d\tau}ds, \qquad t \geq 0, \tag{1.10}$$

and let $H(s) = B_t(s)$. Then, $H(t)$ satisfies the following Volterra equation of the first kind,

$$\int_0^t H(s)e^{-\int_0^{t-s}\mu(\tau)d\tau}ds = F(t), \qquad t \geq 0, \tag{1.11}$$

which, for $\mu \in C^0(I)$ and $[\phi'(\cdot) + \phi(\cdot)\mu(\cdot)] \exp(-\int_0^t \mu(\cdot + \tau)d\tau) \in L^1(\mathbf{R}^+)$, admits a unique solution. Once H is found, $B(t)$ can be determined explicitly in terms of H by integration:

$$B(t) = \phi(0) + \int_0^t H(s)ds. \tag{1.12}$$

If B is then inserted in (1.8), we arrive at an explicit age-distribution of a population of constant size.

The only case when there does not exist a non-equilibrium solution with constant size is when $\phi(s) = \phi(0)\exp(-\int_0^s \mu(\tau)d\tau)$. In this case $F \equiv 0$ in (1.10) (since $\phi' + d\phi \equiv 0$) and so, $H \equiv 0$, which gives $B(t) \equiv B(0) = \phi(0)$. Therefore, (1.8) transforms into (1.5) and, as a result, we see that, from an initial equilibrium distribution, the only possible solution of (1.1) with constant size is the equilibrium distribution p_∞ (1.5).

The examples we give below are constructed by choosing the functions μ and ϕ so that g given by (1.9) has mean value zero. Then, the function F given by (1.10) is computed using the trapezoidal rule for the quadrature of the integral

$$F(t) = \int_0^\infty g(a)\, \Pi(a+t)/\Pi(a)\, da,$$

where the upper limit of integration is replaced by max sup ϕ. Equation (1.11) is then solved for H using the midpoint rule [1] and subsequently it is numerically integrated to obtain B in (1.12). This (discrete) function is substituted in (1.8) to obtain the desired constant size age-distribution. Note that the age specific birth rate b is not specified anywhere in this process and, in fact, it is not clear when a *nonnegative* function b exists as solution of the Fredholm integral equation (1.3). This is an interesting open problem in general which we shall address elsewhere. Therefore, what we obtain are populations with constant size whose birth function B is known but does not necessarily satisfy (1.1). The existence of populations of constant size depends on a relation between births, deaths, and initial distributions, independently of the restriction that a birth rate function b exists. An obvious case when a birth rate function can be found explicitly is when $B(t) \equiv B_0 = u(0,0)$ is constant. Then, it is also true that $b(a) \equiv b_0 = B_0/P_0$ satisfies (1.3). In the next section we make some considerations about the existence of constant birth functions B.

2 Constant birth functions.

We shall establish now some necessary and sufficient conditions for the birth function B to be constant (and, consequently, $B(t) \equiv \phi(0), \quad t \geq 0$) under some restrictions on ϕ and μ. We first note that B is constant if, and only if, $H(s) \equiv 0$ in (1.11), which is in turn equivalent to $F(t) \equiv 0, \ t \geq 0$, in (1.10) (if the conditions for existence and uniqueness of solutions of the Volterra integral equation (1.11) hold). On the other hand, looking at (1.10) and (1.11) we see that, if either ϕ is the equilibrium distribution p_∞ or μ is constant, then $F \equiv 0$.

Our first result basically says that for some class of functions ϕ and μ those are the only conditions under which F will vanish identically.

Theorem 2.1 *Let μ and $\phi \in C^0([0,+\infty))$ be such that $\mu\,\phi \in L^1((0,+\infty))$, μ is nondecreasing, $\phi(0) = \int_0^\infty \mu(a)\phi(a)\,da$, $\phi(0) \geq \phi(a) + \int_0^a \mu(\sigma)\phi(\sigma)\,d\sigma$, $a \geq 0$, and ϕ' is piecewise continuous. Then, $F(t)$ defined by (1.10) vanishes identically on $[0,+\infty)$ if, and only if, there exists $a^* \in [0,+\infty]$ such that $\phi(a) \equiv p_\infty(a)$, $a \leq a^*$, and $\mu(a) \equiv \mu(a^*)$, $a > a^*$ (where the case $a^* = +\infty$ must be read as $\phi(a) \equiv p_\infty(a)$, $a \geq 0$).*

Proof. Let

$$G(a) = \int_0^a g(\sigma)\,d\sigma, \qquad a \geq 0,$$

be a primitive of g, where g is defined by (1.9). Clearly

$$G(0) = 0 \tag{2.1}$$

and, by hypothesis and (1.9) it follows that

$$\lim_{a\to\infty} G(a) = 0. \tag{2.2}$$

Integrating (1.10) by parts and using (2.1) and (2.2) we obtain the following relation,

$$\begin{aligned} F(t) &= G(a)e^{-\int_a^{a+t}\mu(\sigma)d\sigma}\Big|_0^{+\infty} - G(a)e^{-\int_a^{a+t}\mu(\sigma)d\sigma}\big(-\mu(a+t)+\mu(a)\big)\,da \\ &= \int_0^\infty G(a)\big(\mu(a+t)-\mu(a)\big)e^{-\int_a^{a+t}\mu(\sigma)d\sigma}\,da. \end{aligned} \tag{2.3}$$

Note that the last hypothesis on ϕ and μ means exactly that

$$G(a) = \phi(a) - \phi(0) + \int_0^a \mu(\sigma)\phi(\sigma)\,d\sigma \leq 0, \quad a \geq 0. \tag{2.4}$$

Since the last integrand in (2.3) is continuous for $a, t \geq 0$ and, by hypothesis and by (2.4) it does not change sign, it follows that $F \equiv 0$ if, and only if, $G(a)\big(\mu(a+t)-\mu(a)\big) \equiv 0$, $a, t \geq 0$. We now let

$$a^* = \inf\{a \geq 0 : G(a) \neq 0\} \quad (= +\infty \text{ if } G(a) = 0\ \forall a \geq 0).$$

The theorem follows easily.

We can improve this result and show that, for the same class of functions ϕ and μ, the function F vanishes on an unbounded interval if, and only if, $\phi \equiv p_\infty$ until a certain age after which the mortality must be constant.

Theorem 2.2 *Let μ and $\phi \in C^0([0,+\infty))$ be such that $\mu\,\phi \in L^1((0,+\infty))$, μ is nondecreasing, $\phi(0) = \int_0^\infty \mu(a)\phi(a)\,da$, $\phi(0) \geq \phi(a) + \int_0^a \mu(\sigma)\phi(\sigma)\,d\sigma$, $a \geq 0$, and ϕ' is piecewise continuous. Then, $F(t)$ defined by (1.10) vanishes identically on $[t^*,+\infty)$ if, and only if, there exists $a^* \in [0,+\infty]$ such that $\phi(a) \equiv p_\infty(a)$, $a \leq$*

a^*, and $\mu(a) \equiv \mu(a^*)$, $a > a^*$ (where the case $a^* = +\infty$ must be read as $\phi(a) \equiv p_\infty(a)$, $a \geq 0$).

Proof. The proof is the same as that of the previous theorem except that, for $a > a^*$ in a "half-neighborhood" of a^* where G does not vanish, we now have

$$\mu(a+t) = \mu(a), \qquad \forall t \geq t^*,$$

which immediately yields the relation

$$\mu(a) \equiv \mu(a^*), \qquad a \geq a^* + t^*.$$

The theorem follows from the assumption that μ is nondecreasing.

3 Asymptotic behavior of distributions of constant size

We shall prove in this section that, for large times, any distribution of constant size and finite life span tends to the steady state distribution corresponding to its death rate.

Let us consider the case when all individuals die before reaching age A, that is, let us assume that $\Pi(a) = 0$ for $a \geq A$. It follows from (1.7) and (1.8) that the birth function B, for $t \geq A$, satisfies the relation

$$B(t) = \int_0^A B(t-a)\,\mu(a)\,\Pi(a)\,da. \tag{3.1}$$

Furthermore, it follows from (1.4) and the assumption on Π that

$$\int_0^A \mu(a)\,\Pi(a)\,da = 1. \tag{3.2}$$

As a first step let us show that the birth function approaches a finite limit.

Lemma 3.1 *Let ψ be a continuous function which satisfies the equation*

$$\psi(t) = \int_0^A \psi(t-a)\,\theta(a)\,da,$$

where θ is a non-negative function such that

$$\int_0^A \theta(a)\,da = 1. \tag{3.3}$$

Then, there exists $L \in R$ such that

$$\lim_{t\to\infty} \psi(t) = L$$

Proof. Let

$$I_n = \big[(n-1)A, nA\big], \qquad n \in N,$$

and consider the sequences of maxima and minima of ψ on these intervals. We shall show that the maxima decrease and the minima increase towards a common limit. For $n \in N$, let

$$\psi_M^n = \max_{t \in I_n}\{\psi(t)\} = \psi(t_M^n), \qquad t_M^n \in I_n,$$
$$\psi_m^n = \min_{t \in I_n}\{\psi(t)\} = \psi(t_m^n), \qquad t_m^n \in I_n.$$

It follows from the relation

$$\psi(t_M^n) = \int_0^A \psi(t_M^n - a)\,\theta(a)\,da, \tag{3.4}$$

the mean value theorem and (3.3) that the sequence ψ_M^n is decreasing. Similarly, ψ_m^n is increasing and, since each sequence bounds the other, they both converge. Let

$$\lim_{n\to\infty} \psi_M^n = \psi_M, \qquad \lim_{n\to\infty} \psi_m^n = \psi_m.$$

We shall show, by contradiction, that these limits coincide. Hence, let us assume that $\psi_M > \psi_m$. Fix $\varepsilon > 0$ and let $N \in N$ be such that $n > N \Rightarrow \psi_M + \varepsilon > \psi_M^n > \psi_M > \psi_m^n > \psi_m - \varepsilon$. It follows that, for all $n > N+1$ and all $s > (n-2)A$,

$$\psi_M + \varepsilon > \psi_M^n > \psi(s) > \psi_m^n > \psi_m - \varepsilon.$$

Let $0 < \delta < \frac{\psi_M - \psi_m}{2}$ and let I_δ^n be the part of $[0, A]$ on which $\psi(t_M^n - a) < \psi_M - \delta$. We obtain from (3.4) the relation

$$\psi_M^n = \int_{I_\delta^n} \psi(t_M^n - a)\,\theta(a)\,da + \int_{[0,A]\setminus I_\delta^n} \psi(t_M^n - a)\,\theta(a)\,da,$$

from which follows that

$$\psi_M < \psi_M^n < (\psi_M - \delta)\int_{I_\delta^n} \theta(a)\,da + (\psi_M + \varepsilon)\int_{[0,A]\setminus I_\delta^n} \theta(a)\,da.$$

It follows that $\varepsilon > (\varepsilon + \delta)\int_{I_\delta^n} \theta(a)\,da$, which implies that, for $n \to \infty$, the measure of the set I_δ^n tends to zero. Analogously, if we let, $\tilde{I}_\delta^n$ be the part of $[0, A]$ on which $\psi(t_m^n - a) > \psi_m + \delta$, it follows that the measure of $\tilde{I}_\delta^n$ tends to zero as $n \to \infty$.

Next, note that, since the inequalities $\psi_M - \delta < \psi(s) < \psi_m + \delta$ are incompatible, it follows that the intersection of the sets $\mathcal{F}^n = t_M^n - [0, A] \setminus I_\delta^n$ and $\mathcal{G}^n = [0, A] \setminus \tilde{I}_\delta^n$ is empty; that is,

$$\mathcal{F}^n \cap \mathcal{G}^n = \{s^n = t_M^n - q^n = t_m^n - r^n : q^n \in [0, A] \setminus I_\delta^n,\ r^n \in [0, A] \setminus \tilde{I}_\delta^n\} = \emptyset.$$

Hence, $\lim_{n\to\infty} |t_M^n - t_m^n| = A$. Let N be large enough that, for $n > N$, $|t_M^n - t_m^n| > A - \varepsilon$. Without loss of generality, assume that $t_M^n - t_m^n > 0$. Then, for some $\tilde{\varepsilon} < \varepsilon$,

$t_M^n > nA - \tilde{\varepsilon}$. Also, $t_m^n < (n-1)A + \varepsilon$. It follows that, for n sufficiently large, there exists t_*^n such that

$$t_M^n - t_*^n \in [0, A] \setminus I_\delta^n,$$
$$(n-1)A + \varepsilon < t_*^n < (n-1)A + 2\varepsilon,$$

and

$$\psi(t_*^n) > \psi_M - \delta. \tag{3.5}$$

Finally, observe that

$$\begin{aligned}\psi(t_*^n) &= \int_0^{2\varepsilon} \psi(t_*^n - a)\,\theta(a)\,da + \int_{2\varepsilon}^{A} \psi(t_*^n - a)\,\theta(a)\,da \\ &< (\psi_M + \varepsilon)\left[\int_0^{\varepsilon} \theta(a)\,da + \int_{\tilde{I}_\delta^n} \theta(a)\,da\right] \\ &\quad + (\psi_m + \delta)\int_{[2\varepsilon, a]\setminus \tilde{I}_\delta^n} \psi(t_*^n - a)\,\theta(a)\,da.\end{aligned}$$

For n sufficiently large, this contradicts (3.5). Consequently, there does not exist such a δ, that is, $\psi_m = \psi_M$, as we needed.

As a consequence of lemma 3.1 we establish very easily the asymptotic behavior of distributions of constant size.

Theorem 3.1 *Let B be a continuous birth function such that the solution u of (1.1) is of constant size with $\Pi(a) = 0$ for $a \geq A$. Then, for $0 \leq a \leq A$,*

$$\lim_{t\to\infty} u(a,t) = p_\infty(a).$$

Proof. Note that (3.1) and (3.2) say that the lemma can be applied to B with $\theta = \mu\Pi$. Hence, there exists $L \in R$ such that $\lim_{t\to\infty} B(t) = L$. It follows that $B(t-a)$ converges uniformly in a to L on $[0, A]$. Next note that (1.8) implies that $P_0 = \int_0^A B(t-a)\,\Pi(a)\,da$ for $t \geq A$. Hence, $L = P_0 / \int_0^A \Pi(a)\,da$ and (1.8) gives the theorem.

Remark. We have shown that populations of constant size which are not steady state distributions will inevitably tend to these. This is reflected in the simulations we present in the next section.

4 Nontrivial examples

We describe in this section three examples of populations of constant size. The first one has smooth, compactly supported initial data. The second one has a rough initial distribution, and the third one that same initial distribution and a non-compactly supported death rate.

I. We consider a mortality rate

$$\mu(a) = \begin{cases} \rho \tan(\frac{\pi}{4}[1 + \frac{a}{110}]), & 0 \le a < 110, \\ +\infty, & 110 \le a, \end{cases}$$

and an initial age distribution

$$\phi(a) = \begin{cases} \alpha\,[10^{-4}a^2 - 2 \cdot 10^{-2}a + 1], & 0 \le a < 100, \\ 0, & 100 \le a. \end{cases}$$

It follows that

$$\phi'(a) = \begin{cases} \frac{2\alpha}{10}^4(a - 100), & 0 \le a < 100, \\ 0, & 100 \le a, \end{cases}$$

and, (1.4), the survival function is

$$\Pi(a) = \begin{cases} [\sqrt{2}\cos(\frac{\pi}{4}(1 + \frac{a}{110}))]^{\frac{440\rho}{\pi}}, & 0 \le a < 110, \\ 0, & 110 \le a. \end{cases}$$

We choose $\alpha = 30,000$ so that the total population (1.2) satisfies

$$P = \int_0^{100} \phi(a)\,da = \frac{100\alpha}{3} = 1,000,000$$

and $\rho \approx 0.01907$ is chosen as the reciprocal of

$$\int_0^{100} (10^{-4}a^2 - 2 \cdot 10^{-2}a + 1)\tan(\frac{\pi}{4}[1 + \frac{a}{110}])\,da,$$

in order that g given by (1.9) have mean value zero.

Note that the equilibrium distribution (1.5) is compactly supported in the interval $[0, 110]$ in this example (since Π is) and, despite appearances, the function ϕ is quite close to it (see figures 4.1-4.4 below). The value of the factor $p_\infty(0)$ was computed using Simpson's rule to evaluate the integral of Π, and has a value of $30,125.82$. As we have shown that u converges to equilibrium as $t \to +\infty$, we see that the only changes in the age-distribution in this situation are a mild aging (ten more years of age are added to the initial population) and a slight redistribution in the young and middle ages. The numerical algorithm is here very effective in keeping the population constant as shown in Table 4.1. For that table, the total population P is calculated by using Simpson's rule, with $\Delta t = \frac{1}{3}$, in the formula

$$\begin{aligned} P(t) &= \int_0^t B(t-a)\,\Pi(a)\,da + \int_t^\infty \phi(a-t)\Pi(a)/\Pi(a-t)\,da \\ &= \int_0^t B(a)\,\Pi(t-a)\,da + \int_0^\infty \phi(a)\Pi(a+t)/\Pi(a)\,da, \end{aligned}$$

which results from the combination of (1.2) and (1.8).

Table 4.1.

Time(yrs.)	Total Population
0	1,000,000
10	1,000,058
20	1,000,046
30	1,000,023
40	1,000,007
50	999,998
60	999,997
70	999,997
80	999,998
90	1,000,000
100	1,000,000
110	1,000,000

We show in figures 4.1–4.4 the graphs of the initial age-distribution ϕ and the asymptotic distribution p_∞, in figure 4.5 the graph of the mortality function, and in figures 4.6-4.8, respectively, the function F, the birth function B, and the total population function P.

We see in figure 4.8 that the simulator kept the population essentially constant for 115 years, by which time it had reached equilibrium. The computed values of P did oscillate a little bit (from 999,996 to 1,000,058) but their mean is 1,000,000 with a standard deviation of only 20. The oscillation is an almost inevitable consequence of solving a Volterra integral equation. However, the results are very accurate, with a maximum error of only 58 parts per million.

II. We consider now the same death rate μ as in the previous example and use a rough initial distribution (saw-tooth with gap)

$$\phi(a) = \begin{cases} \alpha\,(30-a), & 0 \le a < 30, \\ 0, & 30 \le a < 40, \\ \alpha\,(30-|70-a|), & 40 \le a < 100, \\ 0, & 100 \le a. \end{cases}$$

Then

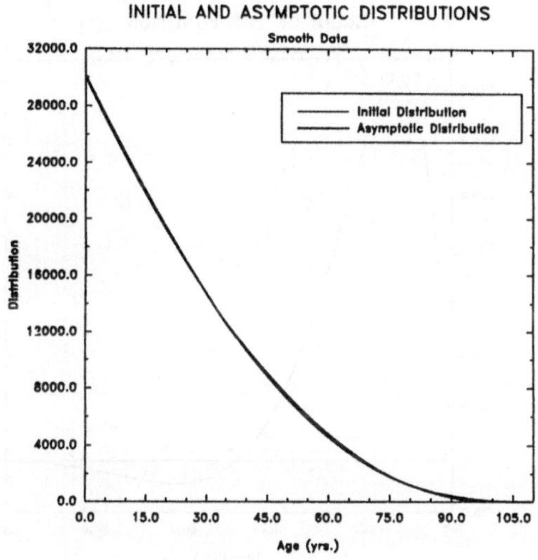

Fig. 4.1. ϕ and p_∞

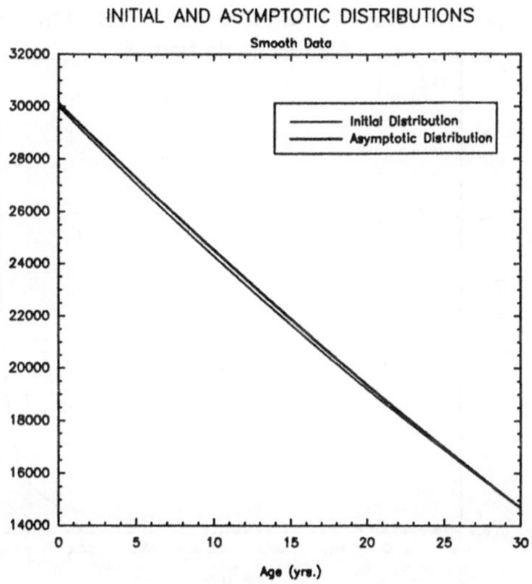

Fig. 4.2. ϕ and p_∞(young ages)

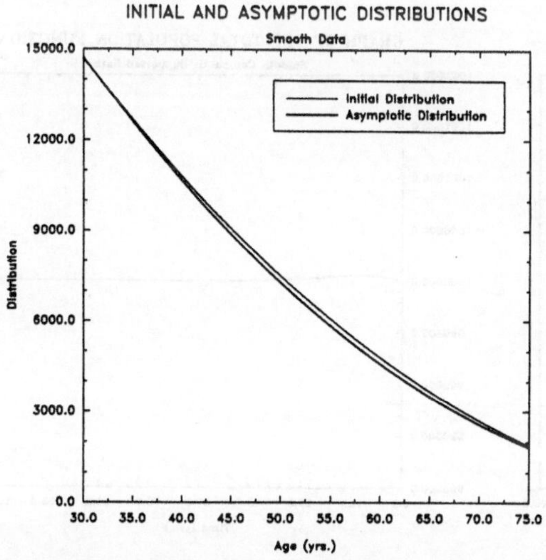

Fig. 4.3. ϕ and p_∞(middle ages)

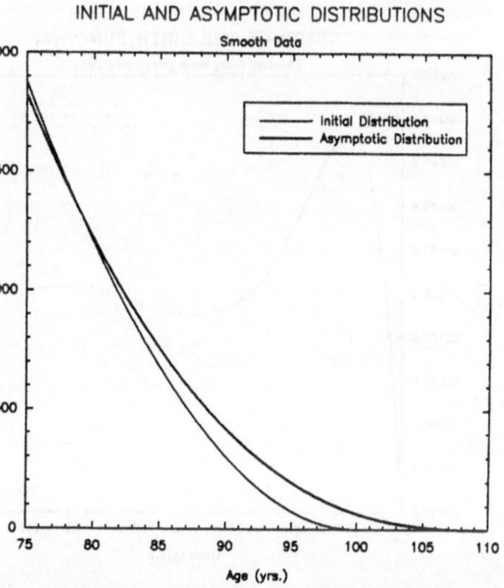

Fig. 4.4. ϕ and p_∞(old ages)

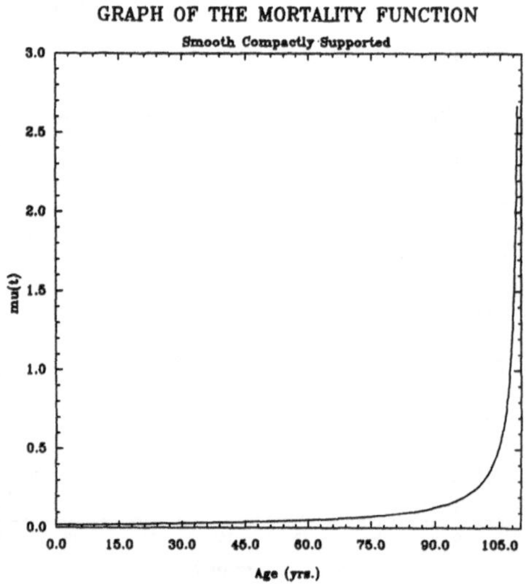

Fig. 4.5. $\mu(a)$

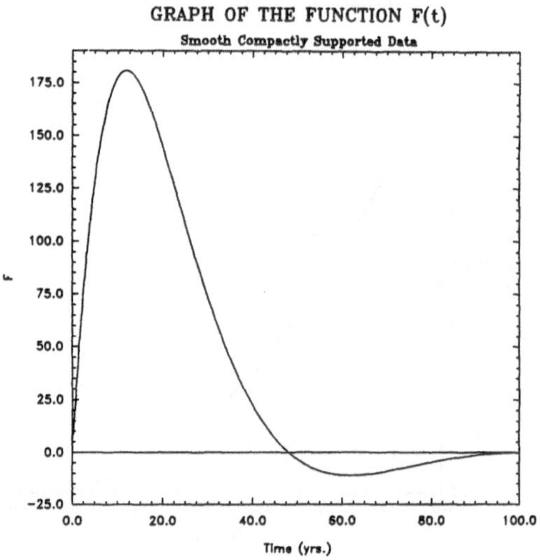

Fig. 4.6. $F(t)$

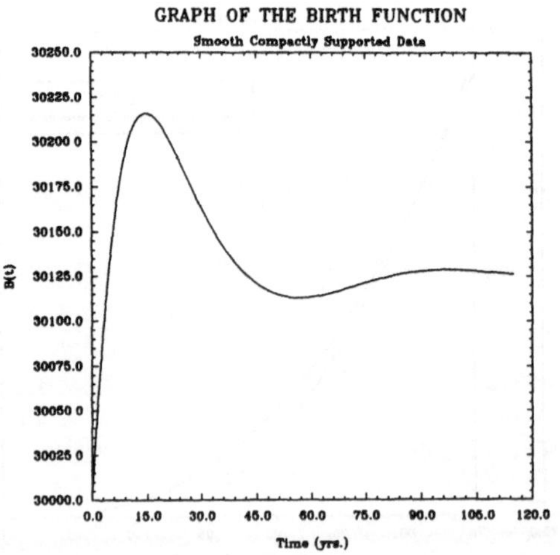

Fig. 4.7. $B(t)$

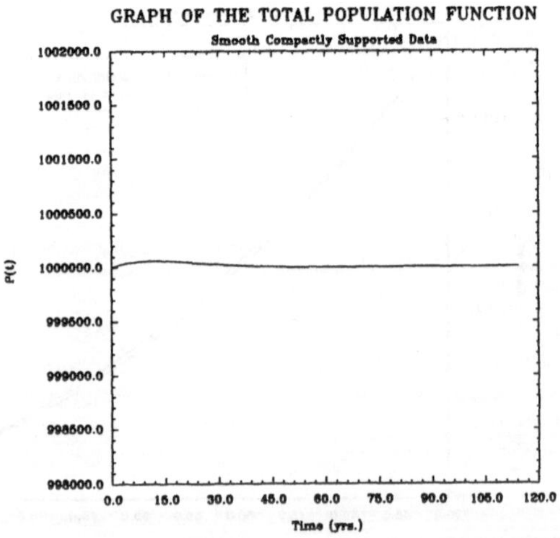

Fig. 4.8. $P(t)$

$$\phi'(a) = \begin{cases} -\alpha & 0 \le a < 30, \\ 0, & 30 \le a < 40, \\ \alpha \operatorname{sgn}(70 - a), & 40 \le a < 100, \\ 0, & 100 \le a. \end{cases}$$

We use again $\alpha = 30,000$, so that the total population (1.2) is

$$P = \int_0^{100} \phi(a)\, da = 1350\alpha = 40,500,000$$

and $\rho \approx 0.0075355$ is chosen as the reciprocal of

$$\int_0^{100} \phi(a) \tan\left(\frac{\pi}{4}\left[1 + \frac{a}{110}\right]\right) da,$$

in order that g given by (1.9) have mean value zero.

The simulator works extremely well in keeping the total population constant, as evidenced in figure 4.12, where the variation was only ± 1 in 40,500,000 in 300 years.

We show in figures 4.9–4.13, respectively, the graphs of the death and survival functions μ and Π, the function F, the birth function B, the total population function P, and the age-distribution at times $t = 0, 30, 80, 130$ and 180, together with the equilibrium distribution (1.5). Note that the small value of ρ used results in an essentially linear survival function Π (figure 4.9). We also observe that the birth function obtained numerically (figure 4.11) approaches its theoretical asymptotic value 721,374 rather slowly and it has even more oscillations than in the preceding example (figure 4.7) although the shape (but not the values) of the function F is rather similar in both cases (figures 4.6 and 4.10). These oscillations turned out to be exactly those necessary for the numerically evaluated total population to be perfectly constant, as shown in figure 4.12. Finally, figure 4.13 shows how the initial distribution changes in time to approach the equilibrium one. As expected, this happens entirely due to the birth process, since only after 80 years does the distribution start to look somewhat like the asymptotic one, that is, when most of the individuals initially present have already died.

III. We finally consider the same initial distribution as in the previous example but allow the individuals to age indefinitely. We use the following death rate:

$$\mu(a) = \rho a, \qquad 0 \le a,$$

which leads to the survival function:

$$\Pi(a) = e^{-\rho a^2/2}, \qquad 0 \le a.$$

We take, as before, $\alpha = 30,000$; $\rho = 4/9000$ is chosen as the reciprocal of

$$\int_0^{100} a\, \phi(a)\, da,$$

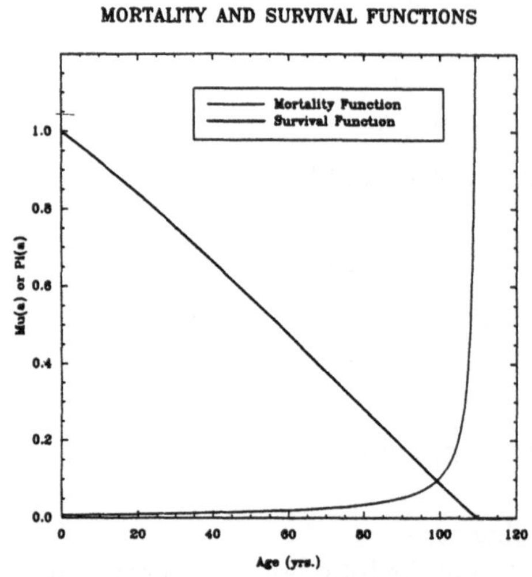

Fig. 4.9. $\mu(a)$ and $\Pi(a)$

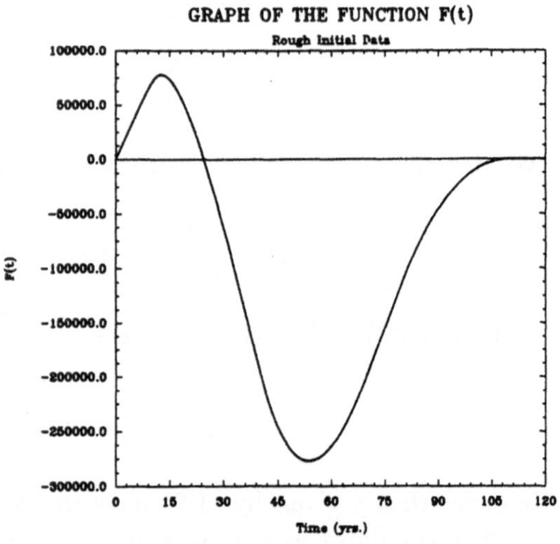

Fig. 4.10. $F(t)$

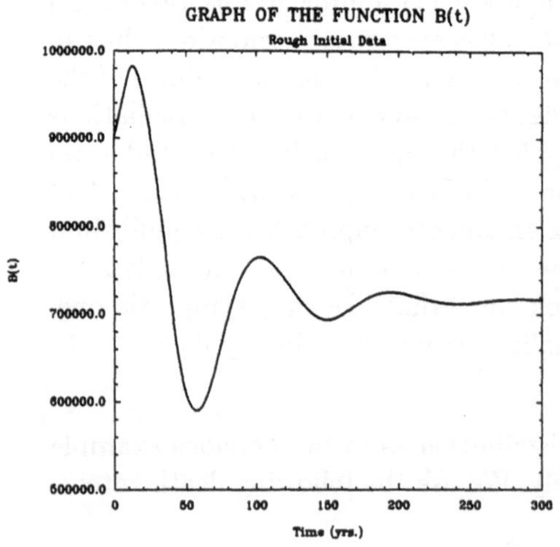

Fig. 4.11. $B(t)$

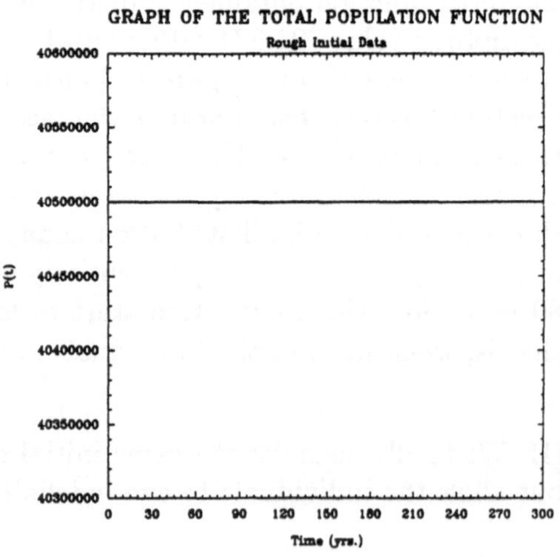

Fig. 4.12. $P(t)$

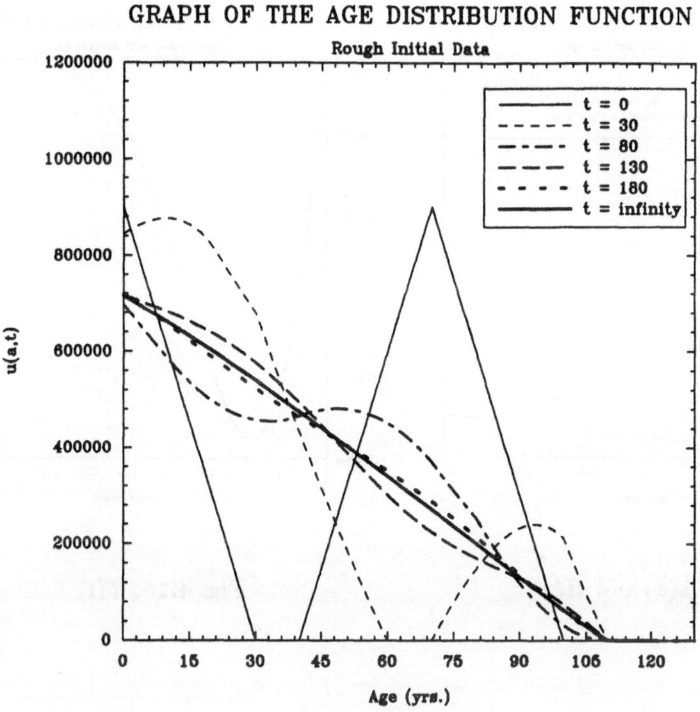

Fig. 4.13. $u(a,t)$

in order that g given by (1.9) have mean value zero. The simulator was run for 300 years. We obtained in this case, just as in the preceding one, a perfectly constant total population.

We present in figures 4.14–4.17, respectively, the graphs of the death and survival functions, of the function F, of the birth function B, and of the age-distribution $u(\cdot,t)$ at times $t = 0, 30, 80, 130$, and ∞. The numerical birth function approaches an asymptotic value a bit short of the theoretical one of 705,964, as seen in figure 4.16.

References

1. Baker, C.T.H. (1977): The Numerical Treatment of Integral Equations. Clarendon Press, Oxford
2. Busenberg, S., Iannelli, M. (1985): Separable models in age-dependent population dynamics. J. Math. Biol. **22**, 145-173
3. Kostova, T., Milner, F.A. (1991): Nonlinear age-dependent population dynamics with constant size. SIAM J. Math. Anal. **22**, 129-137
4. McKendrick, A.G. (1926): Applications of mathematics to medical problems. Proc. Edinburgh Math. Soc. **44**, 98-130
5. von Foerster, H. (1959): Some Remarks on Changing Populations. Grune and Stratton, New York

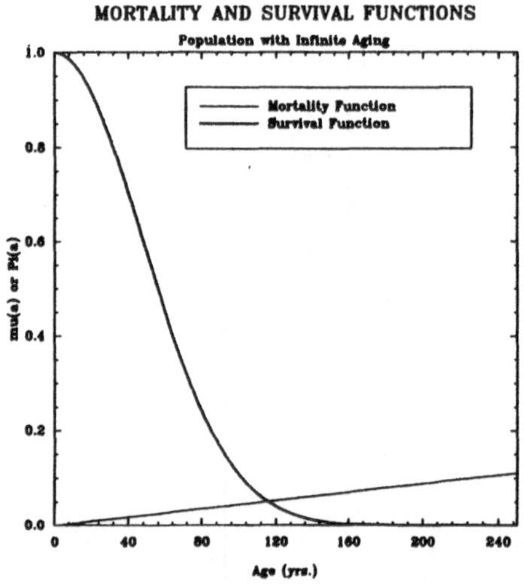

Fig. 4.14. $\mu(a)$ and $\Pi(a)$

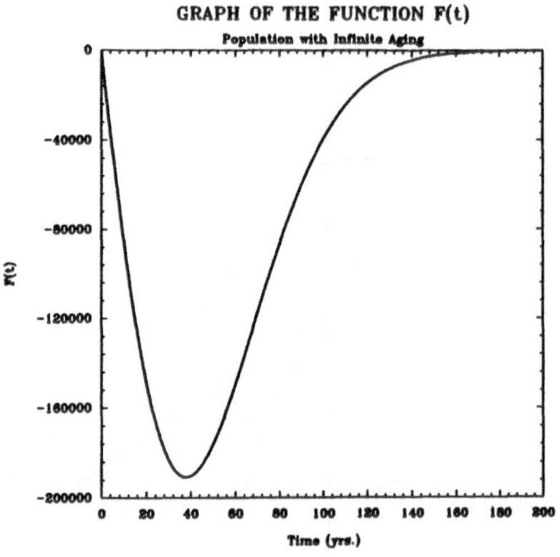

Fig. 4.15. $T(t)$

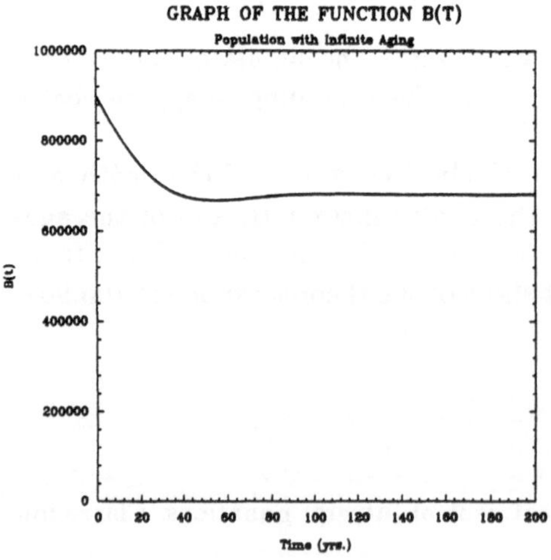

Fig. 4.16. $B(t)$

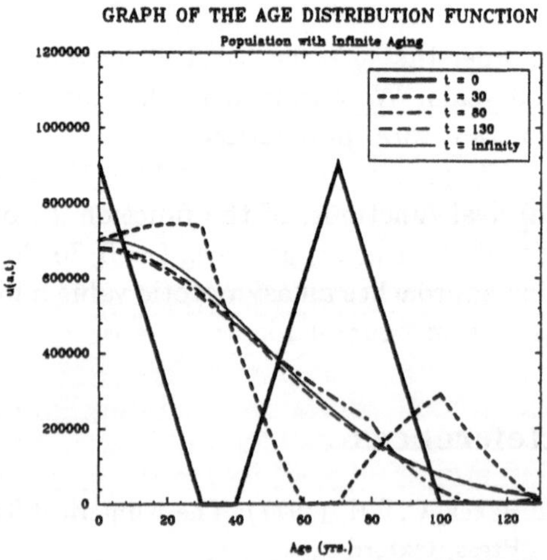

Fig. 4.17. $u(a, t)$

Coexistence in Competition-Diffusion Systems

Masayasu Mimura

Department of Mathematics, Faculty of Science, Hiroshima University, Hiroshima 730, Japan

1 Introduction

It is a central problem in population ecology to understand the mechanism of spatial patterning of ecological communities. In this paper, we will be concerned with regionally segregation of competing species in a homogeneous environment from a theoretical aspect. Suppose the situation where n species are competing with each other and moving by diffusion. Let $u_i(t,x)$ be the population density of the i-th species at time t and position x for $i = 1, 2, ..., n$. Then the dynamics of $u_i(t,x)$ are described by

$$\frac{\partial}{\partial t}u_i = d_i\Delta u_i + (r_i - \sum_{j=1}^{n} a_{ij}u_j)u_i, \quad (t,x) \in (0,\infty) \times \Omega \tag{1.1}$$

$$(i = 1, 2..., n),$$

where Δ is the Laplace operator in R^N, r_j is the intrinsic growth rate, a_{ii} and $a_{ij} (i \neq j)$ are respectively the coefficients of intra- and inter-specific competition and d_i is the diffusion coefficient $(i = 1, 2..., n)$. We assume that a habitat Ω is bounded in R^N. First define a basically homogeneous environment for competing species by the following assumptions:

(A) r_i, a_{ij} and d_i $(i = 1, 2, ..., n)$ are positive constants;

(B) The boundary condition at the boundary $\partial\Omega$ is of zero flux, i.e.

$$\frac{\partial u_i}{\partial \mathbf{n}} = 0 \quad (i = 1, 2, ..., n), \tag{1.2}$$

where $\mathbf{n}$ is the outward normal unit vector at x on $\partial\Omega$. The simplest system of (1.1) is the case when $n = 2$, which is described by

$$\begin{cases} \frac{\partial}{\partial t}u_1 = d_1\Delta u_1 + (r_1 - a_{11}u_1 - a_{12}u_2)u_1 \\ \frac{\partial}{\partial t}u_2 = d_2\Delta u_2 + (r_2 - a_{21}u_1 - a_{22}u_2)u_2. \end{cases} \tag{1.3}$$

One of the main interesting problems for (1.1), (1.2) is to study the possibility of coexistence of competing species in a homogeneous environment. From a mathematical viewpoint, a lot of work have been done on existence or non-existence of stable spatial inhomogeneous, positive periodic as well as equilibrium solutions. First of all, we should note that if all of the diffusion coefficients d_i $(i = 1, 2, ..., n)$ are very large, any non-negative solutions of (1.1), (1.2) tend to be spatially homogeneous for large time [2]. In other words, the asymptotic behavior of solutions of (1.1), (1.2) is qualitatively the same as that of

$$\frac{d}{dt}u_i = (r_i - \sum_{j=i}^{n} a_{ij}u_j)u_i \quad (i = 1, 2, ..., n). \tag{1.4}$$

This result implies that high diffusion effects enhance the spatial homogenization for competing species. Under this situation, we can find by the analysis of the system of equations (1.4) that n species possibly coexist for suitable r_i and a_{ij} even if the interspecific competition rate $a_{ij} (i \neq j)$ is greater than the intraspecific one $a_{ij} (i, j = 1, 2, ..., n)$. However the case when $n = 2$ is an exception, that is, under such strong competition, there is no coexistence of two competing species. More precisely, if

$$\frac{a_{11}}{a_{21}} < \frac{r_1}{r_2} < \frac{a_{12}}{a_{22}} \tag{1.5}$$

holds, then $(u_1(t, x),\ u_2(t, x))$ tends generically to either $(r_1/a_{11}, 0)$ or $(0, r_2/a_{22})$. That is, if the two strongly competing species move by high diffusion, one of the species always becomes extinct. This indicates the competitive exclusion principle which was postulated by Gause. We now set the following problems: If the diffusion coefficients are not necessarily large, is it possible for two strongly competing species to coexist by diffusion? For this problem, we should refer to three fundamental results for the system (1.3). The first one is given by Kishimoto and Weinberger [7]: When Ω is convex, there are no stable spatially inhomogeneous equilibrium solutions. This implies that if r_i and a_{ij} $(i, j = 1, 2)$ satisfy (1.5), there are no stable positive equilibrium solutions. In ecological terms, two competing species never coexist under strong competition even if they move by diffusion. The second is given by Matano and Mimura [10]: There is a appropriate nonconvex domain in $\Omega \subset R^2$ such that there exist stable spatially inhomogeneous, positive equilibrium solutions which exhibit spatially segregating pattern for two competing species. These results show that coexistence for two competing species is possible under strong competition, depending on the shape of domain Ω as well as the magnitude of the diffusion coefficients d_1 and d_2.

Motivated by the above, we study the domain-shape problem by using (1.1), (1.2) with $n = 2$ in two dimensional space. To do it, we specify the habitat Ω to be a dumbbell-shape domain with a small parameter ϵ, say Ω_ϵ, which consists of three disjoint unions

$$\Omega_\epsilon = \Omega_0^L \cup \Omega_0^R \cup R_\epsilon,$$

where Ω_0^L and Ω_0^R are convex domains and R_ϵ is a handle satisfying $|R_\epsilon| \to 0$ as $\epsilon \downarrow 0$, as in Figure 1. We introduce a parameter $\alpha = |\Omega_0^R|/(|\Omega_0^R| + |\Omega_0^L|)$ which is an indicator of the domain-shape of Ω_ϵ. When ϵ is sufficiently small, the limiting case when $\alpha \downarrow 0$ (resp. $\alpha \uparrow 1$) implies the situation that Ω_ϵ is close to the convex domain Ω_0^L (resp. Ω_0^R).

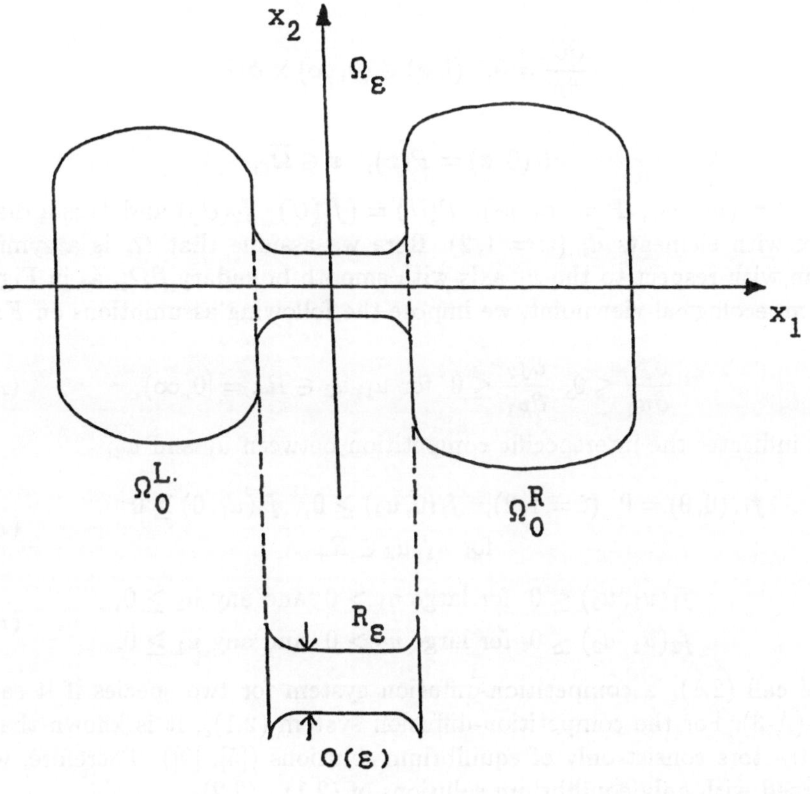

Fig. 1. **Dumbbell-shape domain Ω_ϵ.**

In addition, we assume that the diffusion coefficients d_1 and d_2 are very large. We conveniently introduce positive constants $\overline{d}_1, \overline{d}_2$ and θ such that the relation

$$d_i = \overline{d}_i/\epsilon^\theta \ (i = 1, 2)$$

holds. We are thus interested in the dependence of $\alpha \in (0, 1)$ and $\theta > 0$ on the behavior of (1.1), (1.2) when ϵ is sufficiently small.

Most of the work reported here has been done jointly with S-I. Ei and Q. Fang.

2 Results

We consider the following ϵ-family of reaction-diffusion equations:

$$\frac{\partial U}{\partial t} = D/\epsilon^{\theta}\Delta U + F(U), \quad (t,x) \in (0,\infty) \times \Omega_\epsilon \tag{2.1$_\epsilon$}$$

with

$$\frac{\partial U}{\partial \mathbf{n}} = 0, \quad (t,x) \in (0,\infty) \times \partial\Omega_\epsilon \tag{2.2}$$

and

$$U(0,x) = P(x), \quad x \in \overline{\Omega}_\epsilon, \tag{2.3}$$

where $U = (u_1, u_2)$, $P = (p_1, p_2)$, $F(U) = (f_1(U),\ f_2(U))$ and D is a diagonal matrix with elements d_i $(i = 1,2)$. Here we assume that Ω_ϵ is a symmetric domain with respect to the x_1 axis with smooth boundary $\partial\Omega_\epsilon$, as in Figure 1. From an ecological viewpoint, we impose the following assumptions on F:

$$\frac{\partial f_1}{\partial u_2} \leq 0, \ \frac{\partial f_2}{\partial u_1} \leq 0 \ \text{ for } u_1, u_2 \in R_+ = [0,\infty), \tag{$A-1$}$$

which indicates the interspecific competition between u_1 and u_2;

$$f_i,(0,0) = 0 \ \ (i = 1,2), \ \ f_1(0,u_2) \geq 0, \ \ f_2(u_1,0) \geq 0 \\ \text{for } u_1, u_2 \in R_+; \tag{$A-2$}$$

$$f_1(u_1,u_2) \leq 0 \ \text{ for large } u_1 > 0 \ \text{ and any } u_2 \geq 0, \\ f_2(u_1,u_2) \leq 0 \ \text{ for large } u_2 > 0 \ \text{ and any } u_1 \geq 0. \tag{$A-3$}$$

We call $(2.1)_\epsilon$ a competition-diffusion system for two species if it satisfies (A-1)-(A-3). For the competition-diffusion system $(2.1)_\epsilon$, it is known that stable attractors consist only of equilibrium solutions ([5], [9]). Therefore, we are concerned with only equilibrium solutions of $(2.1)_\epsilon$, (2.2).

We first give some simple remarks ([1]): Define $\Xi \equiv [0,K] \times [0,K]$ for some $K > 0$. Then Ξ is an invariant set of $(2.1)_\epsilon$, (2.2) for large K. Moreover, define

$$\Sigma = \{U \in (H^1(\Omega_\epsilon))^2 \cap (L^\infty(\Omega_\epsilon))^2 \mid U \in \Xi, \ ||U||_{(H^1(\Omega_\epsilon))^2} \leq K'\}$$

for some $K' > 0$. If the initial function $P(x)$ is in Ξ, a solution of $(2.1)_\epsilon$-(2.3) eventually enters Σ for large K' so that we may restrict our discussion to solutions interior to Σ.

Let $0 = \lambda_\epsilon^{(1)} < \lambda_\epsilon^{(2)}$ be the first two eigenvalues of $-\Delta$ in Ω_ϵ with the zero-flux boundary condition and $\omega_\epsilon^{(1)}$, $\omega_\epsilon^{(2)}$ be the corresponding normalized eigenfunctions. For $\alpha = |\Omega_0^R| \ / \ (|\Omega_0^R| + |\Omega_0^L|)$, we define $\phi_\epsilon^{(1)}$ and $\phi_\epsilon^{(2)}$ by

$$\phi_\epsilon^{(1)} = |\Omega_\epsilon|^{1/2}((1-\alpha)\omega_\epsilon^{(1)} - (\alpha(1-\alpha))^{1/2}\omega_\epsilon^{(2)})$$

and

$$\phi_\epsilon^{(2)} = |\Omega_\epsilon|^{1/2}((1-\alpha)\omega_\epsilon^{(1)} - (\alpha(1-\alpha))^{1/2}\omega_\epsilon^{(2)}),$$

respectively. Since Hale and Vegas [4] showed that

$$\omega_\epsilon^{(1)} = |\Omega_\epsilon|^{-1/2}$$

and

$$\lim_{\epsilon\downarrow 0} \omega_\epsilon^{(2)} = \begin{cases} (|\Omega_0^R| \,/\, |\Omega_0||\Omega_0^L|))^{1/2} \text{ in } \Omega_0^L \\ (|\Omega_0^L| \,/\, (|\Omega_0||\Omega_0^R|))^{1/2} \text{ in } \Omega_0^R, \end{cases}$$

we easily find that

$$\lim_{\epsilon\downarrow 0} \phi_\epsilon^{(1)} = \begin{cases} 1 \text{ in } \Omega_0^L \\ 0 \text{ in } \Omega_0^R \end{cases}$$

and

$$\lim_{\epsilon\downarrow 0} \phi_\epsilon^{(2)} = \begin{cases} 0 \text{ in } \Omega_0^L \\ 1 \text{ in } \Omega_0^R \end{cases}$$

Let Q^ϵ be the projection from $(L_2(\Omega_\epsilon))^2$ into $(\text{span}\{\omega_\epsilon^{(1)},\ \omega_\epsilon^{(2)}\})^2$ and $P^\epsilon = Id - Q^\epsilon$, where Id is the identity operator.

Theorem 2.1 ([3]) *There is $\epsilon_1 > 0$ such that for any $\epsilon \in (0, \epsilon_1)$,*

$$M_\epsilon = \{Y_1\phi_\epsilon^{(1)} + Y_2\phi_\epsilon^2 + h_\epsilon(Y_1, Y_2) | Y_1, Y_2 \in R^2\} \cap \Sigma$$

in a 4-dimensional Lipschitz continuous manifold for some $h_\epsilon(Y_1, Y_2) \in (\mathbf{C}(R^4, P^\epsilon((H^1(\Omega_\epsilon))^2))$, where h_ϵ satisfies $h_\epsilon(Y_1, Y_1) = 0$ and $|||h_\epsilon|||_{\epsilon,\infty} = O(\epsilon^{(2\theta+1)/2})$ with the norm $|||h|||_{\epsilon,\infty} = sup\{||h(Y_1, Y_2)||_{(H^1(\Omega_\epsilon))^2} \mid Y_1, Y_2 \in R^2\}$ for $h \in (\mathbf{C}(R^4, P^\epsilon((H^1(\Omega_\epsilon))^2))\}$. Furthermore, it is invariant under the semiflow of $(2.1)_\epsilon$-(2.3) with $P \subset \Sigma$.

Let $(Y_1(t;\epsilon), Y_2(t;\epsilon))$ be the solution to the following 4-dimensional ODEs

$$\begin{cases} \frac{dY_1}{dt} = F(Y_1) + \frac{\lambda_\epsilon^{(2)}}{\epsilon^\theta}\alpha D(Y_2 - Y_1) + R_1^\epsilon(Y_1, Y_2) \\ \frac{dY_2}{dt} = F(Y_2) + \frac{\lambda_\epsilon^{(2)}}{\epsilon^\theta}(1-\alpha) D(Y_1 - Y_2) + R_2^\epsilon(Y_1, Y_2) \end{cases} \tag{2.7$_\epsilon$}$$

with the initial condition

$$(Y_1(0;\epsilon),\ Y_2(0;\epsilon)) = (Y_{10}, Y_{20}), \tag{2.8}$$

where $R_i^\epsilon(Y_j, Y_j) = 0$ $(i, j = 1, 2)$ and $|R_i^\epsilon|_{K_{1,\infty}} = 0(\epsilon^{1/2})$ as $\epsilon \downarrow 0$ with the norm $|R|_{K_{1,\infty}} = sup(|R(Y_1, Y_2)| \mid |(Y_1, Y_2)| \le K_1$ for $R \in C(R^4, R^2))$.

Theorem 2.2 ([3]) *There exist $\epsilon_2 > 0, \mu > 0$ and $C > 0$ such that for any $\epsilon \in (0, \epsilon_2)$, if $U(t;\epsilon)$ is a solution of $(2.1)_\epsilon$-(2.3) and lies in Σ, then there is the initial function (Y_{10}, Y_{20}) for $(2.7)_\epsilon$ such that the solution $Y_1(t;\epsilon)$, $Y_2(t;\epsilon))$ satisfies for $t \ge 0$*

$$||U(t;\epsilon) - Y_1(t;\epsilon)\phi_\epsilon^{(1)} + Y_2(t;\epsilon)\phi_\epsilon^{(2)} + h_\epsilon(Y_1(t;\epsilon, Y_2(t;\epsilon)))||_{H^1(\Omega_\epsilon))^2} \le Ce^{-vt}.$$

Thus, it turns out that the dynamics of solutions of $(2.1)_\epsilon$-(2.3) is asymptotically represented by the 4-dimensional ODE problem $(2.7)_\epsilon$, (2.8).

3 4-dimensional ODEs associated with RD systems

In order to obtain the limiting system of $(2.7)_\epsilon$ as $\epsilon \downarrow 0$, we classify the parameter θ into three cases:

$$1)\ \ 0<\theta<1;\ \ 2)\ \ \theta=1;\ \ 3)\ \ \theta>1.$$

We first note that there is some constant $\tau > 0$ independently of α and ϵ such that ([3])

$$\lim_{\epsilon\downarrow 0}\lambda_\epsilon^{(2)}/\epsilon = \frac{\tau}{\alpha(1-\alpha)}.$$

Therefore, when $\gamma(\theta) = \lim_{\epsilon\downarrow 0}(\lambda^{(2)}/\epsilon^\theta)$, it is obvious that as $\epsilon \downarrow 0$,

$$\gamma(\theta) = \begin{cases} 0, \text{ if } 0<\theta<1, \\ \frac{\tau}{\alpha(1-\alpha)}, \text{ if } \theta = 1, \\ \infty, \text{ if } \theta > 1. \end{cases}$$

For $0 < \theta \le 1$, we define the limiting systems of $(2.7)_\epsilon$ by

$$\begin{cases} \frac{dW_1}{dt} = F(W_1) + \gamma(\theta)\alpha D(W_2 - W_1) \\ \frac{dW_2}{dt} = F(W_2) + \gamma(\theta)(1-\alpha)D(W_1 - W_2). \end{cases} \tag{3.1}$$

Theorem 3.1 *Assume* $0 < \theta \le 1$. *Let* $(\overline{W}_1, \overline{W}_2)$ *be a nondegenerate equilibrium solution of* (3.1). *Then there is* $\epsilon_3 > 0$ *such that the corresponding equilibrium solution* $\overline{U}(\epsilon)$ *of* $(2.1)_\epsilon$, (2.2) *exists for* $0 < \epsilon \le \epsilon_3$, *which satisfies*

$$\lim_{\epsilon\downarrow 0}\overline{U}(\epsilon) = \begin{cases} \overline{W}_1 \text{ in } \Omega_0^L \\ \overline{W}_2 \text{ in } \Omega_0^R \end{cases}$$

with respect to the norm $||.||_{(L^2(\Omega_0))^2}$.

Theorem 3.2 *Assume* $0 < \theta \le 1$. *Let* $\lambda_1, \lambda_2, \lambda_3, \lambda_4 (Re\lambda_1 \ge ... \ge Re\lambda_4)$ *be the eigenvalues of the linearized matrix of* (3.1) *around* $(\overline{W}_1, \overline{W}_2)$ *and* $\mu_\epsilon^{(1)}, \mu_\epsilon^{(2)}, ...,$ $(Re\mu_\epsilon^{(1)} \ge Re\mu_\epsilon^{(2)} \ge ...)$ *be the spectra of the linearized operator of* $(2.1)_\epsilon$, (2.2) *around* $\overline{U}(\epsilon)$. *Then there is* $\epsilon_4 > 0$ *and* $C > 0$ *such that*

$$\lim_{\epsilon\downarrow 0}\mu_\epsilon^{(i)} = \lambda_i \quad (i = 1, 2, 3, 4)$$

and

$$Re\mu_\epsilon^{(i)} < -C/\epsilon^\theta \quad (i \ge 5)$$

for $0 < \epsilon \le \epsilon_4$.

Thus, when $0 < \theta \le 1$, one finds that if one could show the existence and stability (or instability) of nondegenerate equilibrium of (3.1), one could also show the same results on $(2.1)_\epsilon$, (2.2) with sufficiently small $\epsilon > 0$.

On the other hand, when $\theta > 1$, $\gamma(\theta)$ tends to infinity as $\epsilon \downarrow 0$ so that, Theorems 3.1 and 3.2 can not be used. However, we find that the solution $W_1, (t; \epsilon)$, $W_2(t; \epsilon))$ satisfies

$$\lim_{t\to\infty} |W_1(t;\epsilon) - W_2(t;\epsilon)| = 0.$$

This information gives us the following:

Theorem 3.3 *Assume $\theta > 1$. There exist $\epsilon_5 > 0$, $M > 0$ and $\kappa > 0$ such that for $0 < \epsilon < \epsilon_5$,*

$$\|U(t;\epsilon) - \overline{U}(t)\|_{(L^2(\Omega_0)^2)} \leq Me^{-\kappa t}, \; t \geq 0$$

holds, where $\overline{U}(t)$ is the spatial average of $U(t,x;\epsilon)$.

This theorem indicates that any equilibrium solutions of $(2.1)_\epsilon$, (2.2) must be constant when $\theta > 1$.

In the next section, by specifying $F(f_1, f_2)$, we discuss more precisely the dependency of α, $\overline{d}_1$ and $\overline{d}_2$ on the equilibrium solutions of $(2.1)_\epsilon$, (2.2) for sufficiently small ϵ.

4 Global structure of equilibrium solutions

Applying the results in the previous section to a competition-diffusion system with Gause-Lotka-Volterra dynamics, we study the effect of domain-shape on the coexistence of two competing species. The system which we will discuss is

$$\begin{cases} \frac{\partial u_1}{\partial t} = \frac{d}{\epsilon}\Delta u_1 + (r - au_1 - bu_2)u_1, \\ \qquad\qquad (t,x) \in (0,\infty) \times \Omega_\epsilon \\ \frac{\partial u_2}{\partial t} = \frac{d}{\epsilon}\Delta u_2 + (r - bu_1 - au_2)u_2. \end{cases} \tag{4.1$_\epsilon$}$$

with

$$\frac{\partial u_1}{\partial \mathbf{n}} = \frac{\partial u_2}{\partial \mathbf{n}} = 0, \; (t,x) \in (0,\infty) \times \partial\Omega_\epsilon. \tag{4.2}$$

Here we consider the case when $r_1 = r_2 = r$, $a_{11} = a_{22} = a$, $a_{12} = a_{21} = b$, $d_1 = d_2 = d$ in the kinetics of (1.3), and $\theta = 1$ for simplicity only. Let us assume $a < b$ which implies that the interspecific competition is stronger than the intraspecific one. With $W_1 = (w_{11}, w_{12})$ and $W_2 = (w_{21}, w_{22})$, the limiting system resulting form $(4.1)_\epsilon$, (4.2) is described by

$$\begin{cases} \frac{dw_{11}}{dt} = (r - aw_{11} - bw_{12})w_{11} + \frac{rd}{1-\alpha}(w_{21} - w_{11} \\ \frac{dw_{12}}{dt} = (r - bw_{11} - aw_{12})w_{12} + \frac{rd}{\alpha}(w_{22} - w_{12}) \\ \qquad\qquad t \in (0,\infty) \\ \frac{dw_{21}}{dt} = (r - aw_{21} - bw_{22})w_{21} + \frac{rd}{1-\alpha}(w_{11} - w_{21}) \\ \frac{dw_{22}}{dt} = (r - bw_{21} - aw_{22})w_{22} + \frac{rd}{\alpha}(w_{12} - w_{22}). \end{cases} \tag{4.3}$$

For $\alpha = 1/2$, Figure 2(i) shows the global bifurcation diagram of (4.3) when the diffusion coefficient d is globally varied ([8], [14]), where letting (I) = (0,0), (II) = $(r/a, 0)$, (III) = $(0, r/a)$, and (IV) = $(r/(a+b),\ r/(a+b))$, (I,I) means $\overline{W}_1 = 1$ in Ω_0^L and $\overline{W}_2 = I$ in Ω_0^R. Other equilibria (I,II),... are similarly

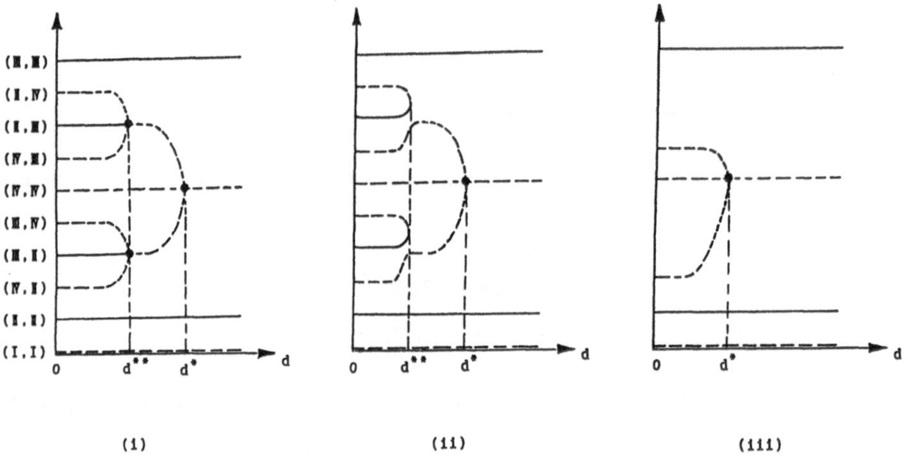

Fig. 2. Global bifurcation picture of equilibria of (4.3) when d is varied. (i) $\alpha = 1/2$. (ii) α is close to $1/2$. (iii) α is small.

defined. We first note that there are 4 trivial equilibria for any $d > 0$; (II, II) and (III,III) are stable while (I,I) and (IV,IV) are unstable. We note that there occurs the primary bifurcation of pitchfork type in the unstable (IV,IV)-solution branch at $d = d_*$, and the bifurcating branches continue to (II,III) or (III,II) as d tends to zero. Further, on each bifurcating branch, there occurs the secondary bifurcation of the same type as the primary one at $d = d_{**}$. As a result, we find that the primarily bifurcating branch is unstable for $d_{**} < d < d_*$, while it is stable for $0 < d < d_*$. On the other hand, for $0 < \alpha < 1/2$, this global bifurcation diagram is drawn in Figure 2(ii) and (iii). When α is close to $1/2$, the secondary bifurcation is deformed as the imperfection of the symmetric case when $\alpha = 1/2$, that is, the primary bifurcation branches continue to (IV,II) or (II,IV) as d tends to zero, and there are knee branches for $0 < d < d_{**}$; the branch which continues to (II,III) or (III,II) is stable. However, when α is small, the knee branches disappear and there is only one bifurcation point at $d = d_*$, and there are no longer stable branches except for the trivial ones (III,III) and (II,II). The case when $1/2 < \alpha < 1$ is similar to the above. So we omit its discussion.

Let us come back to the original competition-diffusion system $(4.1)_\epsilon$, (4.2). We first note that corresponding to the trivial equilibria (I,I), (II,II), (III,III) and

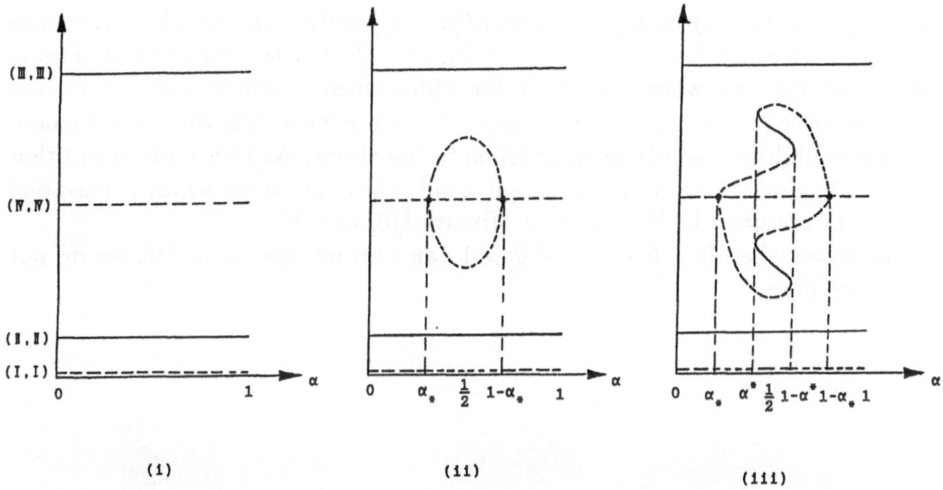

Fig. 3. Global bifurcation picture of equilibria of (4.3) when α is varied. (i) d is large. (ii) d is intermediate. (iii) d is small.

(IV,IV) of (4.3), there exist constant equilibrium solutions of $(4.1)_\epsilon$, which are given by $(0,0), (r/a,0), (0,r/a)$ and $(r/(a+b), r/(a+b))$, respectively. Now we address the following question: Is the global bifurcation picture of equilibria of (4.3) inherited to those of $(4.1)_\epsilon$, (4.2) for sufficiently small ϵ? This has not been completely answered. However, the theorems in the previous section tell us that the branches are inherited except for the ones in a neighborhood of bifurcation and turning points. We thus conclude that when α takes the value near 1/2, there exist stable nonconstant equilibrium solutions for small d. In fact, as d tends to zero, they approximately become

$$(r/a, 0) \text{ in } \Omega_0^L \text{ (or } \Omega_0^R)$$

and

$$(0, r/a) \text{ in } \Omega_0^R \text{ (or } \Omega_0^L)$$

which clearly exhibits spatial segregation for two competing species.

We can also draw the bifurcation diagram of equilibrium solutions of (4.3) when α is globally varied for fixed d. Figure 3(i) shows the case when d is large. It shows that $(4.1)_\epsilon$, (4.2) possess constant equilibrium solutions,

$(0,0), (r/a, 0), (0, r/a)$ and $(r/(a+b), r/(a+b))$ only. This result corresponds to the one given by Conway, Hoff and Smoller [2]. On the other hand, Figure 3(iii) shows the case when d is small, for which when α is near 0 or 1, there are four constant equilibrium solutions, while when α is near $1/2$, there are six non-constant equilibrium solutions in addition to the above. We thus understand that $(4.1)_\epsilon$, (4.2) possess stable nonconstant equilibrium solutions which correspond to the ones obtained in Matano and Mimura [10] and Jimbo [6].

The cases when $0 < \theta < 1$ and $\theta > 1$ can also be discussed, but we do not treat them here.

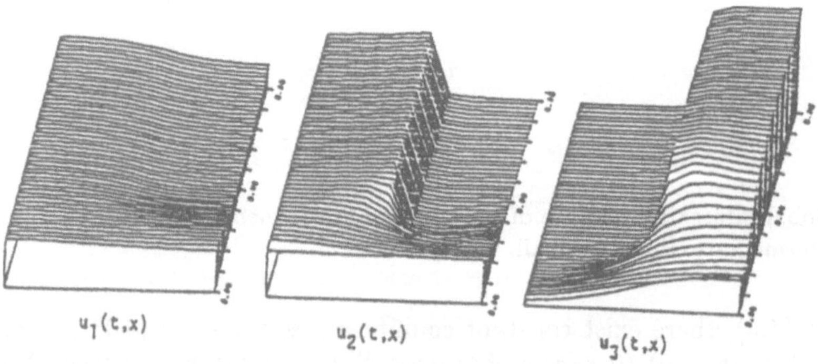

Fig. 4. Spatial segregation among three competing species.

5 Concluding remarks

We have studied the coexistence problem for two competing species by using a competition-diffusion system. Especially, we have been concerned with the influence of diffusion coefficient as well as domain-shape on the stability and existence of equilibrium solutions. We would like to emphasize here that the global bifurcation picture of equilibria of the limiting ODEs gives the answer to the original PDE problem. More general competition-diffusion systems described by

$$\frac{\partial u_i}{\partial t} = d_i \Delta u_i + (r_i - \sum_{j=1}^{n} a_{ij} u_j) u_i \quad (i = 1, 2, ..., n)$$

are been investigated. For n = 3 (i=1,2,···,n), using singular perturbation analysis, Mimura and Fife [13] suggest that there are stable nonconstant stable equilibrium solutions for appropriate values of d_i, R_i and a_{ij} $(i, j = 1, 2, 3)$, even if Ω is convex. Figure 4 numerically shows spatial segregation among three competing

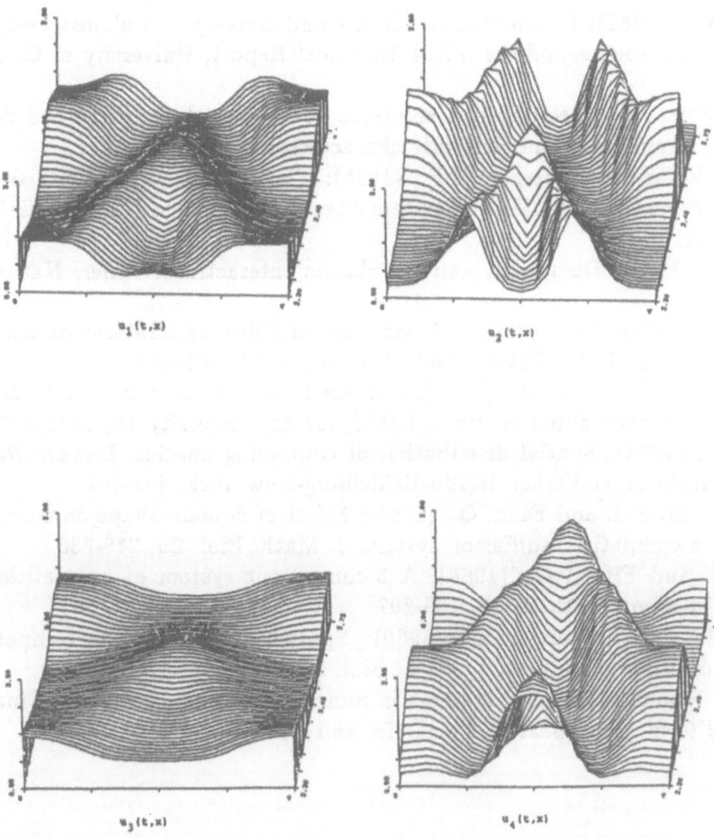

Fig. 5. Spatio-temporal segregation among four competing species.

species in one dimension. Furthermore, for n = 4, bifurcation theory indicates that there are non-constant periodic solutions in one dimension [11], which shows spatio-temporal segregation between 4-species in Figure 5.

References

1. Chueh, K.N., Conley, C.C. and Smoller, J.A., (1977): Positively invariant regions for system of nonlinear diffusion equations. Indiana Univ. Math. J. **26**, 373-392
2. Conway, E., Hoff, D., and Smoller, J.A., (1978): Large time behaviors of solutions of systems of nonlinear reaction-diffusion equations. SIAM J. Appl. Math. **35**, 1-16
3. Fang, Q.: Asymptotic behavior and domain-dependency of solutions to a class of reaction-diffusion systems with large diffusion coefficients. To appear in Hiroshima Math. J.
4. Hale, J.K. and Vegas, J., (1984): A nonlinear parabolic equation with varying domain. Arch. Rat. Mech. Anal. **86**, 99-123

5. Hirsch, M.W., (1982): Differential equations and convergence almost everywhere of strongly monotone semiflows. PAM Technical Report, University of California, Berkeley
6. Jimbo, S., Perturbated equilibrium solutions in the singularly perturbed domain: $L^\infty(\Omega(\zeta))-$ formulation and elaborate characterization. Preprint
7. Kishimoto, K. and Weinberger, H.F., (1985): The spatial homogeneity of stable equilibria of some reaction-diffusion system on convex domains. J. Differential Equations **58**, 15-21
8. Levin, S.A., (1974): Dispersion and population interactions. Amer. Natur. **108**, 207-228
9. Matano, H., (1979): Asymptotic behavior and stability of solutions of semilinear diffusion equations. Publ. RIMS, Kyoto University, **15**, 401-454
10. Matano, H. and Mimura, M., (1983): Pattern formation in competition-diffusion systems in nonconvex domains. Publ. RIMS, Kyoto University **19**, 1049-1079
11. Mimura, M., (1984): Spatial distribution of competing species. Lecture Notes in Biomath. **54**, Springer-Verlag, Berlin-Heidelberg-New York, 492-501
12. Mimura, M., Ei, S.-I. and Fang, Q., (1991): Effect of domain-shape on coexistence problems in a competition-diffusion system. J. Math. Biol. **29**, 219-238
13. Mimura, M. and Fife, P.C., (1986): A 3-component system of competition and diffusion. Hiroshima Math. J. **16**, 189-207
14. Mimura, M. and Kawasaki, K., (1980): Spatial segregation in competitive interaction-diffusion equations. J. Math. Biol. **9**, 49-64
15. Morita, Y.: Reaction-diffusion systems in nonconvex domains: invariant manifold and reduced form. To appear, J. Dynamics and Differential Equations

Population Interactions with Growth Rates Dependent on Weighted Densities

James F. Selgrade[1] *and Gene Namkoong*[2]

[1] Mathematics Department, North Carolina State University, Raleigh, NC 27695
[2] Forest Service, U.S. Department of Agriculture, Southeastern Forest Experiment Station and Departments of Genetics and Forestry, North Carolina State University, Raleigh, NC 27695

Abstract

Differential and difference equation models of interacting populations are presented and analyzed. The per capita growth rates are functions of linear combinations of individual population densities. The asymptotic behavior of these systems is discussed, including the occurrence of strange attractors. For the two dimensional differential equation, a general formula for the stability coefficient of a Hopf bifurcation is derived.

1 Introduction

The complex interactions among different biotic components of an ecosystem determine the development of the ecosystem. Animal and plant populations compete and cooperate with one another in order to obtain sufficient natural resources to sustain their growth and survival. In many cases the densities of the interacting populations determine individual viability and reproduction, and the growth response to density varies among species. Here we attempt to separate a population's response to density from interspecies interactions by assuming that a population's per capita growth rate is a function of its weighted total density. A weighted total density is a linear combination of the densities of the individual populations. An example of this approach is the classical Lotka/Volterra model where per capita growth rates are linear functions of the density variables. More recent studies include May (1974), Comins and Hassell (1976), Hassell and Comins (1976), Hofbauer, Hutson, and Jansen (1987), Cushing (1988), Franke and Yakubu (1990), and Selgrade and Namkoong (1990).

This paper is divided into five sections. Section 2 discusses the differential equation model, including background and motivating examples. Section 3 deals with the 2-dimensional differential equation and derives a formula for the stability coefficient of a periodic solution resulting from a Hopf bifurcation. Section

4 briefly discusses higher dimensional systems. Section 5 presents the difference equation model and describes a 2-dimensional example with strange attractors.

2 Background and model equations

The ecosystem being modeled contains n interacting, continuously reproducing populations. Let M_i, for $i = 1, \cdots, n$, denote the size (density) of the i-th population. For the i-th population define a *weighted total density* variable Z_i by

$$Z_i = \sum_{j=1}^{n} c_{ij} M_j. \tag{1}$$

The c_{ij} are real numbers which may be positive, negative, or zero depending on the effect of the j-th population on the i-th population. If $\mathbf{Z} = (Z_1, \cdots, Z_n)$ and $\mathbf{m} = (M_1, \cdots, M_n)$ are n-dimensional column vectors and $C = (c_{ij})$ is a $n \times n$ matrix, the weighted total density vector is given by

$$\mathbf{Z} = C\mathbf{m}.$$

The matrix C is called the *interaction matrix*, and $|c_{ij}|$ determines the intensity of the effect of the j-th population on the i-th population. The per capita growth rate of the i-th population, f_i, is taken to be a smooth function of the i-th weighted density. Hence $f_i(Z_i)$ measures the response to density of the i-th population and is referred to as a *fitness* function.

The following system of autonomous ordinary differential equations models the interactions among the n populations:

$$dM_i/dt = f_i(Z_i)M_i \qquad i = 1, 2, \cdots, n. \tag{2}$$

In vector form, (2) may be written

$$d\mathbf{m}/dt = \mathbf{F}(\mathbf{m})$$

where the function $\mathbf{F}$ is called a vector field. This vector field is defined on the nonnegative orthant which is invariant because of the form of (2). The derivative of the vector field at any point in the orthant may be expressed in terms of two diagonal matrices and C as

$$D\mathbf{F}(\mathbf{m}) = \begin{bmatrix} f_1(Z_1) & & 0 \\ & \ddots & \\ 0 & & f_n(Z_n) \end{bmatrix} + \begin{bmatrix} M_1 f_1'(Z_1) & & 0 \\ & \ddots & \\ 0 & & M_n f_n'(Z_n) \end{bmatrix} C. \tag{3}$$

The first matrix in (3) vanishes at an equilibrium in the interior of the orthant.

Typically a fitness function will possess certain monotonicity properties as a function of its weighted total density variable. For example, because of intraspecific and interspecific crowding, a fitness f_i may be a monotone decreasing function of Z_i. In a forest ecosystem, a tree population with such a fitness is called a *pioneer* species and we adopt that terminology here. Certain varieties of pine and poplar are considered deprivation intolerant species; they do well at low density but have fitnesses which decrease with increasing density. A more tolerant tree species such as oak or maple is called a *climax* or *successionary* species, and its fitness increases up to a maximum and then decreases with increasing density. At low density the climax species benefits from the presence of additional trees because they provide a more suitable environment for climax reproduction and survival, e.g., more protection and improved soil conditions. Here we refer to any fitness which is first increasing and then decreasing with density as a climax fitness.

In order to obtain Lotka/Volterra systems from (2), we take the f_i*'s* to be linear pioneer fitnesses of the form

$$f_i(Z_i) = r_i - Z_i, \tag{4}$$

where $Z_i = \sum_{j=1}^{n} c_{ij} Z_j$. The sign of c_{ij} determines the effect of the j-th population on the i-th population, i.e., if the j-th population competes with the i-th then $c_{ij} > 0$ or if the j-th cooperates with the i-th then $c_{ij} < 0$. Ricker (1954) concludes that certain fish populations have pioneer fitnesses of the form

$$f(Z) = e^{r(1-Z)} - a. \tag{5}$$

Hassell and Comins (1976) study pioneer fitnesses of the form

$$f(Z) = \frac{r}{(1+bZ)^p} - a. \tag{6}$$

Cushing (1988) in his analysis of age-structured populations and Selgrade and Namkoong (1990) in a forest model suggest climax fitnesses similar to

$$f(Z) = Ze^{r(1-Z)} - a. \tag{7}$$

An equilibrium which is interior to the nonnegative orthant occurs where all the fitness functions vanish. For each fitness f_i, we assume that there is a value $z_i > 0$ where $f_i(z_i) = 0$ and we assume that $f_i'(z_i) \neq 0$. For a pioneer fitness there can be only one such value and $f_i'(z_i) < 0$; for a climax fitness there may be $z_{i,1} < z_{1,2}$ so that $f_i(z_{i,1}) = f_i(z_{i,2}) = 0$ and $f_i'(z_{i,1}) > 0$ but $f_i'(z_{i,2}) < 0$. Equation (2) will always have at least one interior equilibrium if the following linear algebraic system has a solution in the positive orthant:

$$C\mathbf{m} = \mathbf{z} \tag{8}$$

where $\mathbf{z} = (z_1, \cdots, z_n)$. This equilibrium is isolated if $det\, C \neq 0$. Additional isolated interior equilibria exist if (8) has solutions in the positive orthant when some of the components of $\mathbf{z}$ are replaced by other zeroes of the f_i*'s*. Let $\mathbf{e} = (e_1, \cdots, e_n)$ denote an interior isolated equilibrium of (2). The following formulas for the trace and the determinant of $D\mathbf{F}(\mathbf{e})$ are computed from (3):

$$trD\mathbf{F}(\mathbf{e}) = \sum_{i=1}^{n} e_i f_i'(z_i) c_{ii}$$
$$det\, D\mathbf{F}(\mathbf{e}) = det\, C \prod_{i=1}^{n} e_i f_i'(z_i) \tag{9}$$

Stability information about e may be obtained from (9). For example, if all fitnesses are pioneer and each population experiences intraspecific competition, i.e., $c_{ii} > 0$ for $i = 1, \cdots, n$, then $trD\mathbf{F}(\mathbf{e}) < 0$ so the interior equilibrium has at least a one dimensional stable manifold. For two dimensional systems, (9) determines the stability of every interior equilibrium.

3 Two dimensional models

The 2-dimensional system (2) is purely cooperative or competitive if both fitnesses are pioneer and, for $i \neq j$, $c_{ij} < 0$ (cooperative) or $c_{ij} > 0$ (competitive). For such dynamical systems each bounded solution converges to an equilibrium, (see Hirsch (1982) or Selgrade (1980)). However, if one fitness is not pioneer, stable periodic solutions via Hopf bifurcation may be present as seen in a pioneer/climax model studied by Selgrade and Namkoong (1990). In fact, a Hopf bifurcation may occur if the fitnesses have derivatives of opposite signs at their zeroes. Assume that there is intraspecific competition ($c_{ii} > 0$ for $i = 1, 2$) and that $f_1'(z_1) < 0$ and $f_2'(z_2) > 0$ where $f_1(z_1) = f_2(z_2) = 0$. From (9) it is clear that the real parts of the eigenvalues of $D\mathbf{F}(\mathbf{e})$ have the same sign if $det\, C < 0$. Also it may be shown that the $tr\, D\mathbf{F}(\mathbf{e})$ will change sign if either c_{11} or c_{22} is fixed and the other is varied as a parameter. We have the following two results regarding the existence and stability of a periodic orbit from a Hopf bifurcation. In light of assumption (A2) below, assumption (A3) insures that the equilibrium $\mathbf{e} = (e_1, e_2)$ lies in the positive quadrant. Also (A3) implies that $c_{12} > 0$ and $c_{21} > 0$.

Theorem 1 *For the 2-dimensional system* (2), *let $z_1 > 0$ be a zero of f_1 and $z_2 > 0$ be a zero of f_2. Assume $c_{ii} > 0$ for i=1,2 and*
(A1) $f_1'(z_1) < 0$ and $f_2'(z_2) > 0$
(A2) $det C < 0$,
(A3) $c_{22}z_1 - c_{12}z_2 < 0$ and $c_{11}z_2 - c_{21}z_1 < 0$. Then a Hopf bifurcation occurs with respect to the parameter c_{11} as c_{11} decreases through the critical value

$$c_{11} = \frac{c_{21}c_{22}z_1 f_2'(z_2)}{(c_{22}z_1 - c_{12}z_2)f_1'(z_1) + c_{22}z_2 f_2'(z_2)}$$

or with respect to the parameter c_{22} as c_{22} increases through the critical value

$$c_{22} = \frac{c_{11}c_{12}z_2 f_1'(z_1)}{(c_{11}z_2 - c_{21}z_1)f_2'(z_2) + c_{11}z_1 f_1'(z_1)}$$

Proof. The argument is similar to that in Selgrade and Namkoong (1990). □

Theorem 2 *With the same assumptions as in* Theorem 1, *the stability coefficient* α *for the Hopf periodic solution is computed from the following formula. If* $\alpha < 0$ *then the periodic solution is locally asymptotically stable, and if* $\alpha > 0$ *then the periodic solution is unstable.*

$$\begin{aligned} 16\alpha = c_{11}^2 z_2 f_1''(z_1)/(c_{22}e_2) + c_{22}c_{21}z_1 f_2''(z_2)/(c_{12}e_1) \\ + c_{11}c_{21}e_1 f_1'(z_1)\, det\, C\, [f_2''/(c_{12}f_2')]'(z_2) \\ + c_{11}e_2 f_2'(z_2)\, det\, C\, [f_1''/f_1']'(z_1). \end{aligned} \tag{10}$$

Proof. The argument is highly computational and proceeds along the lines discussed in Hassard, Kazarinoff, and Wan (1981) or Guckenheimer and Holmes (1983). Specifically, we use equation (3.4.11) on page 152 in Guckenheimer and Holmes (1983).

The Hopf bifurcation occurs when the eigenvalues of $D\mathbf{F}(\mathrm{e})$ pass through the imaginary axis. At this parameter value the $trD\mathbf{F}(\mathrm{e})$ and the positive imaginary part ω of the eigenvalues are given by:

$$\begin{aligned} trD\mathbf{F}(\mathrm{e}) = c_{11}e_1 f_1'(z_1) + c_{22}e_2 f_2'(z_2) = 0 \\ \omega = (e_1 e_2 f_1'(z_1) f_2'(z_2) \mathrm{det} C)^{\frac{1}{2}}. \end{aligned} \tag{11}$$

An affine change of coordinates is performed so that e is translated to the origin and so that the derivative of the new vector field at the origin is given by:

$$\begin{bmatrix} 0 & -\omega \\ \omega & 0 \end{bmatrix}.$$

The change of coordinates we use is

$$\begin{bmatrix} M_1 \\ M_2 \end{bmatrix} = \begin{bmatrix} e_1 \\ e_2 \end{bmatrix} + \begin{bmatrix} 1 & 0 \\ -\frac{c_{11}}{c_{12}} & \frac{-\omega}{c_{12}e_1 f_1'(z_1)} \end{bmatrix} \begin{bmatrix} x \\ y \end{bmatrix} \tag{12}$$

where the matrix in (12) is obtained from an eigenvector of $i\omega$ and we denote this matrix by P. Under (12), (2) becomes the following system with variables (x, y):

$$\begin{aligned} dx/dt = f(x,y) \\ dy/dt = g(x,y) \end{aligned} \tag{13}$$

where

$$\begin{aligned} f(x,y) &= (e_1 + x) f_1(Z_1) \\ g(x,y) &= -c_{11}e_1 f_1'(z_1)(e_1 + x) f_1(Z_1)/\omega \\ &\quad + [y - c_{12}e_1 e_2 f_1'(z_1)/\omega + c_{11}e_1 f_1'(z_1) x/\omega] f_2(Z_2). \end{aligned}$$

In (13), Z_1 and Z_2 are affine functions of x and y via (1) and (12) with partials given by:

$$\frac{\partial Z_1}{\partial x} = 0, \quad \frac{\partial Z_1}{\partial y} = \frac{-\omega}{e_1 f_1'(z_1)}, \quad \frac{\partial Z_2}{\partial x} = \frac{-det\mathrm{C}}{c_{12}}, \text{ and } \frac{\partial Z_2}{\partial y} = \frac{-\omega c_{22}}{c_{12}e_1 f_1'(z_1)}.$$

Note that the first term in g is a multiple of f.

To find α we need to compute the second and third order partials of f and g and evaluate at $(x,y)=(0,0)$. Since our f and g differ from that of Guckenheimer and Holmes (1983) only by the addition of linear terms, we may compute α by appealing to their formula (3.4.11):

$$\begin{aligned}16\alpha = &f_{xxx}+f_{xyy}+g_{xxy}+g_{yyy}\\ &+[f_{xy}(f_{xx}+f_{yy})-g_{xy}(g_{xx}+g_{yy})-f_{xx}g_{xx}+f_{yy}g_{yy}]/\omega.\end{aligned} \tag{15}$$

The partials of f are straightforward and, in fact, $f_{xx}=f_{xxx}=0$ since $\partial Z_1/\partial x = 0$. The partials of g are much more complicated but propitious use of (11) enable us to cancel and combine various terms leading to the following expressions which are evaluated at $(x,y)=(0,0)$:

$$\begin{aligned}f_{xyy}+g_{xxy}+g_{yyy} = &\,\omega^2 f_1''(z_1)/(e_1 f_1'(z_1))^2 + c_{11}e_1\omega^2 f_1'''(z_1)/(e_1 f_1'(z_1))^2\\ &- c_{21}\,det\,C\,f_2''(z_2)/c_{12} - c_{21}c_{22}e_2\,det\,C\,f_2'''(z_2)/c_{12}\end{aligned}$$

and

$$\begin{aligned}[f_{xy}f_{yy}-g_{xy}(g_{xx}+g_{yy})+f_{yy}g_{yy}]/\omega = &-\omega^2 f_1''(z_1)/(e_1 f_1'(z_1))^2\\ &+c_{11}^2 e_1\,det\,C\,(f_1''(z_1))^2/(c_{22}f_1'(z_1)) + c_{11}^2 z_2 f_1''(z_1)/(c_{22}e_2)\\ &+c_{21}c_{22}z_1 f_2''(z_2)/(c_{12}e_1) + c_{21}\,det\,C\,f_2''(z_2)/c_{12}\\ &+c_{21}c_{22}e_2\,det\,C\,(f_2''(z_2))^2/(c_{12}f_2'(z_2)).\end{aligned}$$

Combining the previous two equations gives (10). This completes the proof. □

Thus the stability of the Hopf periodic solution depends on the interaction matrix C, the coordinates of the equilibrium, the zeroes of the fitness functions, and the first three derivatives of the fitnesses evaluated at their zeroes. Note that the signs of the first two terms in (10) depend on the concavity of the fitnesses at their zeroes. If f_1 is a linear decreasing function like (4) then its second derivative is zero so the first and last terms in (10) are zero. If f_2 is the climax fitness (7) then it is always concave down at its smaller zero (i.e., where $f_2'(z_2)>0$) and $[f_2''/f_2']'$ is always negative there. Hence terms two and three in (10) are negative. Thus if a Hopf bifurcation exists for fitnesses (4) and (7), it always results in a stable periodic solution. Selgrade and Namkoong (1990) study a forest population model where the fitnesses are special cases of (5) and (7). For (5), f''/f' is a constant so the last term in (10) is zero but the first term is positive. Computing α using (10) shows that the periodic solution is stable for this model. In fact, numerical experiments indicate bistable behavior, with solutions asymptotic to a stable equilibrium or to the Hopf periodic solution. If the Hassell/Comins fitness (6) is used as f_1 then $f_1''>0$ and $[f_1''/f_1']'>0$ so that last term in (10) is negative since det C <0. If f_2 is taken to be (7), there is a strong possibility that $\alpha<0$ since three of the four terms in (10) are negative. In fact, for all specific examples we have studied where f_2 is given by (7) and f_1 by (4), (5), or (6), the Hopf periodic solution has been stable.

If either c_{11} or c_{22} is not positive then a Hopf bifurcation may still occur but not under conditions (A1), (A2), and (A3). If Hopf bifurcation occurs in the positive quadrant then (10) may be used to determine stability if the denominators on the right side of (10) are not zero, i.e., $c_{22} \neq 0$, $c_{12} \neq 0$, $f_1'(z_1) \neq 0$, and $f_2'(z_2) \neq 0$.

4 Three dimensional models

Gardini, Lupini, and Messia (1989) report the occurrence of chaotic attractors for 3-dimensional Lotka/Volterra systems, which are included in (2). These attractors arise from repeated period-doubling of the return map of a stable Hopf periodic orbit. We have done some preliminary numerical work on 3-dimensional pioneer/climax models with two pioneer fitnesses like (4) and a climax fitness like (7). Hopf bifurcations occur on the boundaries and in the interior of the positive octant. We have seen an interior stable periodic orbit undergo two period-doublings before disappearing either into an unstable periodic solution or into a union of boundary orbits. We intend to study the onset of chaotic attractors for 3-dimensional pioneer/climax models.

5 Difference equation models

If the interacting populations are reproducing in discrete generations, the following system of n difference equations is analogous to (2):

$$M_i(k+1) = f_i(Z_i)M_i(k) \qquad i = 1, \cdots, n \tag{16}$$

where k is a nonnegative integer and

$$Z_i = \sum_{j=1}^{n} c_{ij} M_j(k).$$

The function $f_i(Z_i) - 1$ measures the per capita change in density per generation for the i-th population. The fitness functions (5), (6), and (7) are appropriate for (16) if the constant terms are deleted so these functions are nonnegative. Because of the form of these fitnesses, chaotic dynamical behavior may be expected for (16) even in one dimension.

An interior equilibrium occurs where each fitness is one. The derivative of the right side of (16) at an interior equilibrium is the same as (3) except the first diagonal matrix in (3) is replaced by the identity matrix. A Hopf bifurcation may occur if a pair of complex eigenvalues of this derivative pass through the unit circle as some parameter is varied. A discussion of Hopf bifurcation for the difference equation (16) in two dimensions parallels that in Sect. 3 for the differential equation (2). However, conditions for bifurcation are more complicated than (A1), (A2), and (A3) because characterizing eigenvalues passing through the unit circle is more difficult than that for passing through the imaginary axis.

Namkoong and Selgrade (1990) study various 2-dimensional examples of pioneer/climax difference equations where Hopf bifurcations result in stable invariant curves. For some examples, varying the bifurcation parameter farther causes

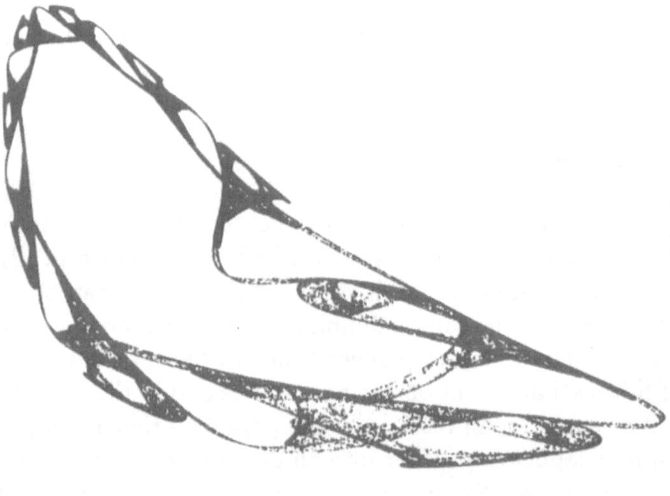

Fig. 1. Interior Attractor for $c_{22} = .6999$.

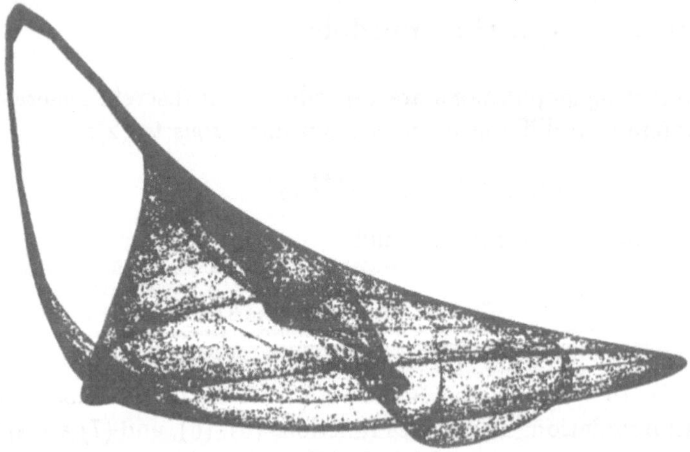

Fig. 2. Interior Attrator for $c_{22} = .78$.

the disintegration of the invariant curve and the onset of a strange attractor as has been observed for other 2-dimensional maps, e.g., Curry and Yorke (1978) or Gumowski and Mira (1980). Take a pioneer fitness f_1 and a climax fitness f_2 given by:

$$
\begin{aligned}
f_1(Z_1) &= 2\,e^{4-20\,Z_1} \\
f_2(Z_2) &= \tfrac{1}{3}\,Z_2 e^{3-2\,Z_2}
\end{aligned}
\tag{17}
$$

where $c_{11} = .33$, $c_{12} = 1$, $c_{21} = 1$, and c_{22} is the bifurcation parameter. A Hopf bifurcation resulting in a stable invariant curve occurs for $c_{22} < .1$. For small values of c_{22}, this curve is oval-shaped, but it becomes irregular as c_{22} increases. For example, when $c_{22} = .6$ the curve has folds and, for larger c_{22}, it crosses itself. The nonnegative quadrant contains three attractors: the invariant curve, a locally stable equilibrium on the M_2-axis, and a locally attracting interval on the M_1-axis within which the dynamical behavior is chaotic. As c_{22} increases beyond .6, the invariant curve attractor is replaced by nine islands which are mapped periodically from one to another. At $c_{22} = .6999$, this interior attractor takes the form of the braided loop in Fig. 1. As c_{22} increase above .7, the braided loop loses all its holes except the center one and assumes the form in Fig. 2 at $c_{22} = .78$. As c_{22} approaches .786, the interior attractor disappears with its domain of attraction engulfed by the domain of attraction of the interval on the M_1-axis. For more details on the dynamical behavior of this example see Namkoong and Selgrade (1990).

Acknowledgment

The authors thank Patrick Hopkins for producing the figures. Research supported by funds provided the USDA-Forest Service, Southeastern Forest Experiment Station, Pioneering (Population genetics of Forest Trees) Research Unit, Raleigh, North Carolina.

References

1. Comins, H.N. and Hassell, M.P. (1976): Predation in multi-prey communities. J. Theor. Biol. **62**, 93-114
2. Curry, J.H. and Yorke, J.A. (1978): A transition from Hopf bifurcation to chaos: Computer experiments with maps on R^2. In The Structure of Attractors in Dynamical Systems, Lecture Notes in Mathematics, **688**, Springer-Verlag, Berlin-Heidelberg-New York, 48-64
3. Cushing, J.M. (1988): Nonlinear matrix models and population dynamics. Natural Resource Modeling **2**, 539-580
4. Franke, J.E. and Yakubu, A.-A. (1990): Extinction of species in systems governed by difference equations of Lotka-Volterra type, preprint
5. Freedman, H.I. (1980): Deterministic Mathematical Models in Population Ecology. Marcel Dekker, New York
6. Gardini, L., Lupini, R. and Messia, M.G. (1989): Hopf bifurcation and transition to chaos in Lotka-Volterra equation. J. Math. Biol. **27**, 259-272
7. Guckenheimer, J. and Holmes, P. (1983): Nonlinear Oscillations, Dynamical Systems, and Bifurcations of Vector Fields. Springer-Verlag, Berlin-Heidelberg-New York
8. Gumowski, I. and Mira, C. (1980): Recurrences and Discrete Dynamic Systems. Lecture Notes in Mathematics, **809**, Springer-Verlag, Berlin-Heidelberg-New York
9. Hassard, B.D., Kazarinoff, N.D. and Wan, Y.-H. (1981): Theory and Applications of Hopf Bifurcation. Cambridge Univ. Press, New York
10. Hassell, M.P. (1976): The Dynamics of Competition and Predation. Edward Arnold Ltd., London
11. Hassell, M.P. and Comins, H.N. (1976): Discrete time models for two-species competition. Theor. Pop. Biol. **9**, 202-221
12. Hirsch, M.W. (1982): Systems of differential equations which are competitive or cooperative. I: Limit sets. SIAM J. Math. Anal. **13**, 167-179

13. Hofbauer, J., Hutson, V. and Jansen, W. (1987): Coexistence for systems governed by difference equations of Lotka-Volterra type. J. Math. Biol. **25**, 553-570
14. Hofbauer, J. and Sidmund, K. (1988): The Theory of Evolution and Dynamical Systems. Cambridge Univ. Press, Cambridge
15. May R. M. (1974): Ecosystem patterns in randomly fluctuating environments. Progress in Theoretical Biology. (ed., Rosen and Snell), Academic Press, New York
16. May R. M. and Oster, G. F. (1976): Bifurcation and dynamic complexity in simple ecological models. Amer. Naturalist **110**, 573-599
17. Namkoong, G. and Selgrade, J.F. (1990): Strange attractors in density dependent succession, preprint.
18. Ricker, W.E. (1954): Stock and Recruitment. J. Fish. Res. Bd. Can. **11**, 559-623
19. Selgrade, J.F. (1980): Asymptotic behavior of solutions to single loop positive feedback systems. J. Diff. Eq. **38**, 80-103
20. Selgrade, J.F. and Namkoong, G. (1990): Stable periodic behavior in a pioneer-climax model. Natural Resource Modeling, **4**, 215-227
21. Waltman, P. (1983): Competition Models in Population Biology. CBMS-NSF 45, SIAM, Philadelphia.

Global Stability in a Population Model with Dispersal and Stage Structure

J.-H. Wu and H.I. Freedman

[1] Department of Mathematics, York University, North York, Ontario M3J 1P3 Canada
[2] Applied Mathematics Institute, Department of Mathematics, University of Alberta, Edmonton T6G 2G1 Canada

1 Introduction

Recently there has been much interest in modeling population growth where the population disperses among patches in a patchy environment [3], [4], [5], [6], [7], [8], [9], [10], [11], [17], [22], [23]. A second group of papers have dealt with models of populations where the life history of the particular species involves two or more stages [1], [2], [13], [18], [24], [25]. In [12], we have combined these two concepts and analyzed a model incorporating stage structure and dispersal.

The idea in [12] was to consider a single-species population whose life history consists of two stages, immature and mature, and where populations in both stages could disperse among patches. In order to carry out our analysis we required that the dispersal be linear.

In the present paper we wish to analyze such a model where there is little or no dispersal for the immature population, but a strong, nonlinear dispersal among patches for the mature population. We have in mind species such as *Malacosoma californicum pluviale* (Dyer), the western tent caterpillar, which has an immature stage as a caterpillar with almost no dispersal, and an adult stage as a fat brown moth with high dispersal rates which are weather dependent.

In the next section we develop our model and state our main results, which give criteria for the global stability of a positive equilibrium. In section 3 we give our proofs.

2 Models and main results

We consider a system composed of n patches connected by dispersion and occupied by a single species population where individual members have a two-stage structure: immature and mature. Let $U(t, s)$ denote the concentration of immatures of age s at time t in the ith patch and $M_i(t)$ denote the concentration of matures at time t in the ith patch, $i = 1, 2, \ldots, n$. The model considered in this paper is described by the following system of functional differential equations:

$$\begin{cases} \left(\frac{\partial}{\partial t} + \frac{\partial}{\partial s}\right) U(t,s) = -\gamma_i U_i(t,s) + \sum_{j \neq i} \delta_{ji}[U_j(t,s) - U_i(t,s)] \\ U_i(t,0) = \alpha_i M_i(t) \\ \frac{d}{dt} M_i(t) = -M_i(t) g_i(M_i(t)) + F_i(M_1(t), \ldots, M_n(t)) + U_i(t, t-\tau), \\ i = 1, \ldots, n, \quad t \geq 0, \end{cases} \tag{2.1}$$

where the notations and biological assumptions are described as follows

(H1): The birth rate into the immature population in the ith patch is proportional to the existing mature population with proportionality constant $\alpha_i > 0$, $i = 1, \ldots, n$.

(H2): The death rate of the immature population in the ith patch is proportional to the existing immature population with proportionality constant $\gamma_i > 0$, $i = 1, \ldots, n$.

(H3): The net exchange rate of the immature population from the jth patch to the ith patch is proportional to the difference of the concentrations with proportionality constants $\delta_{ji} \geq 0$, $i \neq j$, $i, j = 1, \ldots, n$.

(H4): The length of time from birth to maturity is a constant $\tau > 0$, which is uniform for each individual in all patches, and those immature individuals born at time $t - \tau$ and surviving to time t exit from the immature population and enter into the mature population.

Further, $g_i(M_i)$ and $F_i(M_1, \ldots, M_n)$ represent the death rate and the net exchange rate, respectively, of mature population in the ith patch. Detailed assumptions on g_i and F_i will be presented later. According to assumptions (H1) and (H2), we have

$$R_i(t-\tau,\ t-\tau) = \alpha_i M_i(t-\tau), \tag{2.2}$$

$$\frac{\partial}{\partial s} Q_i(s,\ t-\tau) = -\gamma_i Q_i(s,\ t-\tau) + \sum_{j \neq i} \delta_{ji} \left[Q_j(s, t-\tau) - Q_i(s,\ t-\tau)\right], \tag{2.3}$$

where $Q_i(s, t-\tau)$, $t - \tau < s \leq t$, $i = 1, \ldots, n$, denotes the total immature population in the ith patch born at the instant $t - \tau$. Thus

$$\frac{\partial}{\partial s} Q_i(s, t-\tau) = R_i(s, t-\tau), \quad t - \tau < s \leq t. \tag{2.4}$$

From (2.3) and (2.4) it follows that

$$\frac{\partial}{\partial s} R_i(s, t-\tau) = -\gamma_i R_i(s, t-\tau) + \sum_{j \neq i} \delta_{ji} \left[R_j(s, t-\tau) - R_i(s, t-\tau)\right]. \tag{2.5}$$

Solving (2.2) and (2.5), we obtain

$$R_i(t,\ t-\tau) = \sum_{j=1}^{n} b_{ij} \alpha_j M_j(t-\tau) \tag{2.6}$$

Clearly, the integral $I_i(t) := \int_0^{\tau} U_i(t,s) ds$ represents the concentration of immatures at time t in the ith patch, $i = 1, \ldots, n$. From (2.1) we can easily obtain

$$\begin{cases} \frac{d}{dt} I_i(t) = -\gamma_i I_i(t) + \sum_{j\neq i} \delta_{ji}[I_j(t) - I_i(t)] + \alpha_i M_i(t) - U_i(t,\tau) \\ \frac{d}{dt} M_i(t) = -M_i(t) g_i(M_i(t) + F_i(M_1(t), \ldots, M_n(t)) + U_i(t,\tau). \end{cases} \tag{2.7}$$

On the other hand, for $V_i(t,\theta) := U_i(t-\tau+\theta,\theta)$, $0 \le \theta \le \tau$, by (2.1) we have

$$\frac{\partial}{\partial\theta} V_i(t,\theta) = -\gamma_i V_i(t,\theta) + \sum_{j\neq i} \delta_{ji}[V_j(t,\theta) - V_i(t,\theta)], \quad 0 \le \theta \le \tau, \tag{2.8}$$

from which it follows that

$$\begin{pmatrix} U_1(t-\tau+\theta,\theta) \\ \vdots \\ U_n(t-\tau+\theta,\theta) \end{pmatrix} = e^{A\theta} \begin{pmatrix} \alpha_1 M_1(t-\tau) \\ \vdots \\ \alpha_n M_n(t-\tau) \end{pmatrix}, \tag{2.9}$$

in particular, we have

$$\begin{pmatrix} U_1(t,\tau) \\ \vdots \\ U_n(t,\tau) \end{pmatrix} = e^{A\tau} \begin{pmatrix} \alpha_1 M_1(t-\tau) \\ \vdots \\ \alpha_n M_n(t-\tau), \end{pmatrix} \tag{2.10}$$

where

$$A = (a_{ij}) = \begin{pmatrix} -\gamma_1 - \sum\limits_{j\neq 1} \delta_{j1} & \delta_{21} & \ldots & \delta_{n1} \\ \delta_{12} & -\gamma_2 - \sum\limits_{j\neq 2} \delta_{j2} & \ldots & \delta_{n2} \\ \ldots & & \ldots & \ldots \\ \delta_{1n} & \delta_{2n} & \ldots & -\gamma_n - \sum\limits_{j\neq n} \delta_{jn}. \end{pmatrix}$$

Set $B = (b_{ij}) = e^{A\tau}$. Since $a_{ij} = \delta_{ji} \ge 0$ for $i \ne j$, we have $b_{ij} \ge 0$ for $i, j = 1, 2, \ldots, n$. Substituting (2.10) into (2.7), we obtain

$$\begin{cases} \frac{d}{dt} I_i(t) = -\gamma_i I_i(t) + \sum_{j\neq i} \delta_{ji}\left[I_j(t) - I_i(t)\right] + \alpha_i M_i(t) - \sum_{j=1}^n b_{ij}\alpha_j M_j(t-\tau) \\ \frac{d}{dt} M_i(t) = -M_i(t) g_i(M_i(t)) + F_i(M_1(t), \ldots, M_n(t)) + \sum_{j=1}^n b_{ij}\alpha_j M_j(t-\tau). \end{cases} \tag{2.11}$$

for $t \ge 0$ and $i = 1, \ldots, n$.

To specify a solution of the model equation (2.1), we assume that the initial conditions are nonnegative real numbers $M_i(0)$ and nonnegative continuous functions $U_i(0,s)$, $s \in [0,\tau]$, $i = 1, \ldots, n$.

Clearly, once transformed into (2.11), the initial conditions should be nonnegative real numbers $I_i(0)$ and nonnegative continuous functions $M_i(\theta)$, $\theta \in [-\tau, 0]$, $i = 1, \ldots, n$. Moreover, form (2.9) we can obtain the following natural compatibility condition between $I_i(0)$ and $M_i(\theta)$, $\theta \in [-\tau, 0]$, $i = 1, \ldots, n$,

$$\begin{pmatrix} I_1(0) \\ \ldots \\ I_n(0) \end{pmatrix} = \int_0^\tau e^{A\theta} \begin{pmatrix} \alpha_1 M_1(-\theta) \\ \ldots \\ \alpha_n M_n(-\theta) \end{pmatrix} d\theta. \tag{2.12}$$

Mathematical assumptions on the death rate and the net exchange rate of the mature population are described as follows:

(H5): $F_i \in C^2(R^n_+; R)$, $F_i(0,\dots,0) = 0$ for $i = 1,\dots,n$; and for each $i = 1,\dots,n$, we have

$$M_i \le M_j \text{ for } j = 1,\dots,n \quad \text{implies } F_i(M_1,\dots,M_n) \ge 0, \tag{2.13}$$
$$M_i \ge M_j \text{ for } j = 1,\dots,n \quad \text{implies } F_i(M_1,\dots,M_n) \le 0. \tag{2.14}$$

(H6) : $\left(\frac{\partial F_i}{\partial x_j}(x_1,\dots,x_n)\right)$ is cooperative (i.e. $\frac{\partial F_i}{\partial x_j}(x_1,\dots,x_n) \ge 0$ for $i \ne j$) and irreducible

(H7) : $g_i \in C^2(R_+; R)$ and

$$g_i(0) < \sum_{j=1}^{n} b_{ij}\alpha_j < \liminf_{x\to\infty} g_i(x). \tag{2.15}$$

Biologically, the irreducibility of the matrix $\left(\frac{\partial F_i}{\partial x_j}(x_1,\dots,x_n)\right)$ implies that every patch is connected by dispersion directly or indirectly. (2.13) and (2.14) characterizes the dispersion from the jth patch to the ith patch which is usually assumed to be proportional to the difference $M_j - M_i$ of the population densitites in each patch. However our assumption allows a more general nonlinear dispersion form. Assumption (H7) indicates that the death rate of the mature population is of a logistic nature.

We will assume the following "concavity" condition or "sublinearity" condition in order to guarantee the uniqueness of a positive equilibrium

(H8): $\frac{\partial^2 F_i}{\partial x_j^2}(x_1,\dots,x_n) \le 0$ for $i \ne j$,
$-\frac{\partial^2 F_i}{\partial x_i^2}[x_i g_i(x_i)] + \frac{\partial^2 F_i}{\partial x_i^2}(x_1,\dots,x_n) < 0$ for $i = 1,\dots,n$ and $(x_1,\dots,x_n) \in R^n_+$.

(H9): $F_i(ax_1,\dots,ax_n) - aF_i(x_1,\dots,x_n) < ax_i[g_i(ax_i) - g_i(x_i)]$, $j = 1,\dots,n$ for all positive vectors $(x_1,\dots,x_n)$ and constant $a > 1$.

The main result of this paper is described as follows:

Theorem 2.1 *Suppose that* (H1)–(H7) *hold and either* (H8) *or* (H9) *is satisfied. Then there exists a positive vector* $(M_1^*,\dots,M_n^*)$ *such that if* $(I_i(t), M_i(t))$ *is the solution of (2.11) subject to the initial conditions* $I_i(0), M_i(\theta)$, $\theta \in [-\tau, 0]$, $i = 1,2,\dots,n$, *satisfying* (2.12), *then*

(i) $I_i(t) \ge 0,\ M_i(t) \ge 0$ for $i = 1,\dots,n$;

(ii) $\lim_{t\to\infty} M_i(t) = M_i^*$, $i = 1,\dots,n$ and

$$\lim_{t\to\infty}\begin{pmatrix} I_1(t) \\ \dots \\ I_n(t)\end{pmatrix} = A^{-1}(e^{A\tau} - 1)\begin{pmatrix}\alpha_1 M_1^* \\ \dots \\ \alpha_n M_n^*.\end{pmatrix}$$

3 Proof of the main result

We begin by considering the equation

$$\frac{d}{dt}M_i(t) = -M_i(t)g_i(M_i(t)) + F_i(M_1(t), \dots, M_n(t)) + \sum_{j=1}^{n} b_{ij}\alpha_j M_j(t-\tau), \qquad t \geq 0, i = 1, \dots, n, \tag{3.1}$$

which represents the dynamics of the mature populations. Let $C = C([-\tau, 0]; R^n)$ and $C^+ = \{\varphi \in C;\ \varphi \geq 0\}$, where $\varphi = (\varphi_1, \dots, \varphi_n) \geq 0$ means $\varphi_i(\theta) \geq 0$ for $\theta \in [-\tau, 0]$ and $i = 1, \dots, n$. Define $G = (G_1, \dots, G_n) : C \to R^n$ by

$$G_i(\varphi) = -\varphi_i(0)g_i(\varphi_i(0)) + F_i(\varphi_i(0), \dots, \varphi_n(0)) + \sum_{j=1}^{n} b_{ij}\alpha_j\varphi_j(-\tau),\ i = 1, \dots, n,$$

where $\varphi \in C$. Clearly, if $\varphi \in C^+$ and $\varphi_i(0) = 0$ for some $i = 1, \dots, n$, then $G_i(\varphi) \geq 0$. Therefore, by Corollary 2.1 of Smith (1987), C^+ is positively invariant with respect to the semiflow $\{T(t)\}_{t\geq 0}$ defined by

$$T(t)\varphi = M_t(\varphi), t \geq 0, \varphi \in C^+,$$

i.e., if $\varphi \in C^+$, then $M(t, \varphi) \in C^+$ for all $t \geq 0$, where $M(t, \varphi) = (M_1(t, \varphi), \dots, M_n(t, \varphi))$ is the unique solution of the system (3.1) satisfying $M_0(\varphi) = \varphi$.

The following result shows the boundedness and point dissipativeness of the solutions of the system (3.1), here and in what follows, point dissipativeness of the solutions of the system (3.1) means the existence of a bounded subset Q of C^+ such that for any $\varphi \in C^+$, $M(t, \varphi)$ is in Q for sufficiently large t (see [15] for details).

Lemma 3.1 *For any $\varphi \in C^+$, $\sup\{M_i(t, \varphi);\ t \geq 0,\ i = 1, \dots, n\} < \infty$. Moreover, there exists a constant $B > 0$ such that*

$$\limsup_{t\to\infty}\{M_1(t, \varphi) + \cdots + M_n(t, \varphi)\} < B \text{ for any } \varphi \in C^+.$$

Proof. Let $\rho > 1$, $\varepsilon > 0$ and $M > 0$ be given constants such that

$$M_i \geq M \text{ implies that } g_i(M_i) > \rho \sum_{j=1}^{n} b_{ij}\alpha_j + \varepsilon,\ i = 1, \dots, n.$$

The existence of these constants are guaranteed by assumption (H7). Define $M_i(t) = M_i(t, \varphi)$, $W(t) = \max_{1\leq i\leq n} M_i(t)$, $J(t) = \{i = 1, \dots, n;\ W(t) = M_i(t)\}$ and $\dot{W}(t) = \limsup_{h\to 0+} \frac{W(t+h)-W(t)}{h}$. Clearly, at a given instant $t \geq 0$, there exists a sequence $h_m \to 0^+$ as $m \to \infty$ and an integer $k \in J(t)$ such that

$$\dot{W}(t) = \lim_{m\to\infty} \frac{W_k(t+h_m) - W_k(t)}{h_m}$$

$$= -M_k(t)g_k(M_k(t)) + F_k(M_1(t), \ldots, M_r(t) + \sum_{j=1}^{n} b_{ij}\alpha_j M_j(t-\tau)$$

$$\leq -M_k(t)g_k(M_k(t)) + \sum_{j=1}^{n} b_{kj}\alpha_j M_j(t-\tau),$$

since $M_k(t) = W(t) \geq M_j(t)$ for $j = 1, \ldots, n$ implies that $F_k(M_1(t), \ldots, M_n(t)) \leq 0$ by assumption (H5).

If $W(t-\tau) \leq \rho W(t)$ and $W(t) > M$, then $M_j(t-\tau) \leq \rho M_k(t)$ and $M_k(t) > M$ for all $j = 1, \ldots, n$, we have

$$\dot{W}(t) \leq -M_k(t)g_k(M_k(t)) + \sum_{j=1}^{n} b_{kj}\alpha_j \rho M_k(t)$$

$$= -M_k(t)\left[g_k(M_k(t)) - \rho \sum_{j=1}^{n} b_{ij}\alpha_j\right]$$

$$\leq -\varepsilon M_k(t) = -\varepsilon W(t).$$

Therefore our conclusion follows from the classical Liapunov - Razumikhin Theorem for uniform boundedness and uniformly ultimate boundedness (see, e.g., Theorem 4.3 of [14]).

The following result indicates the persistence of system (3.1).

Lemma 3.2 *For any given positive number $\varepsilon > 0$, let*

$$C^+(\varepsilon) = \{\varphi \in C^+;\ \varphi_i(\theta) \geq \varepsilon \text{ for } \theta \in [-\tau, 0] \text{ and } i = 1, \ldots, n\}.$$

Then we have

(i) if $\varepsilon > 0$ is sufficiently small such that $g_i(\varepsilon) < \sum_{j=1}^{n} b_{ij} - \alpha_j$ for $i = 1, \ldots, n$, then $C^+(\varepsilon)$ is positively invariant, i.e. if $\varphi \in C^+(\varepsilon)$, then $M(t,\varphi) \in C^+(\varepsilon)$ for $t \geq 0$;

(ii) if $\varphi \in C^+\backslash\{0\}$, then $\liminf_{t\to\infty} M_i(t,\varphi) > 0$ for $i = 1, \ldots, n$.

Proof. For any $\varepsilon > 0$ such that $g_i(\varepsilon) < \sum_{j=1}^{n} b_{ij} - \alpha_j$, $i = 1, \ldots, n$, and for any $\varphi \in C^+(\varepsilon)$ with $\varphi_i(0) = \varepsilon$ for some i, we have $F_i(\varphi_1(0), \ldots, \varphi_n(0)) \geq 0$ by assumption (H5). Therefore

$$G_i(\varphi) = -\varphi_i(0)g_i(\varphi_i(0)) + F_i(\varphi_1(0), \ldots, \varphi_n(0)) + \sum_{j=1}^{n} b_{ij}\alpha_j\varphi_j(-\tau)$$

$$\geq -\varepsilon g_i(\varepsilon) + \sum_{j=1}^{n} b_{ij} - \alpha_j\varepsilon$$

$$\geq 0.$$

Therefore, by Proposition 1.3 of [19], $C^+(\varepsilon)$, is positively invariant. This proves (i).

It is clear that $G : C^+ \to R^n$ is continuously differentiable. For any $\psi \in C^+$ and $\varphi \in C$, we have

$$dG_i(\psi)\varphi = -[g_i(\psi_i(0)) + \psi_i(0)g_i'(\psi_i(0))]\psi_i(0)$$
$$+\sum_{j=1}^{n} \frac{\partial F_i}{\partial M_j}(\psi_1(0), \dots, \psi_n(0))\varphi_j(0) + \sum_{j=1}^{n} b_{ij}\alpha_j\varphi_j(-\tau),$$

where $i = 1, \dots, n$ and $dG(\psi) = (dG_1(\psi), \dots, dG_n(\psi))$. Hence we have the following observations:

(i) $dG_i(\psi)\phi \geq 0$ if $\phi \in C^+$ and $\phi_i(0) = 0$;

(ii) The matrix

$$(dG(\psi)\hat{e}_1, \dots, dG(\psi)\hat{e}_n) = \begin{pmatrix} A_{11} & A_{12} & \cdots & A_{1n} \\ A_{21} & A_{22} & \cdots & A_{2n} \\ \cdots & \cdots & \cdots & \cdots \\ A_{n1} & A_{n2} & \cdots & A_{nn} \end{pmatrix}$$

where

$$A_{11} = -[g_1(\psi_1(0)) + \psi_i(0)g_1'(\psi_1(0))] + \frac{\partial F_1}{\partial M_1}(\psi_1(0), \dots, \psi_n(0))$$

$$A_{12} = \frac{\partial F_1}{\partial M_2}(\psi_1(0), \dots, \psi_n(0)) + b_{12}\alpha_1$$

$$\cdots$$

$$A_{1n} = \frac{\partial F_1}{\partial M_n}(\psi_1(0), \dots, \psi_n(0)) + b1n\alpha_n$$

$$A_{21} = \frac{\partial F_2}{\partial M_1}(\psi_1(0), \dots, \psi_n(0)) + b_{21}\alpha_1$$

$$A_{22} = -[g_2(\psi_2(0)) + \psi_2(0)g_2'(\psi_2(0))] + \frac{\partial F_2}{\partial M_2}(\psi_1(0), \dots, \psi_n(0)$$

$$\cdots$$

$$A_{2n} = \frac{\{\partial F_n}{\partial M_n}(\psi_1(0), \dots, \psi_n(0)) + b_{2n}\alpha_2$$

$$A_{n1} = \frac{\partial F_n}{\partial M_1}(\psi_1(0)), \dots, \psi_n(0)) + b_{n1}\alpha_1$$

$$A_{n2} = \frac{\partial F_n}{\partial M_2}(\psi_1(0), \dots, \psi_n(0)) + b_{n2}\alpha_2$$

$$\cdots$$

$$A_{nn} = -[g_n(\psi_n(0)) + \psi_n(0)g_n'(\psi_n(0))] + \frac{\partial F_n}{\partial M_n}(\psi_1(0), \dots, \psi_n(0))$$

is cooperative and irreducible, since the matrix (b_{ij}) is nonnegative by assumption (H5), the matrix $\left(\frac{\partial F_i}{\partial M_j}(\psi_1(0), \dots, \psi_n(0))\right)$ is irreducible, where $\{e_1, \dots, e_n\}$

denotes the standard basis in R^n and $\wedge$ denotes the inclusion $R^n \to C$ by $x \to \hat{x}$, $\hat{x}_i(\theta) = x_i$ for $\theta \in [-\tau, 0]$ and $i = 1, \ldots, n$. Therefore, G is cooperative and irreducible in C^+ in the sense of [20], and thus the semiflow $\{T(t)\}_{t\geq 0}$ is monotone and eventually strongly monotone, i.e., $\psi - \varphi \in C^+$ implies $x_t(\psi) - x_t(\varphi) \in C^+$ for $t \geq 0$, and $\psi - \varphi \in C^+\backslash\{0\}$ implies that $x_t(\psi) - x_t(\varphi) \in \text{Int } C^+$ for $t \geq (n+1)\tau$. Therefore if $\varphi \in C^+\backslash\{0\}$, then $M_i(t, \varphi) > 0$ for $t \geq n\tau$ and $i = 1, 2, \ldots, n$, which implies that $M_{(n+1)\tau}(\varphi) \in C^+(\varepsilon)$ for some sufficiently small $\varepsilon > 0$. Now the conclusion that $\liminf_{t\to\infty} M_i(t, \varphi) > 0$ for $i = 1, 2, \ldots, n$ follows from the positive invariance of $C^+(\varepsilon)$. This completes the proof.

We are now in the position to state the proof for the main result.

Proof of Theorem 2.1. According to Lemmas 3.1 and 3.2, under the assumptions (H1) – (H7), the set $C^+(\varepsilon)$ with sufficiently small $\varepsilon > 0$ is positively invariant and the semiflow $\{T(t)\}_{t\geq 0}$ is point dissipative. Therefore by Theorem 4.1.2 of [15], there exists a global attractor G_ε in $C^+(\varepsilon)$ i.e. there exists a compact subset G_ε of $C^+(\varepsilon)$ such that for any bounded subset V of $C^+(\varepsilon), M(t, V)$ approachs G_ε as $t \to \infty$. On the other hand, we have shown that the semiflow $\{T(t)\}_{t\geq 0}$ is monotone and eventually strongly monotone. Therefore Theorem 3.1 of [16] guarantees that G_ε contains at least one equilibrium. This proves the existence of a positive equilibrium.

To consider the uniqueness of a positive equilibrium, we now consider the system of ordinary differential equations

$$\dot{x}_i(t) = \tilde{G}_i(x_1, \ldots, x_n) \tag{3.2}$$

where

$$\tilde{G}_i(x_1, \ldots, x_n) = -x_i g_i(x_i) + F_i(x_1, \ldots, x_n) + \sum_{j=1}^{n} b_{ij}\alpha_j x_j, \quad i = 1, \ldots, n. \tag{3.3}$$

Employing the same argument as that for Lemma 3.1, we can prove that all nonnegative solutions of (3.2) are bounded. By the assumption (H5), the matrix $H(x_1, \ldots, x_n) :\doteq \left(\frac{\partial \tilde{G}_i}{\partial x_j}(x_1, \ldots, x_n)\right)$ is cooperative and irreducible for all $(x_1, \ldots, x_n) \in R^n_+$.

If the assumption (H8) is satisfied, then $H(x_1, \ldots, x_n)$ satisfies the following monotonicity condition

$$(x_1, \ldots, x_n), (\tilde{x}_1, \ldots, \tilde{x}_n) \in R^n_+ \text{ with } (x_1, \ldots, x_n) - (\tilde{x}_1, \ldots, \tilde{x}_n) \in R^n_+\backslash\{0\}$$

implies that

$$H(\tilde{x}_1, \ldots, \tilde{x}_n) - H(x_1, \ldots, x_n) \in R^n_+\backslash\{0\}.$$

Therefore by Theorem 6.1 of [16] either there is no positive equilibrium and the origin is globally asymptotically stable, or there is a unique positive equilibrium which is globally asymptotically stable over $R^n_+\backslash\{0\}$. On the other hand, we have shown that system (3.1) has at least one positive equilibrium. Clearly, any equilibrium of system (3.1) is also an equilibrium for system (3.2) and conversely, therefore, (3.2), and hence (3.1), has one and only one positive equilibrium.

If the assumption (H9) is satisfied, then we can also claim that system (3.2), and hence (3.1), has one and only one positive equilibrium. Indeed, if there are two different positive equilibria $(x_1, \ldots, x_n)$ and $(\hat{x}_1, \ldots, \hat{x}_n)$, then without loss of generality, we may assume that there exists an integer i such that $x_i < \hat{x}_i$. Therefore, there exists a constant $a > 1$ and an integer j such that $ax_j = \hat{x}_j$ and $ax_k \geq \hat{x}_k$ for all $k = 1, \ldots, n$. This implies, by the assumption $\frac{\partial F_i}{\partial x_j}(x_1, \ldots, x_n) \geq 0$ for $(x_1, \ldots, x_n) \in R^n_+$ and $i \neq j$, that

$$\begin{aligned} 0 = \tilde{G}_j(\hat{x}_1, \ldots, \hat{x}_n) &\leq \tilde{G}_j(ax_1, \ldots, ax_{j-1}, \hat{x}_j, ax_{j+1}, \ldots, ax_n) \\ &= \tilde{G}_j(ax_1, \ldots, ax_n) \end{aligned} \tag{3.4}$$

On the other hand, by the definition of $\tilde{G}_j$ and assumption (H9), we have

$$\begin{aligned} &\tilde{G}_j(ax_1, \ldots, ax_n) = -ax_j g_j(ax_j) + F_j(ax_1, \ldots, ax_n) + \sum_{i=1}^{n} b_{ji}\alpha_i a x_i \\ &= -ax_j g_j(ax_j) + F_j(ax_1, \ldots, ax_n) \\ &+ a\Big[-x_j g_j(x_j) + F_j(x_1, \ldots, x_n) + \sum_{i=1}^{n} b_{ji}\alpha_i x_i\Big] + ax_j g_j(x_j) - aF_j(x_1, \ldots, x_n) \\ &= -ax_j\left[g_j(ax_j) - g_j(x_j)\right] + F_j(ax_1, \ldots, ax_n) \\ &- aF_j(x_1, \ldots, x_n) + a\tilde{G}_j(x_1, \ldots, x_n) \\ &= -ax_j\left[g_j(ax_j) - g_j(x_j)\right] + F_j(ax_1, \ldots, ax_n) \\ &- aF_j(x_1, \ldots, x_n) + a\tilde{G}_j(x_1, \ldots, x_n) < 0. \end{aligned}$$

This is contrary to (3.4).

Under either of the assumptions (H8) and (H9), we denote by P the unique positive equilibrium (3.1). P must be in $C^+(\varepsilon)$ for any ε such that $g_i(\varepsilon) < \sum_{j=1}^{n} b_{ij}\alpha_j$ for $i = 1, \ldots, n$. By Theorem 3.3 of [16] $\lim_{t\to\infty} M_t(\varphi) = P$ for any $\varphi \in C^+(\varepsilon)$. This implies that $\lim_{t\to\infty} M_t(\phi) = P$ for any $\phi \in C^+\backslash\{0\}$ since, by Lemma 3.2, for any $\phi \in C^+\backslash\{0\}$ there exists $\varepsilon > 0$ such that $M_t(\phi) \in C^+(\varepsilon)$ for sufficiently large t. Obviously, $P \in C^+$ is of the form $P = (\hat{M}_1^*, \ldots, \hat{M}_n^*)$ for some $(M_1^*, \ldots, M_n^+) \in R^n$. Therefore for any $\varphi \in C^+\backslash\{0\}$, $\lim_{t\to\infty} M_i(t, \varphi) = M_i^*$ for $i = 1, \ldots, n$.

To prove the nonnegative property and the convergence of $I_i(t)$, we rewrite the first equation of the system (2.12) in the following vector form

$$\frac{d}{dt}\begin{pmatrix} I_1(t) \\ \vdots \\ I_n(t) \end{pmatrix} = A\begin{pmatrix} I_1(t) \\ \vdots \\ I_n(t) \end{pmatrix} + \begin{pmatrix} \alpha_1 M_1(t) cr\vdots \\ \alpha_1 M_n(t) \end{pmatrix} - e^{A\tau}\begin{pmatrix} \alpha_1 M_1(t-\tau) \\ \vdots \\ \alpha_n M_n(t-\tau) \end{pmatrix}.$$

This is equivalent to

$$\frac{d}{dt}\left[e^{-At}\begin{pmatrix} I_1(t) \\ \vdots \\ I_n(t) \end{pmatrix}\right] = \frac{d}{dt}\int_{t-\tau}^{t} e^{As}\begin{pmatrix} \alpha_1 M_1(s) \\ \vdots \\ \alpha_n M_n(s) \end{pmatrix} ds,$$

from which and the compatibility condition (2.12) it follows that

$$\begin{pmatrix} I_1(t) \\ \vdots \\ I_n(t) \end{pmatrix} = \int_0^{\tau} e^{A\theta}\begin{pmatrix} \alpha_1 M_1(t-\theta) \\ \vdots \\ \alpha_n M_n(t-\theta) \end{pmatrix} d\theta. \tag{3.5}$$

We notice that $a_{ij} = \delta_{ji} \geq 0$ for $i \neq j$ and $\sum_{j=1}^{n} a_{ij} = -\gamma_i < 0$. Therefore e^{At} is a nonnegative matrix for $t > 0$, and $e^{At} \to 0$ as $t \to \infty$. Since $M_i(t) \geq 0$ for $t \geq -\tau$ and $i = 1, \ldots, n$, from (3.5) we can easily obtain $I_i(t) \geq 0$ for all $i = 1, \ldots, n$, $t \geq 0$, and

$$\begin{aligned} \lim_{t\to\infty}\begin{pmatrix} I_1(t) \\ \vdots \\ I_n(t) \end{pmatrix} &= \int_0^{\tau} e^{A\theta}\begin{pmatrix} \alpha_1 M_1^* \\ \vdots \\ \alpha_n M_n^* \end{pmatrix} d\theta = \int_0^{\tau} A^{-1}\frac{d}{d\theta}e^{A\theta}\begin{pmatrix} \alpha_1 M_1^* \\ \vdots \\ \alpha_n M_n^* \end{pmatrix} d\theta \\ &= A^{-1}(e^{A\tau} - I)\begin{pmatrix} \alpha_1 M_1^* \\ \vdots \\ \alpha_n M_n^* \end{pmatrix}. \end{aligned}$$

This completes the proof.

Acknowledgment

The research of J.-H. Wu was partialy supported by the Gording Kaplan Postdoctoral Fellowship and by the Central Research Fund of the University of Alberta. The research of H.I. Freedman was partially supported by the Natural Sciences and Engineering Research Council of Canada, Grant No. NSERC A4823.

References

1. Aiello, W.G. and Freedman, H.I., (1990): A time-delay model of single-species growth with stage structure. Math. Biosci. **101**, 139-153
2. Barclay, H.J. and van den Driessche, P., (1980): A model for a species with two life history stages and added mortality. Ecolog. Model. **11**, 157–166
3. Beretta, E., Solimano, F. and Takeuchi, Y., (1987): Global stability and periodic orbits for two-patch predator-prey diffusion-delay models. Math. Biosc. **85**, 153–183
4. Beretta, E. and Takeuchi, Y., (1987): Global stability of single-species diffusion models with continuous time delays. Bull. Math. Biol. **49**, 431–448
5. Freedman, H.I., (1989): Persistence and extinction in models of two-habitat migration. Math. Comput. Model. **12**, 105–112

6. Freedman, H.I., Rai, B. and Waltman, P., (1986): Mathematical models of population interactions with dispersal II: Differential survival in a change of habitat. J. Math. and Appl. **115**, 140–154
7. Freedman, H.I., Shukla, J.B. and Takeuchi, Y., (1989): Population diffusion in a two-patch environment. Math. Biosci. **95**, 111–123
8. Freedman, H.I. and Takeuchi, Y., (1989): Global stability and predator dynamics in a model of prey dispersal in a patchy environment. Nonlin. Anal., TMA. **13**, 993–1002
9. Freedman, H.I. and Takeuchi, Y., (1989): Predator survival versus extinction as a function of dispersal in a predator-prey model with patchy environment. Applicable Anal. **31**, 247–266
10. Freedman, H.I. and Waltman, P., (1977): Mathematical models of population interaction with dispersal I: Stability of two habitats with and without a predator. SIAM J. Appl. Math. **32**, 631–648
11. Freedman, H.I. and Wu, J.H., Steady state analysis in a model for population diffusion in a multi-patch environment. Nonlin. Anal., TMA. To appear
12. Freedman, H.I. and Wu, J.H., Persistence and global asymptotic stability of single species dispersal models with stage structure. Quart. Appl. Math. In press
13. Gurney, W.S.C., Nisbet, R.M. and Lawton, J.H., (1983): The systematic formulation of tractable single species population models incorporating age structure. J. Animal Ecol. **52**, 479–495
14. Hale, J.K., (1977): Theory of Functional Differential Equations. Springer-Verlag, Berlin-Heidelberg-New York
15. Hale, J.K., (1988): Asymptotic Behavior of Dissipative Systems, Mathematical Surveys and Monographs, Vol. **25**, Amer. Math. Soc., Providence
16. Hirsch, M.W., (1984): The dynamical systems approach to differential equations. Bull. Amer. Math. Soc. **11**, 1–64
17. Holt, R.D., (1985): Population dynamics in two patch environment: some anomalous consequences of optional habitat selection. Theor. Pop. Biol. **28**, 181–208
18. Koslesov, Yu. S., (1983): Properties of solutions of a class of equations with lag which describe the dynamics of change in the population of a species with the age structure taken into account. Math. USSR Sbornik **45**, 91–100
19. Martin, R.H. and Smith, H.L., Reaction-diffusion systems with time delays: monotonicity, invariance, comparison and convergence. (Preprint)
20. Smith, H.L., (1986): Cooperative systems of differential equations with concave nonlinearities. Nonlin. Anal. TMA. **10**, 1037–1052
21. Takeuchi, Y., (1989): Cooperative systems theory and global stability of diffusion models. Acta Applicandae Math. **14**, 49–57
22. Vance, R.R., (1984): The effect of dispersal on population stability in one-species, discrete-space population growth models. Am. Nat. **123**, 230–254
23. Webb, G.F., (1988): Theory of Nonlinear Age-Dependent Population Dynamics. Marcel Dekker, New York
24. Wood, S.N., Blythe, S.P., Gurney, S.C. and Nisbet, R.M., (1989): Instability in mortality estimation schemes related to stage-structure population models. IMA J. Math. Appl. Med. Biol. **6**, 47–68.

Errata

Lecture Notes in Biomathematics, Vol. 92
S. Busenberg, M. Martelli (Eds.)
Differential Equations Models in Biology,
Epidemiology and Ecology
ISBN 3-540-54283-3

The following typographical errors were noted after this volume was in final production.
In the article "Modelling the Effects of Screening in HIV Transmission Dynamics" by Ying-Hen Shieh, the sums appearing in Eq. (5.1), (5.3), (5.4), (5.5), (5.5a) and (5.7) were incorrectly shown as

$$\sum_{k=1} n \quad \text{and} \quad \sum_{j=1} n$$

and should be read as

$$\sum_{k=1}^{n} \quad \text{and} \quad \sum_{j=1}^{n}$$

In the article "Mathematical Model for the Dynamics of a Phytoplankton Population" by E. Beretta and A. Fasano, the references in Table 1 on page 174 are numbered incorrectly and the name of the biological laboratory should be changed. The correct reference numbers and laboratory name are shown in the following partial reproduction of that table.

Symbol	*Quantity*	*Source*
Y	average nutrient concentration in a cell (Phosphorus P) Y_P. $Y_N = 16Y_P$, $Y_C = 106Y_P$	[3]
D	diffusion coefficient for the limiting nutrient (P)	[5]
σ_w	Light extinction coefficient of water in the absence of biomass	[3]
I_m	the optimum of the light intensity	[1, 2]
L	Michaelis-Menten constant (or half saturation constant) for the phosphorus uptake function	[1]

lab: Daphne - Regione Emilia Romagna - Cesenatico, Italy.

Lecture Notes in Biomathematics

For information about Vols. 1–54
please contact your bookseller or Springer-Verlag

Vol. 55: Modelling of Patterns in Space and Time. Proceedings, 1983. Edited by W. Jäger and J.D. Murray. VIII, 405 pages. 1984.

Vol. 56: H.W. Hethcote, J.A. Yorke, Gonorrhea Transmission Dynamics and Control. IX, 105 pages. 1984.

Vol. 57: Mathematics in Biology and Medicine. Proceedings, 1983. Edited by V. Capasso, E. Grosso and S.L. Paveri-Fontana. XVIII, 524 pages. 1985.

Vol. 58: C.A. Macken, A.S. Perelson, Branching Processes Applied to Cell Surface Aggregation Phenomena. VIII, 122 pages. 1985.

Vol. 59: I. Nåsell, Hybrid Models of Tropical Infections. VI, 206 pages. 1985.

Vol. 60: Population Genetics in Forestry. Proceedings, 1984. Edited by H.-R. Gregorius. VI, 287 pages. 1985.

Vol. 61: Resource Management. Proceedings, 1984. Edited by M. Mangel. V. 138 pages. 1985.

Vol. 62: B.J. West, An Essay on the Importance of Being Nonlinear. VIII, 204 pages. 1985.

Vol. 63: Carme Torras i Genís. Temporal-Pattern Learning in Neural Models. VII, 227 pages. 1985.

Vol. 64: Peripheral Auditory Mechanisms. Proceedings, 1985. Edited by J.B. Allen, J.L. Hall, A. Hubbard, S. Neely and A. Tubis. VII, 400 pages. 1986.

Vol. 65: Immunology and Epidemiology. Proceedings, 1985. Edited by G.W. Hoffmann and T. Hraba. VIII, 242 pages. 1986.

Vol. 66: Nonlinear Oscillations in Biology and Chemistry. Proceedings, 1965. Edited by H.G. Othmer. VI, 289 pages. 1986.

Vol. 67: P.H. Todd, Intrinsic Geometry of Biological Surface Growth. IV, 128 pages. 1986.

Vol. 68: The Dynamics of Physiologically Structured Populations. Edited by J.A.J. Metz and O. Diekmann. XII. 511 pages. 1986.

Vol. 69: S.H. Strogatz, The Mathematical Structure of the Human Sleep-Wake Cycle. VIII, 239 pages. 1986.

Vol. 70: Stochastic Methods in Biology. Proceedings, 1985. Edited by M. Kimura, G. Kallianpur and T. Hida. VI, 229 pages. 1987.

Vol. 71: Mathematical Topics in Population Biology, Morphogenesis and Neurosciences. Proceedings, 1985. Edited by E. Teramoto and M. Yamaguti. IX, 348 pages. 1987.

Vol. 72: Modeling and Management of Resources under Uncertainty. Proceedings, 1985. Edited by T.L. Vincent, Y. Cohen, W.J. Grantham, G.P. Kirkwood, and J.M. Skowronski. VIII, 318 pages. 1987.

Vol. 73: Y. Cohen (Ed.), Applications of Control Theory in Ecology. Proceedings, 1986. VII, 101 pages. 1987.

Vol. 74: A.Y. Yakovlev, A.V. Zorin, Computer Simulation in Cell Radiobiology. VI, 133 pages. 1988.

Vol. 75: H. Knolle, Cell Kinetic Modelling and the Chemotherapy of Cancer. VIII, 157 pages. 1988.

Vol. 76: C.A. Macken, A.S. Perelson, Stem Cell Proliferation and Differentiation. VIII, 113 pages. 1988.

Vol. 77: A. Hastings (Ed.), Community Ecology. Proceedings, 1986. VII, 131 pages. 1988.

Vol. 78: J.J. Linderman, D.A. Lauffenburger. Receptor/Ligand Sorting Along the Endocylic Pathway. VI, 164 pages. 1989.

Vol. 79: P. Prusinkiewlcz, J. Hanan, Lindenmayer Systems, Fractals, and Plants. V, 120 pages. 1989.

Vol. 80: D. House, Depth Perception in Frogs and Toads. VII, 135 pages. 1989.

Vol. 81: C. Castillo-Chavez, S.A. Levin, C.A. Shoemaker (Eds.), Mathematical Approaches to Problems in Resource Management and Epidemiology. Proceedings, 1987. VII, 327 pages.

Vol. 82: A.Yu. Yakovlev, N.M. Yanev, Transient Processes in Cell Proliferation Kinetics. VI, 214 pages. 1989.

Vol. 83: C. Castillo-Chavez (Ed.), Mathematical and Statistical Approaches to AIDS Epidemiology. IX, 405 pages. 1989.

Vol. 84: A. Dress, A. von Haeseler (Eds), Trees and Hierarchical Structures. Proceedings, 1987. III, 137 pages. 1990.

Vol. 85: S. Tuljapurkar, Population Dynamics in Variable Environments. II, 154 pages. 1990.

Vol. 86: J.P. Gabriel, C. Lefèvre, P. Picard (Eds.), Stochastic Processes in Epidemic Theory. Proceedings, 1988. VII, 197 pages. 1990.

Vol. 87: P. Dallas, D. Geisler, J.W. Matthews, M.A. Ruggero, C.R. Steele (Eds.), The Mechanics and Biophysics of Hearing. Proceedings. VII, 418 pages. 1990.

Vol. 88: R. Bradbury (Ed), Acanthaster and the Coral Reef: A Theoretical Perspective. Proceedings, 1988 VI, 338 pages. 1990.

Vol. 89: W. Alt, G. Hoffmann (Eds.), Biological Motion. Proceedings, 1989. IX, 604 pages. 1990.

Vol. 90: R. M. Nisbet, S. N. Wood, Estimation of Mortality Rates in Stage-Structured Populations. VII, 101 pages. 1991.

Vol. 91: F. W. Wiegel, Physical Principles in Chemoreception. IX, 85 pages. 1991.

Vol. 92: S. Busenberg, M. Martelli (Eds.), Differential Equations and Applications to Biology and Population Dynamics. Proceedings, 1990. IX, 267 pages. 1991.

General Remarks

Lecture Notes are printed by photo-offset from the master-copy delivered in camera-ready form by the authors of monographs, resp. editors of proceedings volumes. For this purpose Springer-Verlag provides technical instructions for the preparation of manuscripts. Volume editors are requested to distribute these to all contributing authors of proceedings volumes. Some homogeneity in the presentation of the contributions in a multi-author volume is desirable.

Careful preparation of manuscripts will help keep production time short and ensure a satisfactory appearance of the finished book. The actual production of a Lecture Notes volume normally takes approximately 8 weeks.

For monograph manuscripts typed or typeset according to our instructions, Springer-Verlag can, if necessary, contribute towards the preparation costs at a fixed rate.

Authors of monographs receive 50 free copies of their book. Editors of proceedings volumes similarly receive 50 copies of the book and are responsible for redistributing these to authors etc. at their discretion. No reprints of individual contributions can be supplied. No royalty is paid on Lecture Notes volumes.

Volume authors and editors are entitled to purchase further copies of their book for their personal use at a discount of 33.3 %, other Springer mathematics books at a discount of 20 % directly from Springer-Verlag. Authors contributing to proceedings volumes may purchase the volume in which their article appears at a discount of 20 %.

Commitment to publish is made by letter of intent rather than by signing a formal contract. Springer-Verlag secures the copyright for each volume.

Addresses:

Professor Simon A. Levin, Cornell University
Section of Ecology and Systematics
345 Corson Hall, Ithaca
New York 14853-0239, USA

Springer-Verlag, Mathematics Editorial
Tiergartenstr. 17
W-6900 Heidelberg
Federal Republic of Germany
Tel.: *49 (6221) 487-410